Water, fungi and plants

Water, fungi and plants

SYMPOSIUM OF
THE BRITISH MYCOLOGICAL SOCIETY
HELD AT THE UNIVERSITY OF LANCASTER,
APRIL 1985

EDITED BY

P. G. AYRES & LYNNE BODDY

The right of the
University of Cambridge
to print and sell
all manner of books
was granted by
Henry VIII in 1534.
The University has printed
and published continuously
since 1584.

CAMBRIDGE UNIVERSITY PRESS

CAMBRIDGE
LONDON NEW YORK NEW ROCHELLE
MELBOURNE SYDNEY

CAMBRIDGE UNIVERSITY PRESS
Cambridge, New York, Melbourne, Madrid, Cape Town,
Singapore, São Paulo, Delhi, Tokyo, Mexico City

Cambridge University Press
The Edinburgh Building, Cambridge CB2 8RU, UK

Published in the United States of America by Cambridge University Press, New York

www.cambridge.org
Information on this title: www.cambridge.org/9780521106238

First published 1986
First paperback edition 2011

A catalogue record for this publication is available from the British Library

Library of Congress Cataloguing in Publication data

British Mycological Society. Symposium (1985:
University of Lancaster)
Water, fungi, and plants.
Includes index.
1. Fungi—Congresses. 2. Fungal diseases of
plants—Congresses. 3. Plant–water relationships—
Congresses. 1. Ayres, P. G. (Peter G.) 11. Boddy,
Lynne. 111. Title.
QK600.3.B75 1985 581.2′326 85–26969

ISBN 978-0-521-32294-2 Hardback
ISBN 978-0-521-10623-8 Paperback

Contents

Contributors

A. M. Al-Hamdani, *Department of Botany, University of Sheffield, Sheffield, S10 2TN, UK*

G. Ayerst, *School of Applied Science, The Polytechnic, Wolverhampton, WV1 1LY, UK*

P. G. Ayres, *Department of Biological Sciences, University of Lancaster, Bailrigg, Lancaster, LA1 4YQ, UK*

Lynne Boddy, *Department of Microbiology, University College Cardiff, Newport Road, Cardiff, CF2 1TA, UK*

R. Boyd, *Department of Botany, University of Sheffield, Sheffield, S10 2TN, UK*

D. R. Butler, *Long Ashton Research Station, Long Ashton, Bristol, BS18 9AF, UK*

M. J. Carlile, *Department of Pure and Applied Biology, Imperial College at Silwood Park, Ascot, Berkshire, SL5 7PY, UK*

R. C. Cooke, *Department of Botany, University of Sheffield, Sheffield, S10 2TN, UK*

C. Dennis, *Campden Food Preservation Research Association, Chipping Campden, Gloucestershire, GL55 6LD, UK*

P. Dowding, *Environmental Sciences Unit, Trinity College, Dublin 2, Eire*

J. M. Duniway, *Department of Plant Pathology, University of California, Davis, California 95616, USA*

D. Eamus, *Botany Department, University of Liverpool, Liverpool, L69 3BX, UK*

B. D. L. Fitt, *Rothamsted Experimental Station, Harpenden, Hertfordshire, AL5 2JQ, UK*

T. R. Gordon, *Department of Plant Pathology, University of California, Davis, California 95616, USA*

R. Hall, *Department of Environmental Biology, University of Guelph, Ontario N1G 2W1, Canada*

M. Holderness, *Department of Horticulture, University of Reading, Earley Gate, Reading, RG6 2AU, UK*

D. H. Jennings, *Botany Department, University of Liverpool, Liverpool, L69 3BX, UK*

H. G. Jones, *East Malling Research Station, Maidstone, Kent, ME19 6BJ, UK*

J. Lacey, *Rothamsted Experimental Station, Harpenden, Hertfordshire, AL5 2JQ, UK*

H. A. McCartney, *Rothamsted Experimental Station, Harpenden, Hertfordshire, AL5 2JQ, UK*

D. J. Mulla, *Department of Agronomy and Soils, Washington State University, Pullman, Washington 99164, USA*

R. I. Papendick, *USDA, ARS, Land Management and Water Conservation Research, Washington State University, Pullman, Washington 99164, USA*

N. D. Paul, *Department of Biological Sciences, University of Lancaster, Bailrigg, Lancaster, LA1 4YQ, UK*

G. F. Pegg, *Department of Horticulture, University of Reading, Earley Gate, Reading, RG6 2AU, UK*

A. M. Plumbe, *Department of Biological Science, University of Stirling, Stirling, FK9 4LA, Scotland, UK*

A. D. M. Rayner, *School of Biological Sciences, University of Bath, Claverton Down, Bath, BA2 7AY, UK*

D. J. Read, *Department of Botany, University of Sheffield, Sheffield, S10 2TN, UK*

D. J. Royle, *Long Ashton Research Station, Long Ashton, Bristol, BS18 9AF, UK*

D. F. Schoeneweiss, *Illinois Natural History Survey, 607 East Peabody Drive, Champaign, Illinois 61820, USA*

J. M. Waller, *Commonwealth Mycological Institute, Ferry Lane, Kew, Surrey, TW9 3AF, UK*

C. M. Willmer, *Department of Biological Science, University of Stirling, Stirling, FK9 4LA, Scotland, UK*

Preface

The great majority of fungi enjoy at some stage in their life an intimate relationship with plant tissue, either living or dead. Water has a critical role in this relationship, not just because it is the solvent for metabolic processes and a participant in many of them, but because it is essential for the transport of nutrients within and outside fungal thalli and plants, it has a vital skeletal function in both, and a key role in the behaviour and spread of fungi. This volume includes invited papers from the 3rd General Meeting of the British Mycological Society on the subject of 'Water, Fungi and Plants'. Papers were chosen to provide a broad coverage of this vast subject, but it will be evident to the reader that the germs of several more symposia lie buried within this volume.

The first chapter, by Papendick & Mulla, describes cell and tissue water relations and the methods that may be employed for measuring and controlling water potential in plant and microbial systems; it defines terms and sets the scene for chapters that follow. Papendick & Mulla note that measurements of fungal osmotic potentials are especially problematical, a theme taken up by Eamus & Jennings, who explain the significance of osmotic and turgor gradients in mycelia – the generation of positive turgors being important for both growth and water movement, and for the translocation of solutes. Environmental conditions are, of course, not always suitable for vegetative spread from one substratum to another. Under harsh conditions survival units are often formed. The sclerotium is one such structure and its water relations during formation, survival and germination are reviewed by Cooke & Al-Hamdani. It is clear from the first three chapters that we are still very ignorant of fungal water relations at the cellular level and of the effect of these on physiological and morphogenetic processes.

Both asexually and sexually produced spores are major agents of the spread of decomposer organisms and of disease. In Chapter 4 Lacey reviews how water can affect all stages of fungal reproduction by spores, pointing out that effects on individual fungi may differ widely. Fitt & McCartney then consider in detail one aspect of spore dispersal, namely by rain-splash which is, after wind, the most important agent of dispersal of plant pathogen spores. A further significant mode of dispersal of plant pathogens, particularly of lower fungi invading from the soil, is as zoospores and these are discussed in Chapters 6 and 7. Firstly Carlile considers the problems of zoospores and analogises them with jet aircraft. Next Duniway & Gordon relate soil water relations to pathogen activity by providing a case study of the pathogenic activities of the zoosporic genus *Phytophthora* and contrasting this with non-zoosporic foot-rotting *Fusarium* species. From these chapters it is apparent that reproduction and dispersal of fungi may occur only when particular moisture conditions prevail. Indeed, these are so specific that disease epidemics caused by plant pathogens can often be accurately predicted from known weather patterns. Thus, this section on the spread of disease is concluded by Royle & Butler's discussion of the epidemiological significance of liquid water and its role in disease forecasting.

Plants can be rendered more or less susceptible to invasion as a result of particular environmental pressures, including drought stress or waterlogging. Disease resistance in relation to water is considered by the authors of Chapters 9 to 12. Schoeneweiss provides an overview of water stress predisposition to disease and discusses those diseases most often associated with water stress, root and stalk rots, stem cankers and some vascular wilts. He examines the relationship between the level and duration of water stress, pathogen aggressiveness and plant senescence. These relationships are further emphasised by Waller's appraisal of the extreme conditions experienced by tropical crops subjected to drought and irrigation. The studies reported by Holderness & Pegg on the growth of tomatoes, without soil, in recirculated nutrient-film culture may be regarded as another example of irrigation, but they provide a unique insight into that balance between root loss due to infection and root initiation and growth which is normally hidden in soil. In contrast to the previous three chapters, Willmer & Plumbe (Chapter 12) describe work indicating the inhibitory effects on fungal development of compounds produced by plants·as a result of water stress. They also demonstrate that some phytoalexins possess characteristics of water stress compounds and can affect stomatal behaviour.

The last idea, that fungi can affect host water relations is pursued in Chapters 13 to 16. Jones begins this section by explaining the fundamental principles of water flow in healthy plants, and then outlines the effects of vascular wilts on water flow in xylem. Chapters by Hall and Ayres & Paul examine respectively the effects of root pathogens and foliar pathogens on host water relations. Water stress induced by rhizopathic fungi results largely from increased resistance to movement of water in the roots or lower stem, although reductions in the size of root systems are important, and hormonal changes etc. merit further study. On the other hand foliar pathogens cause an increased flux of water away from infected sites with potentially harmful effect on the whole plant, particularly when associated with reduced root growth. In complete contrast to injurious effects of infection described in these chapters, Read & Boyd's review of the water relations of mycorrhizal plants indicates the possible role of the associated fungi in the enhancement of drought resistance of the plant.

In addition to their pathogenic or mutualistic symbiotic associations with plants, fungi are also important as saprotrophs, decomposing dead or harvested organic matter. This forms the subject of the last four chapters. In Chapter 17 Dowding considers the water status of the surface layer of dead plant material and soil, and how the responses of fungi to the pressures imposed on them by water stress determine, in part, their competitive abilities. Fungi may modify the water status of the environment so influencing their own activity and that of their successors. Rayner continues this theme by relating the aetiology of decay in trees to moisture regime and distinguishes five fundamentally different ecological strategies by which fungi counter water-stress problems imposed by the tree. Chapters 19 and 20 describe the effect of water on, and methods of prevention of, deterioration of agricultural food products: Dennis & Ayerst respectively consider post-harvest spoilage of fruits and vegetables and the biodeterioration of bulk stored products, particularly of grain. In the final chapter, Boddy provides an overview of the effect of water on decomposition processes in terrestrial ecosystems. She considers the effects of water on the individual components and overall process of decomposition and emphasises the need for studying such features in conjunction with the structure and development of fungal communities.

We are grateful to all those contributors who helped to make the symposium a success and in particular to all authors for their cooperation and willingness to follow our suggestions. Finally, the B.M.S. is pleased to record the involvement of the British Society for Plant Pathology in the meeting. Dr Brian Clifford (Programme Secretary) helped to plan the programme and the B.S.P.P. financed half of the speakers in sessions on 'Water, Fungi and Plant Disease'.

Peter G. Ayres
University of Lancaster

Lynne Boddy
University College, Cardiff

June 1985

Note on terminology and abbreviations

There have been significant changes over the years in the way in which the water status of fungal and plant material, of soils, and even of the atmosphere, has been expressed. Measurements of water potential (in units of mega- or kilo-pascals, as recommended by the Royal Society, see *Philosophical Transactions of the Royal Society of London, Series B*, **273** 433 (1976): $1 \text{ MPa} = 10 \text{ bars} = 10^6 \text{J m}^{-3}$; $1 \text{ kPa} = 10 \text{ mbar}$) have largely replaced the water activity (a_w) or equilibrium relative humidity terminologies, although in some fields investigators have clung to the older terminology, most notably in studies of storage fungi. The superiority of the water potential terminology is clearly explained here by Papendick & Mulla (see Chapter 1) and is now being widely adopted in other areas of microbiology (see, for example, D. M. Griffin (1981) Water and microbial stress, in *Advances in Microbial Ecology* V, ed. M. Alexander, pp. 91–136, New York: Plenum Press). It is, therefore, adopted throughout this book. Total water potential, ψ, is subscripted to refer to the system under consideration, e.g. ψ_{leaf}, ψ_{xylem} etc.; ψ_π, ψ_m, ψ_p and ψ_s, refer respectively to the osmotic (= solute), matric, turgor and gravitational components of the total water potential. We have asked authors to convert expressions of water status in the older literature to this terminology – equivalence is based on equation (8) shown in chapter 1 – in order to facilitate comparability. This has meant a lot of tedious work for some authors and we are especially grateful to them for their efforts.

Relative humidity (rh) expressed as a percentage has been used throughout to describe atmospheric moisture content. It is the ratio of the water vapour pressure in air to the saturation vapour pressure at the same temperature. Alternatively, it is the ratio of the mole fraction of water vapour in the moist air sample to the mole fraction in a sample which is saturated at the same temperature.

Following Milburn, J. A. (1979) *Water Flow in Plants*, Longman: London, hydraulic conductance ($m s^{-1} Pa^{-1}$) is represented as L_p. In some texts L_p has been used for hydraulic conductivity ($m^2 s^{-1} Pa^{-1}$), but this is properly represented by L. L depends on pathlength, while L_p is a property of the material through which water flows. The reciprocal of conductance ($1/L_p$) is resistance (R) (see Table 1 in Jones, this volume).

1
Basic principles of cell and tissue water relations

ROBERT I. PAPENDICK[1] and DAVID J. MULLA[2]

[1]*USDA, ARS, Land Management and Water Conservation Research, and*
[2]*Department of Agronomy and Soils, Washington State University, Pullman,*
Washington 99164, USA

Water is essential for cell metabolism and growth in plants and micro-organisms. It hydrates living cell walls and tissues and is the main constituent of cytoplasm and vacuole. In higher plants, appreciable water is also found in intercellular spaces and in conductive tissue. Cell walls may vary from tenths to tens of microns in thickness and generally contain appreciable free space for retention and transfer of water across the cell (Nobel, 1974). The size of the water-holding spaces in the cell wall channels is quite small, e.g. in the order of 0.01 μm diameter or below for plants and probably less for microorganisms. The extracellular water is generally very low in solutes and is subject to strong forces of adhesion because of the nature of the wall matrix. Most of the tissue water is found inside the cells. In contrast to extracellular water, the water inside the cell is concentrated in dissolved inorganic ions, organic acids, sugars, and amino acids.

Most of the effects of water on cellular functions are closely related to the free energy of water in the cells which determines water availability for life processes. Water potential is a fundamental concept now widely used in the biological and soil sciences for quantifying the energy of water in plants, microorganisms, soils, and other related systems (Papendick & Campbell, 1981). Water potential is an abbreviated expression for the 'potential energy of water'. By definition, water potential is the free energy of water in a system relative to the free energy of a reference pool of pure, free water having a specified mass or volume.

The reference state of pure, free water is usually assigned zero water

Contribution from Agric. Res. Serv. US Dep. of Agric., and the College of Agric. and Home Economics, Agric. Res. Center, Washington State Univ., Pullman, WA 99164, USA, Scientific Paper No. 7150.

potential. Water in living cells and tissues is normally subject to certain forces that raise or lower its potential energy relative to the reference state. As in any thermodynamic system, the primary tendency of cells is to achieve water-potential equilibrium with the surrounding environment. Water flows spontaneously from high to low (more negative) potentials and the availability of water for physiological processes decreases as the potential is lowered. On the other hand, energy must be expended to raise the water potential from a negative to a more positive state which makes it more available for plants and microorganisms.

Some basic relationships of water potential

The free energy of water in cells and tissues can be expressed in terms of the chemical potential, μ_w, of the water, which under isothermal conditions is

$$\mu_w = \mu_w^* + RT \ln a_w + V_w P + mgh \tag{1}$$

where μ_w^* is the chemical potential of water in the reference state ($J\,mol^{-1}$), R is the universal gas constant ($8.3143\,J\,mol^{-1}\,K^{-1}$), T is temperature (K), a_w is the activity of water, V_w is the partial molar volume of the water ($1.8 \times 10^{-5}\,m^3\,mol^{-1}$ at 4 °C), P is the hydrostatic pressure (Pa), m is the molecular weight of water ($18 \times 10^{-3}\,kg\,mol^{-1}$), g is the gravitational constant of acceleration ($9.8\,m\,s^{-2}$), and h is the height (m) of the water relative to a specified reference elevation. Chemical potential is a relative quantity, just like electrical or gravitational potential. Its measurement is made relative to a reference chemical potential, μ_w^*, which is by convention taken to be the chemical potential of water at a reference elevation maintained at a temperature of T and an external pressure equal to atmospheric pressure.

It has been most convenient in studies of plant, microbiological, and soil-water relations to express the energy status of water from Eq. (1) in the familiar units of pressure terminology (e.g. bar, atmosphere, cm of water). Hence, the term water potential ψ has been defined as the energy per unit volume of water:

$$\psi = (\mu_w - \mu_w^*)/V_w. \tag{2}$$

Since chemical potential is expressed in units of $J\,mol^{-1}$, dividing by V_w gives units of $J\,m^{-3}$ for water potential, which is dimensionally equivalent to pressure in pascals (Pa). Plant water potentials commonly range from $-0.5\,MPa$ down to $-4\,MPa$, and certain microorganisms such as mycelial and yeast Ascomycotina are known to survive in environments as dry as $-65\,\text{MPa}$ (Harris, 1981).

Dividing Eq. (1) by V_w, and defining $(RT/V_w)\ln a_w$ as the osmotic plus matric component, P as the turgor component, and mgh/V_w as the gravitational component of total water potential gives the following:

$$\psi = (\psi_\pi + \psi_m) + \psi_p + \psi_g \tag{3}$$

where ψ_π, ψ_m, ψ_p, and ψ_g refer to the osmotic, matric, turgor, and gravitational components of the total water potential. The advantage of using the water-potential concept to express the availability of water for cellular processes is that the component potentials arising from the different forces acting on the water can be separated and summed to give the total potential.

The osmotic and matric components of cell water are always negative, meaning that work is expended in moving water from the cell to the reference state, while the turgor pressure is always positive or zero. The gravitational component is either negative, zero, or positive, depending on an arbitrary choice of reference level.

Since cell walls are elastic, changes in turgor pressure will tend to deform the walls and, hence, cause changes in pressure on water confined in the cell wall matrix. This effect, referred to as a deformation potential (Cook & Papendick, 1978), is somewhat analogous to overburden potential in soil caused by weight from overlying soil on water present in the deeper layers (Papendick & Campbell, 1981). Deformation potentials are always positive or zero.

Water in and between cell walls (also in xylem tissue of plants) is subjected primarily to adsorption and capillary effects arising from the wall matrix; these lower the water potential mainly due to a matric potential effect. Inside the intact cells, where the constituents are highly hydrated, the components of water potential are the osmotic potential due to solutes, a small matric component arising from liquid–solid interfaces, and a positive (or sometimes zero) turgor pressure due to the presence of semipermeable membranes and the cell wall. In certain cases, it is necessary to account for the gravitational effect on water potential, e.g. upward movement of water in a tall tree (see also Jones, Ch. 13). However, for water exchange across cell walls and membranes where elevation distances are minute, the gravitational term is neglected.

Differences in water potential across cell walls, which are only a few microns thick, can exist for an extended time only if resistances to flow are extremely high. For this reason, the potential of water inside the cell is likely to be near equilibrium with the potential immediately to the outside, i.e.

$$\psi = (\psi_\pi + \psi_m + \psi_p) = (\psi_m + \psi_d) \tag{4}$$
$$\underset{\text{[inside]}}{} \qquad \underset{\text{[outside]}}{}$$

where ψ_d is a deformation potential. This basic relationship holds at equilibrium even though the components of water potential inside and outside the cells may differ substantially.

Water potential, water activity, and relative humidity

Water activity and relative humidity are widely used parameters for characterising microbial water relations. As shown in Eq. (1), water activity represents only one component of water potential and so it is important to understand the thermodynamic relationships of water activity and relative humidity with water potential, and limitations to use of these parameters for describing cell water relations.

For dilute solutions consisting of n_j moles of j solutes in n_w moles of water, the activity of the water, a_w, is approximately equal to its mole fraction, N_w, i.e.

$$\ln a_w \cong \ln N_w = \ln\left(1 - \frac{\Sigma_j n_j}{n_w + \Sigma_j n_j}\right) \cong -\frac{\Sigma_j n_j}{n_w + \Sigma_j n_j} \cong -\frac{\Sigma_j n_j}{n_w} \tag{5}$$

where an expansion is used in going from step 3 to 4 above. It is also assumed that the number of moles of water is much greater than the number of moles of solutes (Nobel, 1974). To express the osmotic potential in terms of the concentration of solutes, it is only necessary to multiply Eq. (5) by (RT/V_w), thus yielding

$$(RT/V_w)\ln N_w = \psi_\pi = -RT\Sigma_j\left(\frac{n_j}{V_w n_w}\right) = -RT\Sigma_j c_j \tag{6}$$

where the concentration of the jth solute, c_j, is simply the number of moles of solute divided by the volume of water, $V_w n_w$. Equation (6), known as the Van't Hoff Equation, is useful for estimating osmotic potential from measurements of solute concentration of cell contents, provided the solution is dilute and the cell membrane is impermeable to the solutes. Realistically, the solutions present inside cells are not ideal. The activity of water in cells is controlled by its interaction with ions, solutes, and interfaces, and it can be expressed as the product of the mole fraction, N_w, and activity coefficient, γ_w, of the water. Thus, replac-

Table 1.1. *Relationship between water potential and relative humidity at 20°C*

Water potential (MPa)	Relative humidity
−0.005	1.0000
−0.01	0.9999
−0.05	0.9996
−0.1	0.9993
−0.5	0.9963
−1.0	0.9926
−2.0	0.9853
−5.0	0.9637
−10.0	0.9286
−50.0	0.6906
−100.0	0.4769

ing a_w by $N_w \gamma_w$ in Eq. (5) and multiplying by (RT/V_w) yields

$$(RT/V_w) \ln a_w = (RT/V_w) \ln N_w + (RT/V_w) \ln \gamma_w = \psi_\pi + \psi_m. \quad (7)$$

In this form, the term $(RT/V_w) \ln \gamma_w$ represents a correction factor to the osmotic potential $(RT/V_w) \ln N_w$, that term being the matric potential, ψ_m. The major contribution to matric potential, which is generally small inside cells, is from interactions between water and the colloids or solutes it hydrates.

At equilibrium, the sum of osmotic and matric components of water potential of the solution extracted from inside the cell is equal to the water potential of the associated vapour phase, i e

$$\psi_\pi + \psi_m = (RT/V_w) \ln a_w = (RT/V_w) \ln h_r \quad (8)$$

where h_r is the relative humidity (expressed as a fraction) of the air over the cell solution. For water potentials in the range 0 to −1.5 MPa, an approximate relationship between water potential in MPa and relative humidity at 20°C is $\psi \cong 135 \, (h_r - 1)$ (Papendick & Campbell, 1981). A change in relative humidity (or water activity) of 0.01 is therefore equivalent to a change in water potential of about 1.4 MPa. Table 1.1 presents some values of water potential and corresponding relative humidities. As indicated later (p. 19), measurement of equilibrium relative humidity is a very important technique for determining the osmotic plus matric component of water in the cell protoplasm.

Table 1.2. *Equilibrium relative humidities (rh in percent) and corresponding water potentials (ψ in MPa) over saturated solutions at different temperatures. Adapted from Winston & Bates (1960)*

Com-		Temperature (°C)					
pound		10	15	20	25	30	35
KNO$_3$	rh	95.5	—	93.5	92.5	91.0	—
	ψ	−6.0	—	−9.1	−10.7	−13.2	—
KCl	rh	88.0	86.5	85.0	85.0	84.5	83.0
	ψ	−16.7	−19.3	−22.0	−22.4	−23.6	−26.5
NaNO$_3$	rh	—	76.5	76.0	74.0	72.5	71.0
	ψ	—	−35.6	−37.1	−41.1	−45.0	−48.7
Glucose	rh	57.0	—	55.0	55.0	—	55.0
	ψ	−73.5	—	−80.9	−82.3	—	−85.1
LiCl	rh	13.5	13.0	12.5	12.0	11.5	—
	ψ	−252.5	−271.4	−281.4	−291.8	−302.7	—

The equilibrium relative humidity and, hence, water potential of many saturated solutions decreases quite markedly with increase in temperature over the range 10° to 35 °C (Winston & Bates, 1960). Table 1.2 presents relative humidities and corresponding water potentials (computed from Eq. (8)) for several compounds and shows that, for example with KNO$_3$, there is a decrease in relative humidity of 4.5 percentage units when the temperature is increased from 10° to 30 °C. Because of the logarithmic relationship, the numerical and percentage changes in water potential are much greater than those for relative humidity. These data clearly illustrate that the equilibrium relative humidity of solutions does not remain constant with change in temperature and that this change in relative humidity expressed as water potential may be substantial.

It appears from the literature that researchers conducting biological studies over a wide range of humidities, particularly in the dry extremes, often prefer to express the water parameter as water activity or relative humidity instead of water potential. In the wetter range, where more sensitivity is needed, water potential is generally preferred. The importance of measuring small changes in water potential cannot be overemphasised. For example, it is not uncommon to detect certain responses of microorganisms to water potential changes as small as 0.001 MPa (Cook & Duniway, 1981), which in the range of relative humidity between 1.0 and 0.999 corresponds to a change in relative humidity in the order of 0.00001.

Water potential is less sensitive to changes in relative humidity at high humidities than in the drier range, as can be judged from Fig. 1.1. For example, a change in relative humidity from 1 to 0.99 at 20°C results in a change in water potential of 1.36 MPa, which is almost the entire range of available water for higher plants. A change in relative humidity from 0.50 to 0.49 results in a change in water potential of 2.73 MPa. This increases to 6.94 MPa for a change in relative humidity from 0.20 to 0.19.

The perceived usefulness of water activity in describing the energy status of solutions is mostly associated with solution chemistry in systems where the main, if not entire, component of water potential is the osmotic component. The water activity also accounts for a matric effect through lowering of the equilibrium vapour pressure. However, water activity or relative humidity does not account for pressure (or gravitational) potentials and, hence, does not account for turgor pressure inside the cell. Furthermore, at equilibrium, whereas the overall water potential is everywhere equal, the water activity, which contributes to only two components of total water potential, is not necessarily everywhere equal. Hence, the use of water activity for describing the energy relations of water inside and outside microbial and plant cells should be abandoned in favour of the use of water potential.

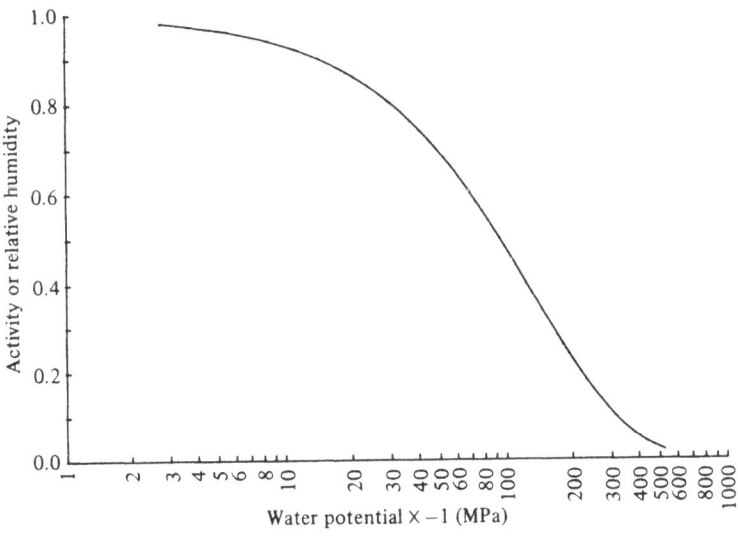

Fig. 1.1. The relationship between water activity or relative humidity and water potential at 20°C.

Relationships between tissue-water content and water potential

Cellular and metabolic processes in plants and microorganisms are sensitive to changes in the water content of cells and tissue, which in turn affect ψ and component potentials. Some rather fundamental theory has been developed to describe physical aspects of water in plant cells and tissue in response to changes in water content. Current models of plant water relations assume that tissue water can be divided into an apoplastic fraction and a symplastic fraction. Apoplastic water, sometimes called 'bound' water, or nonsolvent water, is held in between and within cell walls and in conductive tissue. Symplastic water, on the other hand, is water contained within the intact cell and is more directly involved as an active component in the life processes of the plant. Measurements made by Campbell, Papendick, Rabie & Shayo-Ngowi (1979) show that apoplastic water is essentially solute free and that its ψ_π remains near zero in the presence of living membranes. In contrast ψ_m of symplastic water is very near zero as shown by the fact that with very slight mechanical pressure cell sap tends to exude from thawed tissue that was previously frozen solid. This occurs because freezing causes destruction of the membranes which then results in the cell sap flooding the matrix and mixing freely with the apoplastic water.

Campbell *et al.* (1979) developed a plant-water model which accounted for the relative amount of apoplastic water in plant tissue and its variation with ψ. They showed that neglecting the presence of apoplastic water could cause serious errors in determining tissue-water parameters that are sensitive to plant stress. The theory was used to determine ψ_π of cells at full and zero turgor and the tissue elastic modulus from moisture release curves of winter wheat leaves grown under different moisture and temperature-stress environments. Their equation gives an expression for water potential of plant cells in terms of measurable parameters:

$$\psi = \varepsilon(1 - \alpha)(R_w - R_0) + [\psi_{\pi_1}(1 - B_1)]/[(1 - \alpha)R_w - B_t + \alpha] \quad (9)$$

where ε is bulk elastic modulus (measure of elasticity or stiffness of the cell wall) of the tissue (in MPa), α is a ratio of the change in apoplastic water content to the change in relative water content over a fixed range of apoplastic water content, R_w is the relative water content of plant tissue (water content/water content at full turgor), R_0 is the relative water content at incipient plasmolysis, ψ_{π_1} is the osmotic potential at full turgor, and B_t is the apoplastic water content at full turgor. Estimates of apoplastic water were made by constructing a moisture-release curve with frozen tissue, and assuming the water content of the matrix in intact tissue will equal the water content of the frozen tissue when its ψ is equal to ψ_π of the cell sap. The values for elastic modulus of the tissue over the turgid range are also determined independently by regression of measurements of leaf-water potential vs. leaf weight (Campbell *et al.*, 1979).

Equation (9) is basically the relation $\psi = \psi_p + \psi_\pi$ (actually $\psi_\pi + \psi_m$) where the first group of terms on the right-hand side of the equation is ψ_p, and the next set within the brackets is ψ_π. A plot from equation (9) of ψ, ψ_p, and ψ_π as a function of R_w for a wheat leaf corrected for apoplastic water is shown in Fig. 1.2. At near full turgor, ψ declines steeply as the tissue loses water. The drop in ψ is a result of the sharp decrease in ψ_p as the tissue deflates and its volume decreases. At incipient plasmolysis and at lower water potentials, $\psi_p = 0$ and $\psi = \psi_\pi$.

Field tests with the model showed that there was a significant response of ψ_π to environment, with leaves from a drier environment having potentials 0.5 MPa or more lower than leaves from a more moist site. There was a difference of 0.1 to 0.2 MPa in ψ_π between full and zero

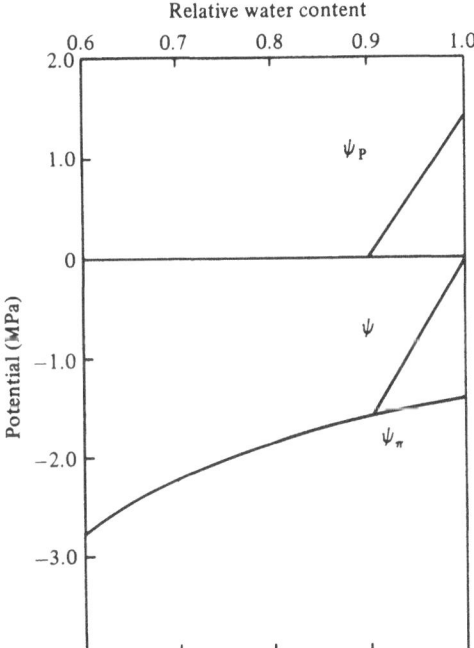

Fig. 1.2. Relation of water potential (ψ) and component osmotic (ψ_π) and turgor (ψ_p) potentials to relative water content for field-grown winter wheat. Calculated from equation (9). (Campbell *et al.*, 1979.)

turgor in leaves at the moist site, and 0.2 to 0.4 MPa in the more stressful environment. Apoplastic water and elastic modulus of the leaves were generally insensitive to environment, averaging around 0.3 relative water content, and 20 MPa, respectively.

Determination of cell volume occupied by osmotically active solutes and proportions of nonsolvent water is much more difficult with fungi than for the higher plants, because of differences in the nature of cell tissue. As with plants, it is generally assumed that the plasma membrane of fungi is the impermeable barrier and the cell wall water is low in solutes. In attempting to account for measured ψ_π on the basis of uptake of specific solutes, Luard (1982a) used electron microscopy to determine wall volume of several fungi for correction of nonsolvent water. In another paper, Luard (1982c) reported that the wall volume of *Phytophthora cinnamomii* was sensitive to sugar osmoticum. The wall volume as per cent of the total cell volume appeared to increase from 16% at 0 MPa to 40% at −3 MPa. In addition, she assumed 25% of the cytoplasmic water had properties altered by macromolecules, making it behave as nonsolvent water. Calculated cell ψ_π values of *P. cinnamomii* based on these assumptions were about 20% greater than values directly measured with thermocouple psychrometry using sucrose as the osmoticum, and over double the values obtained with KCl as the osmoticum (Luard, 1982c). This indicates possible errors with estimations of cell volume or other errors discussed later (p. 20).

Adjustment of water potential in cells

According to Harris (1981) the mechanisms by which cells readjust their ψ to regain equilibrium with an altered external environment are largely biochemical and biophysical in nature. Turgor pressure is thought to be responsible in various ways for the control of cell growth when external ψ is changed from optimum to stressful conditions (Luard & Griffin, 1981; Luard, 1982b). Turgor maintenance is a key process in cell function and is accomplished in part by osmoregulation which, among other things, involves a complex and little understood biofeedback mechanism that appears to control the cell ψ_π. It appears that this control is achieved via an effect on membrane permeability to solutes and ions (Luard, 1982b) or on the electrical properties of the cell membrane (Zimmermann, 1978). In the words of Harris (1981): 'The solute transport properties of microbial membranes and walls play a major role in determining differences in the shock and growth response of microorganisms to water potential stress.' As will be shown, permeability and electrical properties of the membrane are related.

Careful experiments by Luard & Griffin (1981), and Luard (1982a,b,c) with various fungi show that as the ψ external to fungal tissue is lowered by addition of solutes such as sucrose or glucose, or salts such as NaCl and KCl, the internal ψ_π is continually lowered. Internal ψ_π levels are always lower than external ψ_π by a magnitude of 1 to 4 MPa, the difference between the internal and external being attributable to a relatively constant, but increasingly positive, ψ_p with decreasing internal ψ_π (Fig. 1.3). These results indicate that the internal maintenance of ψ_p by water movement is related to changes in ion transport across the cell membrane, or to biochemical synthesis of solutes, or breakdown of macromolecules. The principal ions which appear to be involved in osmoregulation include Na$^+$ and K$^+$, with charge balance being maintained by the movement of Cl$^-$, or solutes such as proline, sucrose, glycerol and mannitol. When external ψ_π was controlled with glucose,

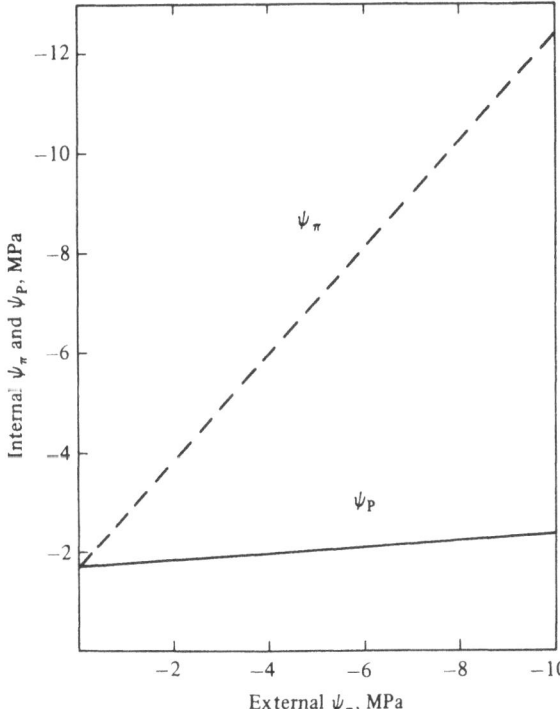

Fig. 1.3. Schematic representation of internal osmotic (ψ_π) and turgor (ψ_p) potentials of fungal cells exposed to different external water (equal to osmotic) potentials.

internal concentrations of K^+ in *Penicillium chrysogenum* were always from 5 to 7 times higher than the concentrations of Na^+, indicating preferential membrane permeability or an active carrier system to maintain the imbalance (Fig. 1.4). Where KCl was used to control ψ of the external medium, K^+ and Cl^- accumulated in the fungi linearly with decrease in ψ but Cl^- was insufficient to balance the cation content (Fig. 1.4). Internal levels of Mg^{2+} and Ca^{2+} were generally from 5 to 15 times lower than those of the Na^+, indicating their lack of transport across the cell membrane (Luard, 1982a).

Luard (1982b,c) induced sudden changes in external ψ by rapidly transferring fungal colonies to an agar medium of different osmotic potential. Hypo-osmotic shock was induced by reducing the external ψ_π, while hyperosmotic shock was induced by increasing it. During the course of the shock, internal ψ_π and ψ_p as well as ion and solute concentrations were monitored. These experiments showed that (i) the internal ψ_π was readjusted within less than 30 min to a level which maintained ψ_p at about 1 MPa, (ii) the changes in ψ_π were accompanied by similar quick changes in internal K^+ and Cl^- concentrations, (iii) the internal level of Na^+ in these specific experiments was relatively constant, and (iv) the changes in internal content of the organic solute proline were relatively slow in occurring but, over a period of hours, were substantial. These results show that osmoregulation in fungi is rapidly induced by

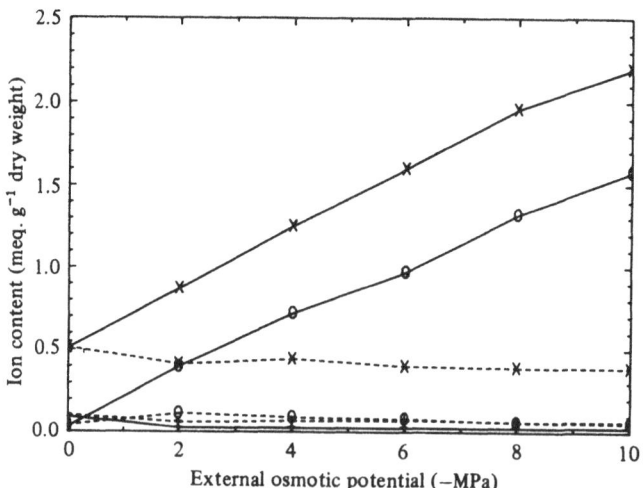

Fig. 1.4. Ion contents of *Penicillium chrysogenum* grown in different osmotic media. ×, K^+; ○, Cl^-; +, Na^+. Solid line, KCl media; broken line, glucose media. (Adapted from Luard, 1982b.)

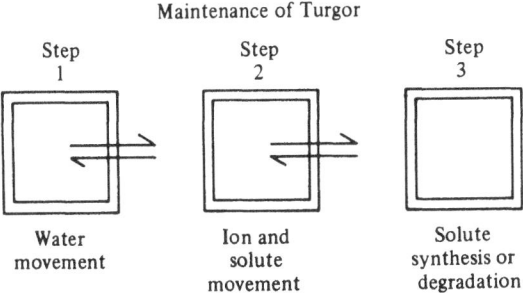

Fig. 1.5. Schematic representation of mechanisms for turgor maintenance in cells.

external changes in water potential, and that the fastest response involves a change in the internal concentration of inorganic ions; this is followed by a slower response involving biochemical synthesis of organic solutes. The net result of osmoregulation is maintenance of nearly constant ψ_p. The advantage of controlling internal ψ_π via changes in solute transport is that ψ_π can remain relatively constant with minimal changes in cell volume.

According to Zimmermann (1978) and Harris (1981), the process of osmoregulation in cells for maintenance of turgor proceeds essentially by three steps: (i) an initial rapid flow of water into or out of the cell wall in response to changes in external ψ; slower steps involving (ii) the transport of ions or solutes across the cell membrane and (iii) the biochemical synthesis or degradation of organic compounds within the cell (Fig. 1.5). The initial rapid response depends largely on the elastic properties of the cell wall governing their ability to withstand volume changes, the hydraulic conductance of the cell wall, the availability of water, and the geometry of the cell. Thus, the effect of the initial response to stress is to regulate internal ψ_π and ψ_p through volume changes of the cell. In the second and third phases, the cell readjusts the cell volume back to its original value by readjustments in the internal concentrations of solutes and ions.

Relationships among osmotic, turgor, and electrochemical potentials

The question in need of answer is, what triggers the second phase of osmoregulation? According to Zimmermann (1978), changes in ψ_p cause changes in cell membrane permeability and potential. These changes in turn cause the required amount of ion transport across the membrane in the appropriate direction. In order to understand this mechanism, it is essential to understand the electrical properties of cell membranes and the conditions governing energy equilibrium of ions on either side of the membrane.

Ions in aqueous solution across a cell membrane are at energy equilibrium if their total chemical potentials, μ_j, are equal. Since ions are charged species, their total chemical potential is affected by an electrical potential term, $z_j F \psi_E$ (a term not in Eq. (1) because water is an uncharged species), where z_j is the valence of the jth ion, F is the Faraday constant $(96487 \, J \, mol^{-1} \, volt^{-1})$, and ψ_E is the electric potential of the ion:

$$\mu_j = \mu_j^* + RT \ln a_j + V_j P + z_j F \psi_E + m_j g h. \tag{10}$$

In Eq. (10), μ_j^*, a_j, V_j, and m_j are the reference state chemical potential, activity, partial molal volume, and molecular weight of the jth ion, in that order. Normally, the gravitational and $V_j P$ terms can be neglected for ion movement across membranes and we can define the electrochemical potential of the ion as $\mu_j^* + RT \ln a_j + z_j F \psi_E$. According to Nobel (1974), the magnitude of the difference in electrical potential across a membrane is often in the order of $100 \, mV$. For a monovalent ion having a partial molal volume of $30 \, cm^3$ per mole, this potential difference is equivalent to a change in hydrostatic pressure across the membrane of $320 \, MPa$. Since ψ_p usually ranges from 1 to $4 \, MPa$ (Luard & Griffin, 1981), the energy stored in the electrical potential is far greater than that stored in the turgor component.

At equilibrium, the electrochemical potentials inside and outside the cell membrane must be equal, i.e.

$$RT \ln a_j^o + z_j F \psi_E^o = RT \ln a_j^i + z_j F \psi_E^i \tag{11}$$

where the superscripts o and i refer to outside and inside the cell membrane, and the μ_j^* term drops out because the reference chemical potentials are equal inside and outside the cell. Solving for the difference in electrical potential across the membrane yields the Nernst equation for membrane potential, $\Delta\psi$:

$$\Delta\psi = \psi_E^i - \psi_E^o = (RT/z_j F) \ln (a_j^o/a_j^i). \tag{12}$$

At $25 \, °C$, the Nernst equation (12) predicts that increasing the activity of a monovalent ion by a factor of 10 inside the membrane leads to a $59 \, mV$ increase in magnitude of the membrane potential. A plot of Eq. (12) shows that $\Delta\psi$ increases markedly for $0 < a^o/a^i < 0.1$ and becomes asymptotic thereafter (Fig. 1.6). In addition to describing the

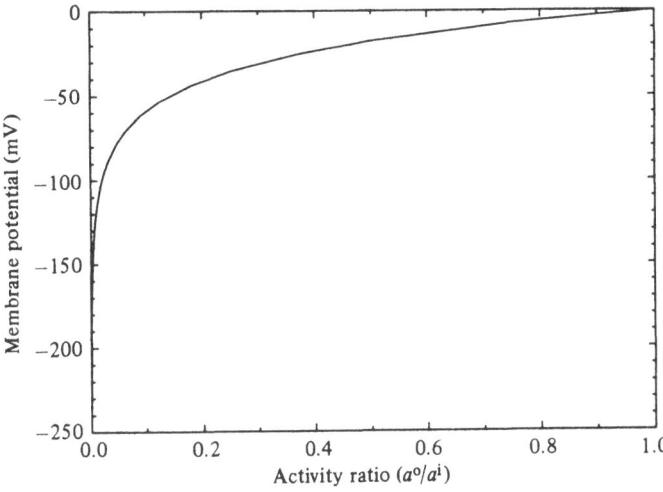

Fig. 1.6. A plot of Nernst equation (Eq. (12)) for cell membrane potential for different ratio activities of ions inside (i) and outside (o) the cell.

energy relations of ions free to move across a membrane, the Nernst equation also describes the energy difference resulting from an unequal cation distribution arising from a Donnan phase potential. The Donnan phase refers to fixed negative charges within a membrane (Baker & Hall, 1975), which are electrically balanced by a swarm of mobile cations outside the membrane. Together, the swarm of mobile cations and fixed anions constitute an electric double layer.

In general, the membrane potential is larger in magnitude (more negative) than that predicted from the Nernst equation. Ample evidence exists for fungi (Harold, 1977; Poole, 1978; Slayman, 1985) showing that this extra contribution results from an electrogenic proton-transport pump which is driven by ATP. This process involves active transport of protons out through the cell membrane in conjunction with inward flux of potassium, or simultaneous transport inward of protons and sugars. Thus, the total membrane potential, called the proton-motive potential difference across the membrane, Δp, is the sum of a Nernst membrane potential for the co-transported substance and a pH difference term (calculated as the internal pH minus the external pH):

$$\Delta p = \Delta \psi + (2.3RT/F)\Delta pH. \tag{13}$$

The latter expression for the proton-motive potential is extremely important to our discussion, as it is directly related to the amount of work that can be performed by the organism to carry out its mechanical, chemical, and osmotic functions (Harold, 1977). It is generally accepted that the storage of energy in the total membrane potential is accomplished by the hydrolysis of ATP (Harold, 1977). For *Neurospora crassa*, the reported values of the proton-motive potential, membrane potential, and pH potential terms are -240, -220, and $-20\,mV$, respectively (Harold, 1977). If, in the latter example, the energy stored in the proton-motive potential is used to drive the co-transport of protons and sugars, i.e.

$$\Delta p = (RT/F) \ln (a_{\text{sugar}^\circ}/a_{\text{sugar}^i}) + (2.3\,RT/F)\,\Delta pH \qquad (14)$$

a simple calculation shows that the ratio of internal to external sugar concentration which could be stored at room temperature using the membrane potential is about 5250:1.

Various researchers have studied the relation between membrane potential and uptake by fungi of different ions and solutes. Ohsumi & Anraku (1981, 1983) have found that basic amino acids and calcium ions are actively transported across the cell membrane of the yeast *Saccharomyces cerevisiae* in response to a membrane potential; Hauer, Uhlemann, Neumann & Hofer (1981) found that K^+ uptake was driven by an active proton pump in *Rhodotorula gracilis*, and van Brunt, Caldwell & Harold (1982) found that K^+ uptake in *Blastocladiella emersonii* was driven by a membrane potential. Hence, evidence is accumulating to show that the transport of ions and solutes across fungal cell membranes is often driven by the energy in the proton-motive potential or the membrane potential.

To answer the original question concerning the interrelationship between the cell permeability, the membrane electrical potential, and the internal ψ of the cell, consider the following analysis. For steady-state diffusion of charged ions across cell membranes having a nonzero membrane potential, the flux into the cell, J, of a cation is given by the Goldman–Hodgkin–Katz (GHK) equation (Baker & Hall, 1975; Schultz, 1980)

$$J = -\frac{PF\Delta\psi}{RT}\left[\frac{a^\circ - a^i\exp(F\Delta\psi/RT)}{1 - \exp(F\Delta\psi/RT)}\right] \qquad (15)$$

where P is the cell permeability to the cations. An equation differing from Eq. (15) in signs only:

$$J = \frac{PF\Delta\psi}{RT}\left[\frac{a^\circ - a^i\exp(-F\Delta\psi/RT)}{1 - \exp(-F\Delta\psi/RT)}\right] \qquad (16)$$

can be used to calculate the flux of anions across the membrane. Equations (15) and (16) indicate that ion fluxes and activities, cell permeabilities, and membrane potential are all intimately related. According to Zimmermann (1978), the changes ψ_p during cell stress cause either changes in cell permeability or direct changes in the membrane potential due to membrane depolarisation. In either case, the transport of ions across the cell membrane is affected and internal osmotic potential is adjusted to maintain constant turgor.

For active transport processes, which are believed to be common in fungi, the relationship between proton-motive potential and ion permeabilities has not been worked out exactly. For *Paramecium*, active transport processes appear to be sensitive to changes in turgor pressure or mechanical stimuli (Harold, 1977); in response to mechanical stimuli, the proton electrical potential is transiently depolarised (perhaps in a fashion analogous to the piezoelectric effect). Action potentials such as those observed for *Paramecium* have been induced in *Neurospora crassa* (Harold, 1977; Poole, 1978) in response to glucose uptake across the membrane. As glucose is transported inward, protons are pumped into the cell membrane. The effect of co-transport of glucose and protons is the depolarisation of the membrane potential. It is known that conditions other than mechanical stimuli depolarise the proton-motive potential. van Brunt *et al.* (1982) showed that the membrane potential of the water mould *Blastocladiella emersonii* is depolarised by addition of external K. Also, low external pH and low external Ca are known to cause depolarisation of the membrane potential (Poole, 1978). Poole (1978) discusses the possible reasons for these effects.

Measurement and control of water potential

Various methods have been developed to measure and control ψ and its components in plants, soils, microorganisms, and growth media for microorganisms. Though the individual techniques may vary, most make use of principles of equilibrium relative humidity, solute content, or a means to apply a force which directly balances the force on the water in the media being measured. Measurements are usually made in bulk samples, small enough so that the water potential can be assumed to be homogeneous throughout.

Control in the external medium

In most experiments with plants and microorganisms, control of ψ is achieved through adjustment of the ψ_π and ψ_m of the growth media. For some experiments involving the effects of fungal disease on plants grown in soil or sand, ψ_m can be controlled best by the suction or tension plate method (Cassell & Hering, 1982). This method involves attaching a hanging water column to the bottom of a soil moisture tension table, and adjusting the distance of the meniscus from the bottom of the table to the desired tension in centimetres of water (1024 cm correspond to 1 bar or 0.1 MPa). Another application of the suction method, which is used for control of ψ_m of soil in which seedling plants are grown for several weeks or less, consists of a ceramic tube secured within a Plexiglass container which is maintained water-filled and connected to a hanging water column (Howie, 1985). The ceramic tube is filled with soil, equilibrated with a desired water tension, and maintained at the given potential through equilibration with water in the reservoir surrounding the tube.

Suction methods are limited to ψ_m of -0.08 MPa or above, which are relatively high for growth of most fungi. Lower ψ_m can be controlled down to about -1.5 MPa with porous ceramic plates where air pressure instead of suction is applied to balance the soil water (Lifshitz & Hancock, 1983). Much lower potentials can be achieved using isopiestic techniques where the medium is allowed to equilibrate with the water vapour held constant above a solution of known water potential. The key concern with this technique is the uncertainty of whether moisture equilibration has been achieved.

Commonly, in microbiological studies, the external ψ_π is controlled by addition of known concentrations of solutes and salts to an agar medium. Various types of solutes and salts have been used to control external ψ_π, including glycerol (Luard, 1983; Magan, Cayley & Lacey, 1984), sucrose (Stevens, Dix & Thompstone, 1983), glucose, fructose, or arabinose (Kushner, Rosenzweig & Stotzky, 1979), NaCl (Edgley & Brown, 1983), KCl (Fisher, Marasas & Toussoun, 1983; Wong, 1983; Stapper, Lyda & Jordan, 1984), as well as a wide variety of other inorganic salts (Jaffee & Zehr, 1983). (Specific solute effects are discussed later.) The conversion of solute or salt concentration to ψ_π can be calculated by using a modified Van't Hoff equation, but this method is subject to several limiting assumptions. It is more accurate to calculate ψ_π from a calibration curve and a measurement of the output from a thermocouple psychrometer as described by Papendick & Campbell (1981) or Luard & Griffin (1981).

Researchers should be aware that certain fungi respond differently to matric and osmotic-induced stresses. For example, it is not uncommon to obtain a stimulatory effect in growth as ψ_π of conventional media is decreased from near 0 MPa to some optimum usually in the range −0.5 to −1 MPa, after which further reductions in potential will result in decreased growth (Cook & Duniway, 1981). This stimulation does not occur with reductions in ψ_m. Moreover, these fungi may cease growth at ψ_m 0.5 to 1 MPa higher than their lowest ψ_π for growth. These effects need to be kept in mind in experiments where effects of ψ on fungal growth are being studied.

Measurement of potentials in tissues and cells

The pressure bomb is probably one of the most popular methods for obtaining plant ψ and is well adapted for field use (Boyer, 1967). Thermocouple psychrometry has not been well developed for *in situ* field use because of problems with temperature control. Its use for measurement of ψ of excised tissues is also questionable because of problems with equilibration periods and unknown changes in metabolic processes associated with the excision itself. The psychrometer is a useful method for obtaining cell ψ_π using frozen tissue and correcting for nonsolvent water. The method lacks sensitivity at high ψ (> -0.1 MPa) and has a sensitivity of 0.03 to 0.05 MPa in the range −0.2 to −6 MPa. The rapid-exchange psychrometric method of Campbell & Wilson (1972) has reportedly been used to measure ψ as low as −100 MPa.

Measurements of ψ of fungal hyphae or cells and their ψ_π are best made with psychrometry (see Brown & van Haveren (1972) and Slavik (1974) for methods). Measurements of equilibrium relative humidity with a thermocouple psychrometer or a condensation dewpoint hygrometer can be used to determine the total water potential of the water vapour over intact cells and, hence, the combined value of the osmotic, matric, and turgor components of the cell water potential. Techniques for direct measurement of ψ_m and ψ_π are either not available or are extremely difficult, so their values are usually estimated indirectly. The osmotic plus matric potential of the cell can be obtained psychrometrically by measurement of the equilibrium relative humidity of cell contents expressed from thawed tissue and correcting for nonsolvent water. The ψ_π of the cell can be obtained independently from measured concentration of solutes of the cell contents and Eq. (6). The ψ_m is the difference between the psychrometer reading and that determined from solute concentration of the thawed tissue. Turgor pressure is calculated as the difference between the total potential of the external medium and the psychrometer reading of the thawed tissue.

Luard & Griffin (1981) calculated ψ_p of several fungi as the difference between medium ψ (the total potential) and hyphal ψ_π. Hyphal ψ_π were determined from samples of colonies grown on Cellophane. These were placed in a psychrometer sample holder, frozen on dry ice, and allowed to thaw and equilibrate, after which the readings were corrected for nonsolvent water.

There are serious limitations to calculating the ψ_π from the measured solute concentrations and the Van't Hoff equation. First, as noted by Schultz (1980), the cell membrane is seldom strictly impermeable to solutes. The derivation of the Van't Hoff equation implicitly assumed, however, that such was the case. Hence, the solutes will always be somewhat free to move across the cell membrane, causing the effective ψ_π to be somewhat higher than the values predicted from the solute concentrations. A correction factor to the ψ_π, called the reflection coefficient, is used to account for this problem (Nobel, 1974). The value of the reflection coefficient ranges from zero when the membrane is equally permeable to solutes and water to unity when the membrane is strictly impermeable to solutes but permeable to water. Hence, the effective osmotic potential of the solution inside the cell, $\psi_\pi(\text{eff})$ is

$$\psi_\pi(\text{eff}) = -RT\Sigma_j \sigma_j c_j \tag{17}$$

where σ_j is the reflection coefficient of the jth solute.

A second problem with the Van't Hoff equation is that real solutes are usually neither ideal nor dilute, and the concentration term in Eq. (17) must be replaced by the activity of the jth ion, a_j. Since a_j is equal to the product $\gamma_j c_j$, where γ_j is the activity coefficient of the jth ion, Eq. (17) is often written as:

$$\psi_\pi(\text{eff}) = -RT\Sigma_j \phi_j c_j \tag{18}$$

where the osmotic coefficient of the jth ion, ϕ_j, is equal to the product of the reflection coefficient and activity coefficient of the jth ion. Tables are available which list values of ϕ_j for various solutes and salts (Papendick & Campbell, 1981), but it must be realised in using most of these tables that the reflection coefficients are assumed to be unity. In practice, the reflection coefficient is an important parameter to measure for a complete understanding of cellular water relations as the following example illustrates.

Consider Luard's (1982c) experiment with the fungus *Phytophthora cinnamomi*. One culture was grown on a sucrose osmoticum at an external ψ_π of -2 MPa, and a second culture was grown on a KCl osmoticum

Table 1.3. *Comparison of measured hyphal osmotic potentials with those calculated from activity coefficients (γ) with and without use of the reflection coefficient (σ). Adapted from Luard (1982c)*

	−2 MPa Sucrose			−2 MPa KCl			
	molal conc.	γ	σ	ψ_π (effective)	molal conc.	γ	σ	ψ_π (effective)
Na^+	0.131	0.93	1.0	−0.605	0.108	0.93	1.0	−0.498
K^+	0.269	0.91	0.8	−0.971	1.349	0.90	0.4	−2.409
Proline	0.402	1.00	0.5	−0.499	0.490	1.00	0.5	−0.608
Sucrose	0.420	1.00	1.0	−1.042	—	—	—	—
Calculated ψ_π(effective)				−3.12				−3.51
Measured ψ_π				−3.15				−3.43
Luard's (1982c) calculated ψ_π				−3.86				−7.74

at −2 MPa. The cell solution was analysed by thermocouple psychrometry for ψ_π and the ion concentrations of the cell, corrected for nonsolvent water, were determined analytically. Ion concentrations and activity coefficients for each ion as reported by Luard (1982c) are given in Table 1.3. If it is assumed, as Luard (1982c) did, that the reflection coefficient of each ion is always unity, the calculated internal ψ_π do not compare well to the measured ψ_π. The discrepancy between calculated and measured ψ_π can be diminished by recalculating using reasonable values for the reflection coefficients. Normally, the reflection coefficient for sucrose is unity, while the reflection coefficients of amino acids average about 0.5 (Nobel, 1974). Furthermore, assuming that cell membranes are always more permeable to K^+ than Na^+, and that the extent of K^+ permeability depends on the external concentration of K^+, we will assign K^+ and Na^+ reflection coefficients of 0.8 and 1.0 in the sucrose osmoticum and 0.4 and 1.0 in the KCl osmoticum. It should be noted that, as discussed by Eamus & Jennings (Ch. 2), errors can be introduced into measurements of solute concentration because ions are released from the (extracellular) wall during extraction of cell sap. Nevertheless, using the information in Table 1.3 and Eq. (18), the calculated ψ_π approximately equals the measured ψ_π regardless of the osmoticum used to control external ψ_π. This example illustrates the advantage of directly measuring the internal ψ_π with thermocouple psychrometry, as well as the importance of the reflection coefficient in Eq. (18).

As already discussed, there is little reason to expect that ψ_π of the external agar medium, which is usually precisely known, is equal to ψ_π inside the cell. Moreover, as external ψ_π is lowered sufficiently, the growth and vigour of fungi become increasingly inhibited. For several fungal species, however, at equal external ψ_π, the inhibiting effect is always greater for salts such as KCl and NaCl than for sugars such as sucrose and glucose (Kushner *et al.*, 1979; Jaffe & Zehr, 1983; Stapper *et al.*, 1984). These results can be explained by a specific solute effect on membrane potential or cell wall permeability rather than as a strictly osmotic effect. Thus, it is important to measure internal ψ_π, and its readjustments under changing conditions of external osmotic stress, in order completely to understand cell water relations.

Measurement of membrane potential and pH gradients

Various methods are available for direct measurement of the salient biophysical factors that control osmoregulation – namely, membrane potential, the pH gradient across the membrane, and membrane permeabilities to selected cations and anions (van Brunt *et al.*, 1982). In larger fungal hyphae or cells, such as those of *Neurospora crassa* (Harold, 1977) and *Blastocladiella emersonii* (van Brunt *et al.*, 1982), direct measurement of membrane potential with microelectrodes is possible; the external osmotic potential can be altered with the electrode in place and subsequent changes in membrane potential can be recorded (Rona, Pitman, Luttge & Ball, 1980). When hyphae or cells are too small for the use of microelectrodes, the membrane potential can be estimated from the Nernst equation using a substance whose transport across the membrane occurs without mediation by a carrier or a specific transport channel. In experiments with the yeast *Rhodotorula gracilis*, Hauer *et al.* (1981) and Hofer, Huh & Kunemund (1983) monitored the decrease in external concentration of triphenylphosphonium cation (TPP^{2+}) with liquid scintillation in order to calculate the internal concentration of TPP^{2+}. Then the membrane potential was calculated using the Nernst equation and the measured internal and external concentrations of TPP^{2+}. A disadvantage of this technique is that an assumption must be made concerning the water content of the cell. Kakinuma, Ohsumi & Anraku (1981) have measured the membrane potential in a similar approach using the Nernst equation and the partition ratio of the radioisotope [^{14}C]KSCN.

Measurement of the pH gradient across the cell membrane by three methods has been described for *Rhodotorula gracilis* by Hauer *et al.*

(1981). The first method involves the use of a spectrophotometer to measure changes in concentration of the indicator dye bromophenol blue, which is partitioned across the cell membrane in response to the pH gradient. A second method involves repeated freezing and thawing of cells until the pH of the cell-free extract remains constant. In the third method, a small amount of nystatin is added to a 2.5% unbuffered cell suspension, and the pH is monitored until it ceases to change. Thus, a variety of methods are available for measuring both the membrane and pH components of proton-motive potential in fungal cells.

References

Baker, D. A. & Hall, J. L. (1975). Introduction and general principles. In *Ion Transport in Plant Cells and Tissues*, ed. D. A. Baker & J. L. Hall. Amsterdam: North-Holland Publishing Co.

Boyer, J. S. (1967). Leaf water potential measured with pressure chamber. *Plant Physiology*, **42**, 133–7.

Brown, R. W. & van Haveren, B. P. (1972). *Psychrometry in Water Relations Research*. Logan, Utah: Utah Agricultural Experiment Station.

Campbell, G. S., Papendick, R. I., Rabie, E. & Shayo-Ngowi, A. J. (1979). A comparison of osmotic potential, elastic modulus, and apoplastic water in leaves of dryland winter wheat. *Agronomy Journal*, **71**, 31–6.

Campbell, G. S. & Wilson, A. M. (1972). Water potential measurements of soil samples. In *Psychrometry in Water Relations Research*, ed. W. Brown & B. P. Van Haveren, pp. 142–9. Logan, Utah: Utah Agricultural Experiment Station, Utah State University.

Cassell, D. & Hering, T. F. (1982). The effect of water potential on soil-borne diseases of wheat seedlings. *Annals of Applied Biology*, **101**, 367–75.

Cook, R. J. & Duniway, J. M. (1981). Water relations in life-cycles of soilborne plant pathogens. In *Water Potential Relations in Soil Microbiology*, Soil Science Society of America Special Publication No. 9, Madison, Wisconsin: Soil Science Society of America.

Cook, R. J. & Papendick, R. I. (1978). Role of water potential in microbial growth and development of plant disease with special reference to postharvest pathology. *HortScience*, **13**, 11–16.

Edgley, M. & Brown, A. D. (1983). Yeast water relations: physiological changes induced by solute stress in *Saccharomyces cerevisiae* and *Saccharomyces rouxii*. *Journal of General Microbiology*, **129**, 3453–63.

Fisher, N. L., Marasas, W. F. O. & Toussoun, T. A. (1983). Taxonomic importance of microconidial chains in *Fusarium* section Liseola and effects of water potential on their formation. *Mycologia*, **75**, 693–8.

Harold, F. M. (1977). Ion currents and physiological functions in microorganisms. *Annual Review of Microbiology*, **31**, 181–203.

Harris, R. F. (1981). Effect of water potential on microbial growth and activity. In *Water Potential Relations in Soil Microbiology*, Soil Science Society of America Special Publication No. 9. Madison, Wisconsin: Soil Science Society of America.

Hauer, R., Uhlemann, G., Neumann, J. & Hofer, M. (1981). Proton pumps of the plasmalemma of the yeast *Rhodotorula gracilis*: their coupling to fluxes of potassium and other ions. *Biochimica et Biophysica Acta*, **649**, 680–90.

Hofer, M., Huh, H. & Kunemund, A. (1983). Membrane potential and cation permeability: a study with nystatin-resistant mutant of *Rhodotorula gracilis* (*Rhodosporidium toruloides*). *Biochimica et Biophysica Acta*, **735**, 211–14.

Howie, W. J. (1985). Factors affecting colonization of wheat roots and suppression of take-all by pseudomonads antagonistic to *Gaeumannomyces graminis* var. *tritici*. Ph.D. Thesis. Pullman: Washington State University.

Jaffee, B. A. & Zehr, E. I. (1983). Effect of certain solutes, osmotic potential, and soil solution on parasitism of *Criconemella xenoplax* by *Hirsutella rhossiliensis*. *Phytopathology*, **73**, 544–6.

Kakinuma, Y., Ohsumi, Y. & Anraku, Y. (1981). Properties of H+-translocating adenosine triphosphatase in vacuolar membranes of *Saccharomyces cerevisiae*. *Journal of Biological Chemistry*, **256**, 10859–63.

Kushner, L., Rosenzweig, W. D. & Stotzky, G. (1979). Effects of salts, sugars, and salt–sugar combinations on growth and sporulation of an isolate of *Eurotium rubrum* from pancake syrup. *Journal of Food Protection*, **41**, 706–11.

Lifshitz, R. & Hancock, J. G. (1983). Saprophytic development of *Pythium ultimum* in soil as a function of water matric potential and temperature. *Phytopathology*, **73**, 257–61.

Luard, E. J. (1982*a*). Accumulation of intracellular solutes by two filamentous fungi in response to growth at low steady-state osmotic potential. *Journal of General Microbiology*, **128**, 2563–74.

Luard, E. J. (1982*b*). Effect of osmotic shock on some intracellular solutes in two filamentous fungi. *Journal of General Microbiology*, **128**, 2575–81.

Luard, E. J. (1982*c*). Growth and accumulation of solutes by *Phytophthora cinnamomi* and other lower fungi in response to changes in external osmotic potential. *Journal of General Microbiology*, **128**, 2583–90.

Luard, E. J. (1983). The problems encountered in preparation of fungi grown at low osmotic potential for microscopy. *Transactions of the British Mycological Society*, **80**, 529–33.

Luard, E. J. & Griffin, D. M. (1981). Effect of water potential on fungal growth and turgor. *Transactions of the British Mycological Society*, **76**, 33–40.

Magan, N., Cayley, G. R. & Lacey, J. (1984). Effect of water activity and temperature on mycotoxin production by *Alternaria alternata* in culture and on wheat grain. *Applied and Environmental Microbiology*, **47**, 1113–17.

Nobel, P. S. (1974). *Introduction to Biophysical Plant Physiology*. San Francisco: W. H. Freeman and Co.

Ohsumi, Y. & Anraku, Y. (1981). Active transport of basic amino acids driven by a proton motive force in vacuolar membrane vesicles of *Saccharomyces cerevisiae*. *Journal of Biological Chemistry*, **256**, 2079–82.

Ohsumi, Y. & Anraku, Y. (1983). Calcium transport driven by a proton motive force in vacuolar membrane vesicles of *Saccharomyces cerevisiae*. *Journal of Biological Chemistry*, **258**, 5614–17.

Papendick, R. I. & Campbell, G. S. (1981). Theory and measurement of water potential. In *Water Potential Relations in Soil Microbiology*, Soil Science Society of America Special Publication No. 9. Madison, Wisconsin: Soil Science Society of America.

Poole, R. J. (1978). Energy coupling for membrane transport. *Annual Review of Plant Physiology*, **29**, 437–60.

Rona, J., Pitman, M. G., Luttge, U. & Ball, E. (1980). Electrochemical data on compartmentation into cell wall, cytoplasm, and vacuole of leaf cells in the CAM genus *Kalanchoe. Journal of Membrane Biology*, **57**, 25–35.

Schultz, S. G. (1980). *Basic Principles of Membrane Transport*. New York: Cambridge University Press.

Slavik, B. (1974). *Method of Studying Plant Water Relations*. New York: Springer-Verlag.

Slayman, C. L. (1985). Plasma membrane proton pumps in plants and fungi. *BioScience*, **35**, 34–7.

Stapper, M. F., Lyda, S. D. & Jordan, W. R. (1984). Temperature × water potential interactions on growth and sclerotial germination of *Phymatotrichum omnivorum. Phytopathology*, **74**, 509–13.

Stevens, L., Dix, N. J. & Thompstone, A. (1983). Effects of high water activity on growth and metabolism in *Aspergillus sejunctus. Transactions of the British Mycological Society*, **80**, 527–9.

van Brunt, J., Caldwell, J. H. & Harold, F. M. (1982). Circulation of potassium across the plasma membrane of *Blastocladiella emersonii*: K^+ channel. *Journal of Bacteriology*, **150**, 1449–61.

Winston, P. W. & Bates, D. H. (1960). Saturated solutions for the control of humidity in biological research. *Ecology*, **41**, 232–7.

Wong, P. T. W. (1983). Effect of osmotic potential on the growth of *Gaeumannomyces graminis* and *Phialophora* spp. *Annals of Applied Biology*, **102**, 67–78.

Zimmermann, U. (1978). Physics of turgor- and osmoregulation. *Annual Review of Plant Physiology*, **29**, 121–48.

2
Water, turgor and osmotic potentials of fungi

D. EAMUS and D. H. JENNINGS

Botany Department, University of Liverpool, Liverpool, L69 3BX, UK

There have been few investigations into the water potential (ψ) of fungal cells or mycelia and its individual components, turgor (ψ_p) and osmotic potential, sometimes referred to as solute potential, (ψ_π). There are two major reasons for the relative neglect of water relations of fungal mycelium. First, it has been assumed mostly implicitly but sometimes explicitly (see p. 29) that ψ of the medium and of the growing mycelium are the same within the closed system of the Petri dish culture. Second, there are often technical difficulties in studying the water relations of fungi. Some of these difficulties are indicated below. Nevertheless it is clear that a better understanding of mycelial water relations is very important for our understanding of mycelial functioning (Jennings, 1984*a*). It is fortunate that increasingly satisfactory approaches to fungal water relations are being developed. These approaches will be discussed in this Chapter.

Any attempt to describe water flow through, or to, a hypha requires knowledge of the driving force, and the resistance to flow imposed by the pathway itself. Since these parameters will change according to the pathway, it is necessary to state exactly the pathway under consideration.

Water and turgor potentials
Flow of water between hyphae and their environment
Theoretical considerations. In the first instance, we will not distinguish between tip and basal regions of the hyphae. Thus we will consider water fluxes as being between environment and all regions of a hypha. Nevertheless, caution must be exercised in assuming that the resistance to water flow is the same all along the length of a hypha; droplet production in apical parts of hyphae of *Serpula lacrimans* shows how resistance

to water flow may vary over very short distances (Coggins, Jennings & Clarke, 1980; Brownlee & Jennings, 1981).

For water to move from one point to another a force must exist to drive the movement, the driving force being the gradient of water potential ($\Delta\psi$) between those two points. When dealing with volume flow, the conductance of the pathway (L_p) (the inverse of resistance) must be measured. The pathway from environment to inside the hypha involves traversing the cell wall and hyphal membrane. Since the cell wall will be fully hydrated and the water in the wall in continuous contact with water in the surrounding environment, and the cell wall is porous in structure, with interstices typically of 10 nm diameter, the cell wall presents little resistance to water movement. The hydrophobic membrane, however, presents a very large activation energy barrier to water flow of $100\,\mathrm{kJ\,mol^{-1}}$ (Eamus & Wilson, 1985). It is therefore the membrane hydraulic conductance (L_p) that principally determines the magnitude of water flow. The derivation of L_p has been dealt with extensively by Zimmermann & Steudle (1978). We will deal with the experimental determinations of $\Delta\psi$ and L_p and show how they influence water fluxes, growth, and turgor.

Experimental considerations. When considering water movement from the environment (agar) into a hypha, we are dealing with the $\Delta\psi$ between these two points. The environmental ψ is open to experimental manipulation. Thus ψ of agar may be varied by the addition of solutes, thereby decreasing its ψ_π and hence ψ. This technique has been used extensively to investigate the influence of substrate ψ upon fungal growth. When making studies using agar, there is a need to remember that it can contribute appreciably to the final ψ of the substratum (Gardner, Dalton & Harris, 1972). Further, we must warn that it is absolutely necessary to measure ψ of any agar after it has been poured and set.

External water potential and linear extension. Much work has been performed to elucidate the influence of substrate ψ upon linear extension of mycelium growing on agar. Generally, there appear to be four typical relationships between ψ of the substratum and rate of linear extension (Fig. 2.1). It can be seen that as the external ψ is decreased slightly below that of the basal medium, there may or may not be a stimulation of growth. The reason for this stimulation is discussed later (see p. 36). As ψ is lowered further there is a progressive inhibition of extension.

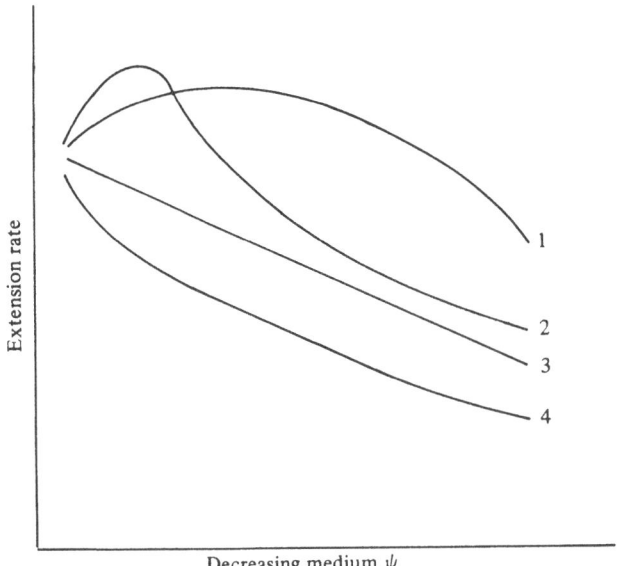

Fig. 2.1. Characteristic relationships between the extension rate of mycelium and water potential of the medium. Relationships 2, 3 and 4 have
been observed by Boddy (1983); 3 by Eamus & Jennings (1985); 1, 2 and
4 by Hocking & Pitt (1979); 1, 2 and 3 by Luard & Griffin (1981); 2 and
4 by Magan & Lacey (1984); 1 and 2 by Pitt & Hocking (1977); 3 by
Stevens, Dix & Thompstone (1983) and 2, 3 and 4 by Wilson & Griffin
(1979).

Extension and turgor. One would anticipate that reduced growth with
reduced ψ is the result of reduced turgor. Cell turgor provides the internal
force for cellular expansion. Any reduction in this driving force will
result in a decrease in the rate of cell expansion and hence growth (except
in circumstances where a change in volumetric elastic modulus occurs;
scc p. 35). Thus, if growth and turgor in fungi were found to be correlated, a possible explanation becomes apparent.

Both Adebayo, Harris & Gardner (1971) and Luard & Griffin (1981)
were unable to find any clear relationship between mycelial turgor and
growth rate. However, in both studies it was assumed wrongly that mycelial ψ was equal to the substrate ψ. From the above, it is clear that
for water uptake to occur, a gradient of ψ between agar and hypha
is necessary. Therefore hyphal ψ cannot equal external ψ. Since this
basic assumption is manifestly incorrect, the observed lack of relationship
becomes explicable. On the other hand, Eamus & Jennings (1985) found
a linear relationship between linear extension and turgor in *Serpula lacri-*

mans and *Phallus impudicus* (Fig. 2.2). Clearly from these data, it may be concluded that decreasing turgor at the hyphal apex is a major determinant of decreasing extension rate with decreasing external ψ. For higher plants such a relationship has been accepted for many years.

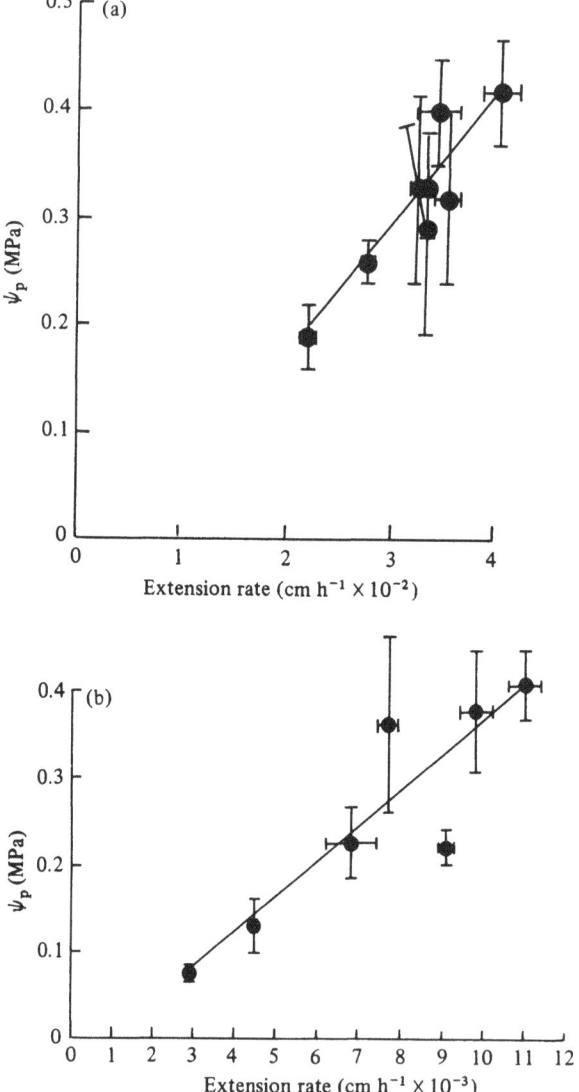

Fig. 2.2. The relationship between mycelial turgor potential and extension rate for (a) *Serpula lacrimans* and (b) *Phallus impudicus* grown on agar the water potential of which had been adjusted with sucrose (Eamus & Jennings, 1985).

There is further experimental support for this relationship from two other studies. Eamus & Jennings (1985), using a Petri dish with a central compartment, also studied the growth of *S. lacrimans* and *P. impudicus* from a medium of low ψ to one of higher potential and vice versa. Though the growth rate of *S. lacrimans* hardly changed as it moved to the different medium, *P. impudicus* could grow only for a very short distance onto a medium of low ψ from one of higher potential. Under such conditions the mycelium at the growing front exhibited virtually zero turgor (Fig. 2.3). Under all other conditions where growth rate was significant there was positive turgor. In another study of growth of *S. lacrimans*, this time from wood blocks over Perspex, growth still occurred when the mycelium was subjected to an external ψ of a magnitude to cause a flow of water out of the mycelium associated with the wood (Thompson, Eamus & Jennings, 1985). There was positive turgor at the growing mycelial front in spite of a reversal of the turgor gradient across the mycelium, such that turgor was higher at the mycelial front than at the nutrient base. Under normal conditions of growth the gradient is in the reverse direction.

Intra-hyphal flow of water and translocation
Theoretical considerations. Previously we have stated that $\Delta\psi$ is the driving force for water flow between the environment and inside the hyphae. We have also shown that by altering the external ψ it is possible to influence water uptake, turgor, and hence growth. However, within a hypha a somewhat different situation prevails. Flow of a solution, J_v, is equal to the gradient of the driving force, divided by the resistance of the pathway. Since resistance is the inverse of conductance (L_p), this can be written

$$J_v = L_p (\Delta\psi) \tag{1}$$

When considering flow across a membrane, the fact that not all solutes cross the membrane with equal ease must be taken into account. Consequently, the term 'reflection coefficient' must be incorporated. When the reflection coefficient (σ) equals one, the membrane is impermeable to the solute. A σ of zero, indicates that the membrane causes no impedance of solute movement relative to water. Typically, sucrose is often taken to have a σ of 1, and alcohols such as methanol have $\sigma = 0$. Studies on chloroplast membranes from pea (Nobel, 1974) have shown that amino acids can have a range from 0.03 (glycine) to 0.51 (leucine).

Dividing ψ into its component parts, and including σ, Eq. (1) can be rewritten as:

$$J_v = L_p \left(\Delta\psi_p - \sigma\Delta\psi_\pi\right) \qquad (2)$$

Within a hypha, movement from the tip to the base does not necessitate crossing a membrane. Consequently $\sigma = 0$ for all solutes, and the term

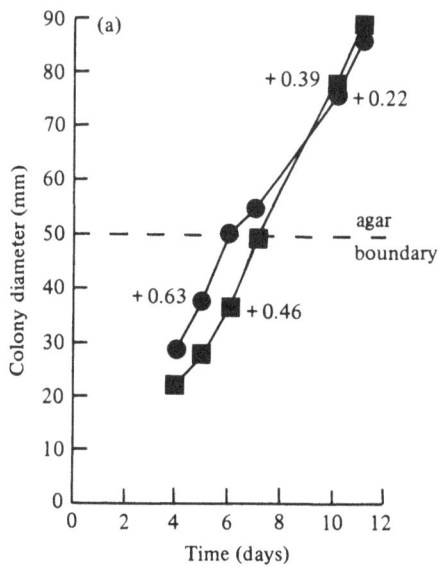

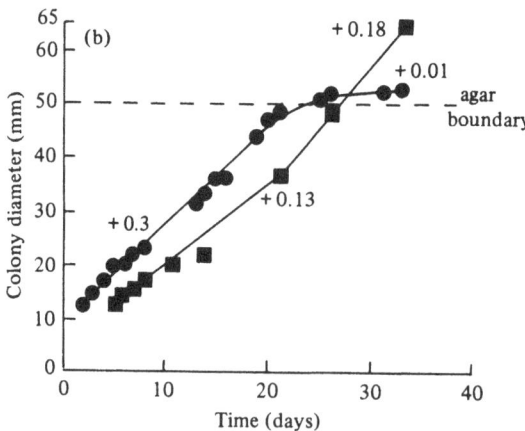

Fig. 2.3. The growth of (a) *Serpula lacrimans* and (b) *Phallus impudicus* from a sucrose-adjusted agar of water potential −2.24 MPa on to agar of −0.4 MPa (■) and vice-versa (●). Figures on the graphs indicate the turgor potentials at the mycelial front of replicate samples (from Eamus & Jennings, 1985).

$\sigma\Delta\psi_\pi$ equals zero. Thus, $J_v = L_p$ (ψ_p); it is a gradient of turgor that drives water flow along hyphae. From this it is clear that a likely mechanism for long-distance translocation in mycelium (and in particular in rhizomorphs and cords) is a turgor-driven bulk flow, as proposed first by Jennings, Thornton, Galpin & Coggins (1974), with additional evidence in favour of this mechanism reviewed by Jennings (1982, 1984a). Most of the evidence published in those reviews has been indirect. Three criteria must be fulfilled for pressure-driven flow to be an acceptable mechanism for translocation in plants (Zimmermann, 1971), namely: (i) turgor gradients between source and sink must be demonstrated, (ii) the conducting channel must be very permeable to the flow of water and solutes in a longitudinal direction, and (iii) the conducting channel must be relatively impermeable to the flow of water in a lateral direction.

Experimental considerations. The evidence for turgor gradients in fungal mycelium is as follows. Robertson & Rizvi (1968) calculated from isopiestic measurements turgor gradients of 0.5 MPa between tip and more basal regions of *Neurospora crassa*. More recently, Eamus & Jennings (1984) have shown the existence of turgor gradients in a range of structures (mycelial mats, rhizomorphs, cords), in *Armillaria mellea*, *Phallus impudicus*, *Phanerochaete velutina* and *Serpula lacrimans* under a variety of growing conditions both in the laboratory and in the field (see Table 2.1). Thompson, Eamus & Jennings (1985) also observed turgor gradients in *S. lacrimans*, and demonstrated that the addition of 2 M sucrose solution to the mycelium at the food source reversed the direction of the gradient under conditions in which translocation was inhibited. It seems likely that turgor gradients in mycelia are of very wide occurrence.

As shown above, a measure of the L_p of the pathway of water movement is needed for a fuller understanding of water flow and such a measure is also required to satisfy criterion (ii). While values for such conductances for hyphae in undifferentiated mycelia have yet to be obtained, Eamus, Thompson, Cairney & Jennings (1985) have provided values for the translocating structures, rhizomorphs and cords of respectively, *A. mellea* and *P. impudicus*; these were determined by measuring the volume flow of water through short lengths, induced by the application of known pressure gradients. This study also demonstrated the presence of wide (10–20 μm diameter) vessel hyphae running longitudinally along these structures similar to those found in mature cords of *Serpula lacrimans* (Clarke, 1979), *Phanerochaete laevis*, *P. velutina*, *Steccherinum fimbriatum*, *Tricholomopsis platyphylla* (Thompson & Rayner,

Table 2.1. Water (ψ), osmotic (ψ_π) and turgor (ψ_p) potentials (MPa) of basal and tip regions of hyphae, cords and rhizomorphs of four species of basidiomycetes, grown under different conditions (Eamus & Jennings, 1984)

Species, tissue	Growing conditions	Basal			Tip		
		ψ	ψ_s	ψ_p	ψ	ψ_s	ψ_p
Armillaria mellea							
rhizomorphs	Field grown	−2.1	−3.1	1.0	−3.1	−3.6	0.5
Phallus impudicus							
hyphae	Over agar	−1.1	−1.3	0.2	−3.3	−3.4	0.1
cords	Over agar	−1.5	−1.9	0.4	−1.6	−1.9	0.3
cords	Field grown	−0.85	−1.2	0.35	−1.7	−1.9	0.2
Phanerochaete velutina							
cords	Over agar	−0.7	−0.95	0.25	−2.1	−2.3	0.2
cords	Field grown	−0.9	−1.3	0.4	−1.6	−1.7	0.1
Serpula lacrimans							
hyphae	Over agar	−0.5	−1.3	0.8	−2.1	−2.2	0.1
cords	Over perspex	−1.0	−1.5	0.5	−1.8	−2.0	0.2

1982, 1983) and ectomycorrhizal rhizomorphs of *Suillus bovinus* (Duddridge, Malibari & Read, 1980). Significantly, the observed L_p, and those calculated on the assumption that the vessel hyphae were the major conducting channels for water flow, were very similar. Furthermore, from measurements of the osmotic potential of these structures, and assuming that trehalose is the principal translocated carbohydrate, it was possible to show that the observed L_p, and the observed volume flow for typical pressure gradients in these structures, was sufficient to account for the observed rate of growth. Further, the L_p obtained was far greater (by a factor of approximately 10^4—10^5) than typical membrane L_p values measured in algae and higher plants (Zimmermann & Steudle, 1980; Eamus & Wilson, 1985). From this it was concluded that flow in a longitudinal direction did not traverse a membrane, and that the conducting pathway is suited to longitudinal flow.

With respect to the third criterion, there is as yet no quantitative information; we have no values for the L_p of the walls and associated plasma membranes to compare with values for L_p in the longitudinal direction.

Future work. From the above, it should be clear that much more information is required concerning water flow both between mycelia and their external media and through mycelia. At the very least, data are required for fungi other than those already studied. But substantial progress will come only when mycologists consider the water relations of fungi at a level of complexity similar to that now present in the conceptual framework of those working in the field of water relations of algae and higher plants. It is now becoming clear that turgor regulation in the cells of those organisms is a major aspect of homeostasis. Thus, turgor has been shown to influence membrane hydraulic conductance directly (Zimmermann & Steudle, 1974; Zimmerman, 1978), an aspect of considerable importance to a plant in a fluctuating environment where water availability and water loss vary through space and time. Furthermore, a pressure dependency of E (volumetric elastic modulus) has been observed (Zimmermann & Steudle, 1978). E is a measure of the elasticity of the cell wall, and relates instantaneous changes in cell volume with changes in cell pressure. From the direct relationship between E and cell volume, it may be concluded that E may have a regulatory role in cell growth.

Thus, much more work is needed on the water relations of fungi. In particular, values of E and membrane L_p are required. More specifically, we need to know about differences in E and L_p along a hypha and how these are affected by time and environment. E and L_p can be measured using the micropressure probe, which requires the insertion of a capillary tip into a cell. Cells of $10\,\mu$m diameter or greater are required, which will limit the application of this technique in fungi, but hyphae of this size are to be found in many species. With respect to the homeostatic aspect of turgor, it has been recently shown that in the alga *Acetabularia mediterranea* cell turgor regulation is associated with transient changes in membrane potential and the efflux of chloride (Wendler, Zimmermann & Bentrup, 1983). Since the ionic relations of *Neurospora crassa*, in particular the membrane potential and H^+ and K^+ fluxes, have been so extensively studied by Slayman and co-workers (Slayman & Sanders, 1984), this fungus would appear to present an ideal system to investigate the role of turgor, E, and L_p in relation to ion and water fluxes.

Osmotic potential

Protoplasmic water content

Before we deal with published data concerning the nature of the solutes contributing to the osmotic potential of a fungus, there is a need to say something about methodology, namely the determination of the water content of protoplasm and the amounts of individual solutes dissolved in that water. It should be noted first that a value for the protoplasmic water content cannot be obtained simply by subtracting the dry weight from the fresh weight. The water in the wall is likely to make a significant contribution to the fresh weight but, being outside the plasma membrane, it must be excluded in any calculation of ψ_π of the protoplasm. As can be seen from Table 2.2, the contribution of wall water to the total water content of the mycelium is as high as 16.7% in *N. crassa*. For that reason, determination of the total water content by weighing should be accompanied by some procedure for determining the water in the wall or more correctly in the free space, that volume of fungus external to the plasmalemma.

Table 2.2 shows that there is considerable variation in the values for the water content of fungal mycelium. It is not clear whether the variation is related to the morphology of the different species studied, or the amount of dry matter within their mycelium, or to the different techniques used. There is need for a thorough study of the methods for determining protoplasmic water content. Nevertheless, if we accept the data in Table 2.2 at face value, it is very interesting to see, when considering the data for *Chaetomium globosum* and the marine fungus *Asteromyces cruciatus*, that with decreasing growth medium ψ brought about by salinisation, the water content first increases and then decreases. Because there is no information about the water content of the other species listed in Table 2.2 after growth on basal media, we cannot yet say how far one can generalise from results for the two fungi above. At least it is clear that, when the two marine species *Dendryphiella salina* and *Thraustochytrium aureum* are grown in media of which ψ is decreased by salinisation, the water content of the mycelium and cells is decreased with decreasing medium ψ from around $-1\,\mathrm{MPa}$.

These few data suggest that hyphae can behave rather like higher plant cells with respect to increased salinity in the external medium, namely that hyphae and cells can become more succulent with increased salinity (Jennings, 1976). The situation underlying the observed change in water content with growth on media of increasing salinity must be akin to that described by Oertli (1975) with respect to elongation of cells in media of different salinities. He postulated that solute requirements for cell elongation at constant turgor increase linearly with decreasing external ψ. Solute absorption must be such that it obeys saturation kinetics. Therefore, at zero solute supply a finite requirement has to be fulfilled with no solutes, with the elongation rate approaching zero, and at the other extreme an infinite requirement must be satisfied with a supply which is finite because of the rate of functioning of the transport system. In the latter case the elongation rate must again decrease towards zero. It follows therefore that there must be an optimum concentration of external solute at which cell elongation occurs. The changes of growth rate with decreasing medium ψ (increasing external solute concentration) described on p. 28 can be explained in a similar manner.

The data for *Penicillium chrysogenum* concerning water content of the mycelium in relation to decreasing medium ψ generated by the addition of glucose does not fit with what has been just said. That is probably due in part to the presence of a carbon substrate for growth already available in the basal medium but it is also probably due to special features of glucose metabolism which, as will be described later, prevent the sugar from having a major direct effect after entry on ψ_π.

Components of osmotic potential

Given that the protoplasmic water content can be determined, the question remains as to whether or not the concentration of individual solutes calculated from values of that water content can be validly used in estimating the total ψ_π of the protoplasm. Such calculations are made on the basis of extracts from known amounts of mycelium. The usual assumption is that all ions and molecules thus extracted had resided in the protoplasm. For organic solutes, of which polyols are a good example, this assumption is warranted but for ions, particularly cations, it is not. Thus, Wethered, Metcalf & Jennings (1985) have shown that for *Dendryphiella salina* grown in the presence of 0.8 osmol (kg water)$^{-1}$ NaCl, as much as 50% of the extracted Na was in the wall. Presumably the ion was associated with fixed negative charges, but that needs to be demonstrated. In contrast to the situation for green plant cells, there is a dearth of information about the ability of the fungal cell walls to bind ions. This situation needs to be rectified.

Information about the major components of the protoplasmic ψ_π for a variety of fungi is given in Table 2.3. With each set of percentage values for a particular species, there are also values for the protoplasmic ψ_π calculated on the basis of the values for protoplasmic solute concentrations. In the case of ions, it is only for *D. salina* that some account has been taken of the cations which might be bound to the wall. Further, the values for the contribution of individual solutes to ψ_π are for those selected for analysis; in most instances these do not represent the full

Table 2.2. *Fresh weight : dry weight ratio and mycelial and protoplasmic water contents for various fungi*

Organism Medium (water potential)	Fresh weight : dry weight	Mycelial water content $g\,g^{-1}$ d.wt	Protoplasmic water content $cm^3\,g^{-1}$ d.wt	Reference
Asteromyces cruciatus				
Basal (−0.3 MPa)	*2.53*	*1.53*		*Troke (1976)*
+0.2 M NaCl (−1.8 MPa) (1)[a]	2.82	1.82		
+1.0 M NaCl (−5.4 MPa) (1)	2.63	1.63		
+0.2 M NaCl (16)	4.53	3.53		
+0.1 M NaCl (16)	3.94	2.94		
Chaetomium globosum				
Basal (−0.3 MPa)	*3.04*	*2.04*		*Troke (1976)*
+50 mM NaCl (−1.15 MPa) (1)[a]	3.67	2.67		
+300 mM NaCl (−2.2 MPa) (1)	3.83	2.83		
+50 mM NaCl (16)	3.4	2.4		
+300 mM NaCl (16)	2.89	1.89		
Dendryphiella salina				
0.4 osmol (kg water)$^{-1}$ NaCl (−0.96 MPa)	3.8	2.8		Wethered, Metcalf & Jennings

	3.5	2.5	0.8 osmol (kg water)⁻¹ NaCl (−1.92 MPa)	
Neurospora crassa				Slayman & Tatum (1964)
Basal	4.27	3.27	2.54[b]	
Penicillium chrysogenum				Luard (1982a)
Basal	3.87	2.87		
+glucose (−2 MPa)	3.20	2.20		
+glucose (−10 MPa)	2.91	1.91		
Penicillium ochro-chloron				Gadd, Chudek, Foster & Reed (1984)
Basal			1.26[c]	
+0.5 M NaCl			0.72	
Thraustochytrium aureum				Wethered & Jennings (1985)
Basal + 50% seawater (−1.24 MPa)	3.45	2.45	2.45[d]	
+75% seawater (−1.85 MPa)	3.15	2.15	2.15	
+100% seawater (−2.46 MPa)	2.69	1.69	1.69	

[a] Number in second parentheses refers to number of culture cycles before mycelium was harvested for analysis.
[b] Free space correction determined with inulin.
[c] Free space correction determined with [14C]sorbitol.
[d] Wall-less sporangia; intercellular water corrected for in determination of mycelial water content.

Table 2.3. *Percentage contribution, where known, to the osmotic potential (ψ_π) of major classes of solute within the protoplasm of various fungi together with values for the osmotic potential either calculated from the concentrations of the known solutes or determined directly from the mycelium*

Organism Osmoticum added to basal medium (water potential)	Contribution as % ψ_π					Identity of major organic solute	ψ_π		Reference
	Carbohydrate	α-amino nitrogen	Inorganic ions	Organic acid	Major organic		Calculated	Measured (−MPa)	
Chrysosporium fastidium[b]									
Glucose (−10 MPa)	53	—	11	—	24.5	glycerol	7.6	11.85	Luard (1982a)
Dendryphiella salina[a]									
NaCl (−0.96 MPa)	28	31	41	0	14	glycerol	1.63	—	Wethered, Metcalf & Jennings (1985)
(−1.92 MPa)	27	14	59	0	15	glycerol	3.30	—	
MgCl$_2$ (−0.96 MPa)	29	20	51	0	19	mannitol	1.88	—	
(−1.92 MPa)	19	11	66	4	8	mannitol	4.32	—	
Inositol (−0.96 MPa)	26	34	32	8	14	inositol	1.58	—	
(−1.92 MPa)	33	26	33	8	23	inositol	2.30	—	
Penicillium chrysogenum[b]									
Glucose (−10 MPa)	44	—	15	—	24	glycerol	7.8	13.4	Luard (1982a)
KCl (−10 MPa)	20	—	74	—	14	glycerol	11.9	12.7	
Phytophthora cinnamomi[b]									
Sucrose (−2 MPa)	33	32	58	—	33	sucrose	3.9	3.15	Luard (1982b)
KCl (−2 MPa)	0	35	187	—	35	proline	7.7	3.43	
Thraustochytrium aureum[a]									
Seawater 50% (−1.24 MPa)	1	9	90	0	0.5	proline	1.36	—	Wethered & Jennings (1985)
75% (−1.85 MPa)	1	11	88	0	0.5	proline	1.78	—	
100% (−2.46 MPa)	2	15	83	0	6	proline	2.45	—	

[a] Percentage of calculated ψ_π.
[b] Percentage of measured ψ_π.

complement of solutes and in certain instances not necessarily all those making a major contribution to the protoplasmic ψ_π.

For *D. salina* and *T. aureum*, one can argue that estimated ψ_π are realistic. This statement can be made with some degree of certainty for *T. aureum* because the sporangia are essentially wall-less and therefore there will be little binding of cations external to the wall. Also, the protoplasmic ψ_π would be expected to be equal to the medium ψ; this was found to be the case. For *D. salina*, the calculated ψ_π is lower than the external ψ. This is necessarily so if turgor is to be generated, though we have no precise knowledge of how far the observed difference is realistic in terms of the mycelial ψ_p.

In view of the above, and the fact, pointed out by Papendick in this volume, that calculations of effective ψ_π have generally not taken into account the reflection coefficients of membranes towards different solutes, one must have doubts about the correctness of some of the data indicating the extent of the contribution of particular classes of solute to the protoplasmic ψ_π of the other fungi listed in Table 2.3. Most confidence can be placed in those for *Phytophthora cinnamomi* grown in the presence of a sucrose medium ($\sigma = 1$) with ψ -2 MPa. Greatest doubts must centre on the data for *Chrysosporium fastidium* and *Penicillium chrysogenum* grown in glucose medium of ψ -10 MPa since the calculated ψ_π is higher than both that determined directly and the medium ψ. On the other hand, the ψ_π calculated for *P. cinnamomi* grown in -10 MPa KCl medium is so much less than that determined directly as to suggest that, at least, there are errors in the determination of the protoplasmic concentrations. This last possibility was essentially acceded to by Luard (1982*a*) who suggested that a considerable proportion of the K extracted from mycelium of *P. cinnamomi* had been bound to the wall.

In spite of the fact that some of the information in Table 2.3 must be treated with caution, certain generalities emerge. A seemingly obvious point, but important nevertheless, is that solutes present in the medium in high concentrations can themselves be assimilated into the protoplasm and thus can make a substantial contribution to its ψ_π. These solutes may be inorganic or organic, such as inositol or sucrose respectively in the cases of *D. salina* and *P. cinnamomi*. The chemical nature of these solutes, including those produced metabolically such as glycerol and proline, is such that high cytoplasmic concentrations are unlikely to perturb enzyme activity greatly (Jennings, 1984*a*). On the other hand, inorganic solutes might accumulate to concentrations which, if they were cytoplasmic and if the enzymes had a sensitivity to ions similar to the sensitivities of similar enzymes in higher plant halophytes (Jennings, 1976), would have serious metabolic consequences. With *D. salina* and *T. aureum* growing in the presence of -1.92 MPa NaCl medium, the Na concentration in the protoplasm of each fungus was no more than 250 mM, which from higher plant studies may be metabolically tolerable (Jennings, 1976). On the other hand, the data for *P. chrysogenum* indicate that K could be at very high concentration in the protoplasm, even taking into account binding of the ion to the wall. Luard (1982*a*) gives a value of 2.0 molal for the mycelial concentration of K and it is hard to see the extent of wall binding reducing this value to a level which is metabolically tolerable. One must assume that significant proportions of the cation and the associated anion (probably Cl^-) are sequestered in vacuoles.

The role of vacuoles in fungal solute relations urgently needs further study. There is extensive information indicating that vacuoles in fungi are storage compartments for amino acids, particularly in yeasts (Matile, 1978), but there is very little information about whether they function in a similar manner to the vacuoles of higher plants with respect to inorganic ions. There is now good evidence for the presence of a proton-stimulated ATPase at the tonoplast of fungi (see Comerford, Spencer-Phillips & Jennings, 1985) which suggests that significant ion movements must take place across the tonoplast. Well defined vacuoles can be seen in the older parts of the mycelium of *D. salina* and tracer flux studies indicate that significant amounts of Na might be sequestered in vacuoles.

Table 2.3 indicates that a variety of organic solutes can contribute to ψ_π of fungi. This view is confirmed for *D. salina*; Table 2.4 shows the concentration of individual polyols after 48 h growth at two different osmolalities generated in four different ways. It is very striking to find that the values for total polyol concentration at any one osmolality are similar, though the proportions of individual polyols may differ. Jennings (1984*b*) has argued that polyols must be seen as compounds involved in buffering the cytoplasm with respect to its redox balance and pH, and that the extent to which one polyol is accumulated and another is depleted depends on the nutrients available in the external environment and the metabolic consequences of the absorption of such nutrients. The essential point here is that many fungi have to be opportunists and their metabolism must be capable of coping with this mode of existence. Polyols can be considered as being part of the necessary metabolic flexibility. While glycerol is clearly an important polyol in fungi, particu-

Table 2.4. *Concentration* (*mol* l^{-1}) *of polyols in mycelium of* Dendryphiella salina *after 48 h growth in media containing various osmotica. The media were initially buffered to pH 7.2 with 50 mM Tris/HCl. Each result is the mean of four replicates* (*Wethered, Metcalf & Jennings, 1985*)

External osmoticum	Sodium chloride		Magnesium chloride		Sodium sulphate		Inositol	
osmol kg^{-1} H$_2$O	0.4	0.8	0.4	0.8	0.4	0.8	0.4	0.8
Glycerol	0.098	0.197	0.038	0.121	0.039	0.062	–	–
Erythritol	–	0.006	0.006	–	0.002	0.002	–	–
Arabitol	0.007	0.048	0.034	0.065	0.033	0.051	–	0.013
Mannitol	0.086	0.102	0.142	0.128	0.126	0.129	0.082	0.082
Inositol	–	–	–	–	–	–	0.091	0.214
Total	0.191	0.353	0.220	0.314	0.200	0.244	0.173	0.309

– = not present.

larly in certain fungi under certain conditions (Brown, 1978), we must be cautious about placing too central a role on this particular polyol with respect to the generation of low ψ_π in stressful environments.

Turgor regulation

We have emphasised already that a growing hypha needs to generate an appropriate turgor for extension to continue and it is almost certain that turgor is regulated. It is likely that the process of regulation will be such as to respond only to small changes of turgor. It is well known from experiments involving abrupt decreases in the external ψ bringing about an increased flow of water into hyphae, that too great ψ_p can lead to bursting of hyphal tips. It seems that turgor changes of large amplitude are of sufficient magnitude to override any regulation in the apical compartment that might be brought about via changes in ψ_π. Under such circumstances bursting of tips is unavoidable. However, it seems likely in septate fungi that loss of cytoplasm from the damaged hypha is reduced by rapid occlusion. In the case of *Neurospora crassa* this is brought about by hexagonal crystals (Trinci & Collinge, 1974). It seems probable that Woronin bodies will act in a similar manner in other fungi (Reichle & Alexander, 1965). Once the septal pore is plugged, growth can recommence either by the production of side branches just below the septa, or by the production of branches from lateral walls, or by the formation of intra-hyphal branches from the plugged septa themselves (Trinci & Collinge, 1974).

If we turn to those situations where there might be changes in $\Delta\psi$ between the protoplasm and the external medium of a magnitude not great enough to cause bursting of the hyphal tips, the question which has to be answered is, how does turgor regulation take place? At the moment, we have little idea; there is little information available to match that for green plants (Gutknecht, Hastings & Bisson, 1978; Zimmermann, 1978). Essentially, regulation of turgor in plants is via regulation of transport processes across the tonoplast and the outer membrane. Though there is a considerable amount of information about how particular transport processes might be regulated in fungi, e.g. H^+ extrusion and K^+ influx in *N. crassa* (Slayman & Slayman, 1970; Gradman & Slayman, 1975; Gradman, Hansen & Slayman, 1982; Slayman & Sanders, 1984) and amino acid transport in general (Eddy, 1980, 1982; Wolfinbarger, 1980), there is little empirical information about the relationship between transport of solutes across the plasmalemma and either concentrations of solutes in the protoplasm or, more particularly, turgor regulation.

Finally, the maintenance of an appropriate ψ_π within the protoplasm of a fungus depends not only on transport processes across the plasmalemma but also upon the regulation of metabolism such that there are not large changes in the concentrations of metabolites. It has been pointed out already how, at any one osmolality of the growth medium, the total concentration of polyols within the mycelium of *D. salina* is relatively constant irrespective of solute composition of the growth medium but that the concentrations of individual polyols may differ. When non-growing mycelium accumulates the non-metabolised sugar 3-*O*-methyl glucose, while there is a stage when the soluble carbohydrate concentration rises above that initially present, the final concentration achieved is the same irrespective of the amount of the sugar absorbed (Jennings & Austin, 1973). This constancy of total carbohydrate concentration is brought about in part by inhibition of glycogen breakdown (McDermott & Jennings, 1976) and loss of low molecular weight carbon (probably three-carbon) compounds into the external medium (Metcalf & Jennings, 1982). The regulation of metabolism is not directly through turgor since the non-metabolised sugar L-sorbose does not elicit the same metabolic response (Jennings & Thornton, 1984).

References

Adebayo, A. A., Harris, R. F. & Gardner, W. R. (1971). Turgor pressure of fungal mycelia. *Transactions of the British Mycological Society*, **57**, 145–51.

Boddy, L. (1983). Effects of temperature and water potential on growth rates of wood rotting basidiomycetes. *Transactions of the British Mycological Society*, **80**, 141–9.

Brown, A. D. (1978). Compatible solutes and water stress in eukaryotic microorganisms. *Advances in Microbial Physiology*, **17**, 181–242.

Brownlee, C. & Jennings, D. H. (1981). Further observations on tear or drop formation by mycelium of *Serpula lacrimans*. *Transactions of the British Mycological Society*, **77**, 615–19.

Clarke, R. W. (1979). *A study of the Ecology and Physiology of Growth of the True Dry Rot Fungus* Serpula lacrimans (*Wulf ex Fr.*) *Schroet*. Ph.D. Thesis, University of Liverpool.

Coggins, C. R., Jennings, D. H. & Clarke, R. W. (1980). Tear or drop formation by mycelium of *Serpula lacrimans*. *Transactions of the British Mycological Society*, **75**, 63–7.

Comerford, J., Spencer-Phillips, P. T. H. & Jennings, D. H. (1985). Membrane-bound ATPase activity, the properties of which are altered by growth in saline conditions, isolated from the marine yeast *Debaryomyces hansenii*. *Transactions of the British Mycological Society*, **85**, 431–8.

Duddridge, J. A., Malibari, A. & Read, D. J. (1980). Structure and function of mycorrhizal rhizomorphs with special reference to their role in water transport. *Nature*, **287**, 834–6.

Eamus, D. & Jennings, D. H. (1984). Determination of water, solute and turgor potentials of mycelium of various basidiomycete fungi causing wood decay. *Journal of Experimental Botany*, **35**, 1782–6.

Eamus, D. & Jennings, D. H. (1985). Turgor and fungal growth – studies on the water relations of mycelia of *Serpula lacrimans* and *Phallus impudicus*. *Transactions of the British Mycological Society* (in press).

Eamus, D., Thompson, W., Cairney, J. W. G. & Jennings, D. H. (1985). Internal structure and hydraulic conductivity of basidiomycete translocating organs. *Journal of Experimental Botany*, **36**, 1110–16.

Eamus, D. & Wilson, J. M. (1985). The influence of temperature upon the water relations parameters of *Rhoeo discolor*. *Plant Science Letters* (in press).

Eddy, A. A. (1980). Some aspects of amino acid transport in yeast. In *Microorganisms and Nitrogen Sources*, ed. J. W. Payne, pp. 35–62. Chichester: John Wiley & Son Ltd.

Eddy, A. A. (1982). Mechanisms of solute transport in selected eukaryotic microorganisms. *Advances in Microbial Physiology*, **23**, 1–78.

Gadd, G. M., Chudek, J. A., Foster, R. & Reed, R. H. (1984) The osmotic responses of *Penicillium ochro-chloron*: changes in internal solute levels in response to copper and salt stress. *Journal of General Microbiology*, **130**, 1969–75.

Gardner, W. R., Dalton, F. N. & Harris, R. F. (1972). Thermocouple psychrometry for the study of water relations of soil micoorganisms. In *Psychrometry in Water Relations Research*, ed. R. W. Brown & B. P. Van Haveren, pp. 150–3. Logan, Utah: Utah Agricultural Experimental Station.

Gradmann, D., Hansen, U-P & Slayman, C. L. (1982). Reaction-kinetic analysis of current–voltage relationships for electrogenic pumps in *Neurospora* and *Acetabularia*. *Current Topics in Membranes and Transport*, **16**, 257–76.

Gradmann, D. & Slayman, C. L. (1975). Oscillations of an electrogenic pump in the plasma membrane of *Neurospora*. *Journal of Membrane Biology*, **23**, 181–212.

Gutknecht, J., Hastings, D. F. & Bisson, M. A. (1978). Ion transport and turgor pressure regulation in giant algal cells. In *Membrane Transport in Biology*, ed. G.

Giebisch, D. C. Tosteson & H. H. Ussing, pp. 125–74. New York: Springer-Verlag.

Hocking, A. D. & Pitt, J. I. (1979). Water relations of some *Penicillium* species at 25 °C. *Transactions of the British Mycological Society*, **73**, 141–5.

Jennings, D. H. (1976). The effects of sodium chloride on higher plants. *Biological Reviews*, **51**, 453–6.

Jennings, D. H. (1982). The movement of *Serpula lacrimans* from substrate to substrate over nutritionally inert surfaces. In *Decomposer Basidiomycetes: their Biology and Ecology*, ed. J. C. Frankland, J. N. Hedger & M. J. Swift, pp. 91–108. Cambridge: Cambridge University Press.

Jennings, D. H. (1984a). Water flow through mycelia. In *The Ecology and Physiology of the Fungal Mycelium*, ed. D. H. Jennings & A. D. M. Rayner, pp. 143–64. Cambridge: Cambridge University Press.

Jennings, D. H. (1984b). Polyol metabolism in fungi. *Advances in Microbial Physiology*, **25**, 149–93.

Jennings, D. H. & Austin, S. (1973). The stimulatory effect of the non-metabolised sugar 3-O-methylglucose on the conversion of mannitol and arabitol to polysaccharide and other insoluble compounds in the fungus *Dendryphiella salina*. *Journal of General Microbiology*, **75**, 287–94.

Jennings, D. H. & Thornton, J. D. (1984). Carbohydrate metabolism in the fungus *Dendryphiella salina*. VII. The effect of L-sorbose on ethanol-soluble carbohydrate. *New Phytologist*, **98**, 399–403.

Jennings, D. H., Thornton, J. D., Galpin, M. F. J. & Coggins, C. R. (1974). Translocation in fungi. In *Transport at the Cellular Level*, 28th Symposium of the Society for Experimental Biology, August 1973, ed. M. A. Sleigh & D. H. Jennings, pp. 139–56. Cambridge: Cambridge University Press.

Luard, E. J. (1982a). Accumulation of intracellular solutes by two filamentous fungi in response to growth at low steady-state osmotic potential. *Journal of General Microbiology*, **128**, 2563–74.

Luard, E. J. (1982b). Growth and accumulation of solutes by *Phytophthora cinnamomi* and other lower fungi in response to changes in external solute potential. *Journal of General Microbiology*, **128**, 583–90.

Luard, E. J. & Griffin, D. M. (1981). Effect of water potential on fungal growth and turgor. *Transactions of the British Mycological Society*, **76**, 33–40.

Magan, N. & Lacey, N. (1984). Effect of water activity, temperature and substrate on interactions between field and storage fungi. *Transactions of the British Mycological Society*, **82**, 83–93.

Matile, P. (1978). Biochemistry and function of vacuoles. *Annual Review of Plant Physiology*, **29**, 193–213.

McDermott, J. C. B. & Jennings, D. H. (1976). The relationship between uptake of glucose and 3-O-methyl glucose and soluble carbohydrate and polysaccharide in the fungus *Dendryphiella salina*. *Journal of General Microbiology*, **97**, 193–209.

Metcalf, E. & Jennings, D. H. (1982). Carbohydrate metabolism in the fungus *Dendryphiella salina*. VI. Increased loss of metabolites from mycelium brought about by 3-O-methyl glucose. *New Phytologist*, **92**, 243–9.

Nobel, P. S. (1974). Chloroplast reflection coefficients: influence of partition coefficients, carriers and membrane phase transitions. In *Membrane Transport in Plants*, ed. U. Zimmermann & J. Dainty, pp. 289–95. Berlin, Heidelberg, New York: Springer-Verlag.

Oertli, J. J. (1975). Effect of absorbable and non-absorbable solutes on elongation of barley coleoptiles. *Zeitschrift fur Pflanzenphysiologie*, **75**, 287–95.

Pitt, J. I. & Hocking, A. D. (1977). Influence of solute and hydrogen ion concentration on the water relations of some xerophilic fungi. *Journal of General Microbiology*, **101**, 35–40.

Reichle, R. E. & Alexander, J. V. (1965). Multiperforate septations, Woronin bodies and septal plugs in *Fusarium*. *Journal of Cell Biology*, **24**, 489–96.

Robertson, N. F. & Rizvi, S. R. H. (1968). Some observations on the water relations of the hyphae of *Neurospora crassa*. *Annals of Botany*, **32**, 279–91.

Slayman, C. L. & Sanders, D, (1984). Electrical kinetics of proton pumping in *Neurospora*. In *Electrogenic Transport: Fundamental Principles and Physiological Implications*, ed. M. P. Blaustein & M. Lieberman, pp. 307–22. New York: Raven Press.

Slayman, C. W. & Slayman, C. L. (1970). Potassium transport in *Neurospora*. Evidence for a multisite carrier at high pH. *Journal of General Physiology*, **55**, 758–86.

Slayman, C. W. & Tatum, E. L. (1964). Potassium transport in *Neurospora*. 1. Intracellular sodium and potassium concentrations and cation requirements for growth. *Biochemica et Biophysica Acta*, **88**, 578–92.

Stevens, L., Dix, N. J. & Thompstone, A. (1983). Effects of high water activity on growth and metabolism in *Aspergillus sejunctus*. *Transactions of the British Mycological Society*, **80**, 527–9.

Thompson, W., Eamus, D. & Jennings, D. H. (1985). Water flux through mycelium of *Serpula lacrimans*. *Transactions of the British Mycological Society*, **84**. 601–8.

Thompson, W. & Rayner, A. D. M. (1982). Structure and development of mycelial cord systems in *Phanerochaete laevis* in soil. *Transactions of the British Mycological Society*, **78**, 193–200.

Thompson, W. & Rayner, A. D. M. (1983). Extent, development and functioning of mycelial cord systems in soil. *Transactions of the British Mycological Society*, **81**, 333–45.

Trinci, A. P. J. & Collinge, A. J. (1974). Occlusion of the septal pores of damaged hyphae of *Neurospora crassa* by hexagonal crystals. *Protoplasma*, **80**, 57–67.

Troke, P. F. (1976). *Aspects of the Physiology and Biochemistry of Salt Tolerance in Fungi*. Ph.D. Thesis, University of Sussex.

Wendler, S., Zimmermann, U. & Bentrup, F.-W. (1983). Relationship between cell turgor pressure, electrical membrane potential, and chloride efflux in *Acetabularia mediterranea*. *Journal of Membrane Biology*, **72**, 75–84.

Wethered, J. M. & Jennings, D. H. (1985). The major solutes contributing to the solute potential of *Thraustochytrium aureum* and *T. roseum* after growth in media of different salinities. *Transactions of the British Mycological Society*, **85**, 439–46.

Wethered, J. M., Metcalf, E. C. & Jennings, D. H. (1985). Carbohydrate metabolism in the fungus *Dendryphiella salina*. VIII. The contribution of polyols and ions to the mycelial solute potential in relation to the external osmoticum. *New Phytologist*, **101**, 631–49.

Wilson, J. M. & Griffin, D. M. (1979). The effect of water potential on the growth of some soil basidiomycetes. *Soil Biology and Biochemistry*, **11**, 211–12.

Wolfinbarger, Jr. L. (1980). Transport and utilization of amino acids by fungi. In *Microorganisms and Nitrogen Sources*, ed. J. W. Payne, pp. 63–87. Chichester: John Wiley & Sons Ltd.

Zimmermann, M. H. (1971). Transport in the phloem. In *Trees, Structure and Function*, ed. M. H. Zimmermann & C. L. Brown, pp. 221–75. New York: Springer-Verlag.

Zimmermann, U. (1978). Physics of turgor- and osmoregulation. *Annual Review of Plant Physiology*, **29**, 121–48.

Zimmermann, U. & Steudle, E. (1974). The pressure dependence of the hydraulic conductivity, the membrane resistance and membrane potential during turgor regulation in *Valonia utricularia*. *Journal of Membrane Biology*, **16**, 331–52.

Zimmermann, U. & Steudle, E. (1978). Physical aspects of water relations of plant cells. *Advances in Botanical Research*, **6**, 45–117.

Zimmermann, U. & Steudle, E. (1980). Fundamental water relations parameters. In *Plant Membrane Transport: Current Conceptual Issues*, ed. R. M. Spanswick, W J. Lucas & J. Dainty, pp. 113–27. Amsterdam: Elsevier/North-Holland Biomedical Press.

3
Water relations of sclerotia and other infective structures

R. C. COOKE and A. M. AL-HAMDANI
Department of Botany, University of Sheffield, Sheffield, S10 2TN, UK

The important, and changing, role of water at all stages of sclerotium morphogenesis is self-evident, yet water relations have been studied largely in general terms and for the most part without the precision afforded to similar investigations on mycelia and spores. This situation is puzzling since their size, shape and ease of handling confer on sclerotia numerous advantages as experimental materials. In addition many sclerotium-forming species are serious and intractable plant pathogens for which a clearer understanding of water relations has direct relevance to the development of control measures. In this chapter attention is given to those aspects of water relations which, as far as is known, determine the effectiveness of sclerotia as survival and infection units. Some other kinds of multicellular infective structures are also considered briefly. In doing so it is not always possible or desirable to ignore the involvement of water in wider physiological processes. In view of the paucity of detailed information, those areas within which critical investigations might be rewarding are indicated.

Sclerotia
Form and function
Modes of development and structural features of sclerotia have been reviewed to the point of exhaustion (Willetts, 1972, 1978; Chet & Henis, 1975; Cooke, 1983). In brief, sclerotia arise in one of four distinct ways: from loose primordia formed through irregular branching of neighbouring hyphae; from terminal primordia produced by dense branching of a single hyphal tip or the tips of adjacent hyphae; from strand primordia built up via interweaving of parallel lateral branches of one or more leader hyphae; and from the differentiation of an exten-

49

sive, pre-existing mycelium or asexual fruiting structure. Sclerotia produced from loose primordia tend to be poorly organised but the remainder commonly possess a rind of flattened, dead, pigmented cells and, in some, additional differentiation may give rise to several tissue types constituting a distinct cortex and medulla. A notable feature of some sclerotia both with and without a rind is the presence of a prominent, extracellular polysaccharide matrix which fills the interhyphal spaces within the body of the sclerotium.

Germination can be myceliogenic, where the sclerotium gives rise to vegetative hyphae or hyphal aggregates, sporogenic, where asexual spores are produced, or carpogenic, where sexual fruit-bodies arise from the sclerotium (Coley-Smith & Cooke, 1971). These modes of germination are not mutually exclusive although most species exhibit only one of them.

Particular aspects of sclerotium function will be discussed in detail later but a few useful generalisations may be made here. Sclerotium formation facilitates maintenance of a quiescent viable state in the absence of either a suitable host or of conditions which favour saprotrophic growth. This developmental phase may either be transient or be continued for many years, with germination, if successful, resulting in either the resumption of vegetative growth or rapid dissemination via spores. Sclerotia of root-infecting fungi, some of which are good competitive saprotrophs, tend to be myceliogenic while those of parasites of aerial structures are sporogenic or carpogenic. As an infective structure, the minimum effective size of a sclerotium is determined by the volume of fungal material which is necessary to support germination. For this reason myceliogenic sclerotia are commonly very small whilst carpogenic forms are correspondingly large. It might be expected that carpogenic sclerotia, in which germination involves concerted morphogenetic processes, would exhibit a greater degree of internal differentiation than sporogenic or myceliogenic sclerotia where hyphae arise simply from individual cells. However, this is not the case and some highly differentiated forms are typically sporogenic or myceliogenic. Furthermore there seems to be no relationship between degree of differentiation or mode of germination and durability (Table 3.1).

Formation

The growth phase of sclerotia, that is the period during which there is rapid influx of materials into primordia, is often remarkably brief and is marked by copious exudation of fluid. In some fungi, notably

Table 3.1. *Survival under field conditions of sclerotia with different degrees of differentiation and modes of germination* (*based on data from Coley-Smith & Cooke, 1971; Cooke, 1983*)

	Degree of differentiation[a]	Survival period
Myceliogenic		
Phymatotrichum omnivorum	R	7–9 years
Rhizoctonia tuliparum	0	10 years
Sclerotium cepivorum	R	4 years
Sclerotium delphinii	R, D	2 years
Sporogenic		
Botrytis tulipiae	R, D	15 months
Colletotrichum coccodes	R, D	7 months
Carpogenic		
Claviceps microcephala	R, D	1 year
Sclerotinia sclerotiorum	R	>2 years
Sclerotinia trifoliorum	R	>7 years
Typhula idahoensis	R, D	6 months

[a] 0, no differentiation; R, rind present; D, differentiation into tissue types.

Claviceps species, production of sugar-rich honeydew decreases the water potential of the developing sclerotium and so maintains a flow of nutrient-rich sap towards it (Mower & Hancock, 1975). However, in other fungi the exudate is not a honeydew and the reasons for its production are still a matter for debate.

Typically such low-sugar exudates are found in *Sclerotinia* and *Sclerotium* species and contain small amounts of sugars, sugar alcohols, amino acids, organic acids, inorganic salts and enzymes. The nature and amounts of these substances fluctuate during sclerotium development, there being also some evidence for their selective movement into exudate. These observations have prompted some rather vague speculations as to the possible role of exudate constituents in maintaining internal physiological balance during sclerotium growth (Cooke, 1969; Colotelo, 1973, 1978; Christias, 1980). On reflection it seems more likely that the primary function of exudation is expulsion of water *per se* from the sclerotium interior, a process which inevitably would be accompanied by some accumulation of materials within the external fluid. The selective movement of some substances into exudate need not imply a major physiological role for them. This then raises two important, and as yet unanswered, questions as to the mechanisms of low-sugar exudation and its function.

In *Sclerotinia sclerotiorum* exudation begins as soon as the primordium is recognisable, rises to a maximum within three days and terminates after a further two, cessation coinciding with the sclerotium achieving its final dry weight (Fig. 3.1). Over the same period, labelling of exudate from sclerotia developing on ^{14}C-labelled media is initially relatively high but falls rapidly throughout the growth phase (Fig. 3.2). By contrast the ability of tissues to incorporate label increases until the fourth day, only then to decline markedly (Fig. 3.3). That label per unit volume of exudate should decrease, while both the volume of exudate and the amount of labelling in tissues are increasing, strongly indicates that water movement to the exterior does not depend on the presence of solutes within exudate. Furthermore, although quantitative data on exudate composition are severely limited, there is good reason to suppose that solute concentration is too low, and hence water potential (ψ) too high, to facilitate water movement via osmosis (Christias, 1980). It seems likely that exudation results from the generation of internal hydrostatic pressure which forces water to the sclerotium surface. Such pressure-driven

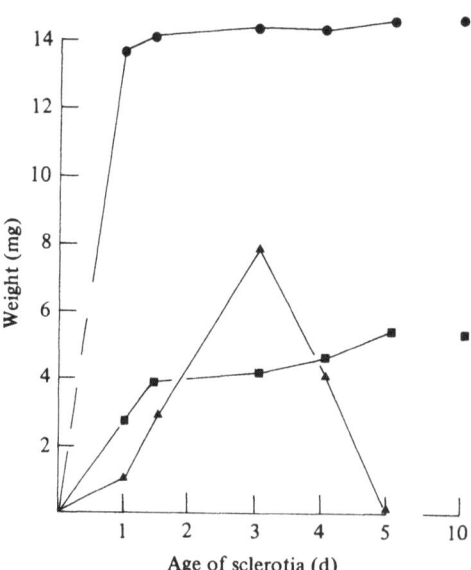

Fig. 3.1. Sclerotia of *Sclerotinia sclerotiorum*. Average fresh weight (●) exclusive of exudate, dry weight (■) and amount of exudate (▲) present at various stages of development. Based on data from Cooke (1971*a*).

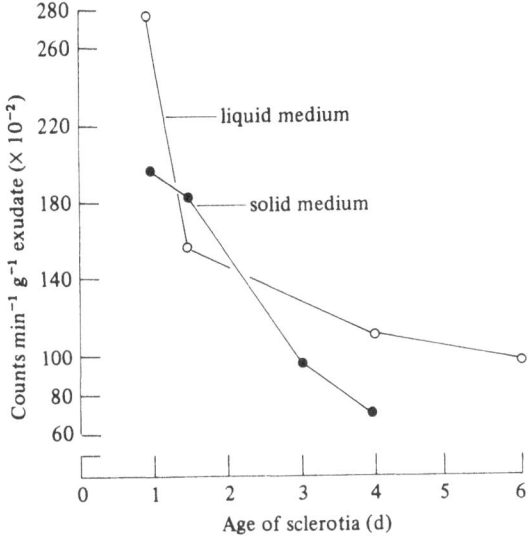

Fig. 3.2. *Sclerotinia sclerotiorum*. Amounts of radioactivity detected in exudate collected from sclerotia developing on solid or liquid media containing [^{14}C]glucose. Based on data from Cooke (1971*a*).

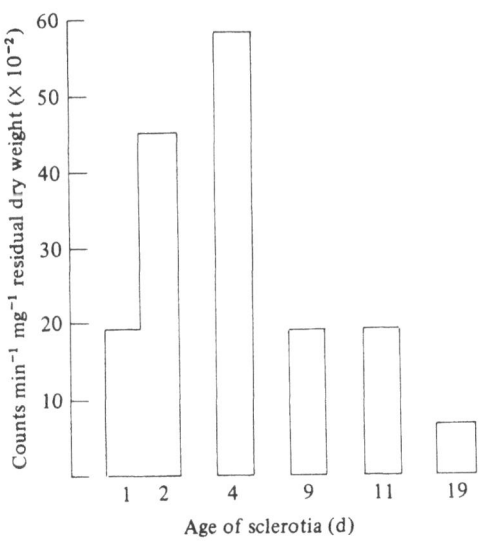

Fig. 3.3. *Sclerotinia sclerotiorum*. Levels of radioactivity in slices of sclerotia of different ages after 4 h incubation in medium containing [^{14}C]glucose. Based on data from Cooke (1971*b*).

water flow occurs in hyphae and hyphal strands, and probably in other kinds of organs, during their growth phase where it is associated with rapid translocation of materials to growth sites (Jennings, Thornton, Galpin & Coggins, 1974; Coggins, Jennings & Clarke, 1980; Brownlee & Jennings, 1981).

In addition to its probable but unproven translocatory role, exudation is linked to another important process. In sclerotia of *Sclerotinia sclerotiorum* and *Sclerotinia trifoliorum* fresh weight and dry weight increase during growth, but percentage water content of tissues steadily falls and can continue to do so long after the attainment of maximum dry weight and the disappearance of liquid from the sclerotium surface (Fig. 3.4). It would seem that a reduction in hydration, associated with deposition of insoluble intra- and extracellular reserves, is an essential preliminary to quiescence, and that it is rapidly achieved at first via exudation and later by evaporation through the rind.

Physiological studies on sclerotium formation have generally ignored the possible effects of ψ on development. This is despite the fact that, in nature, sclerotia are produced either on host tissues, where osmotic potential (ψ_π) may be of importance, or on plant residues lying upon or within soil where matric potential may have a greater influence. The

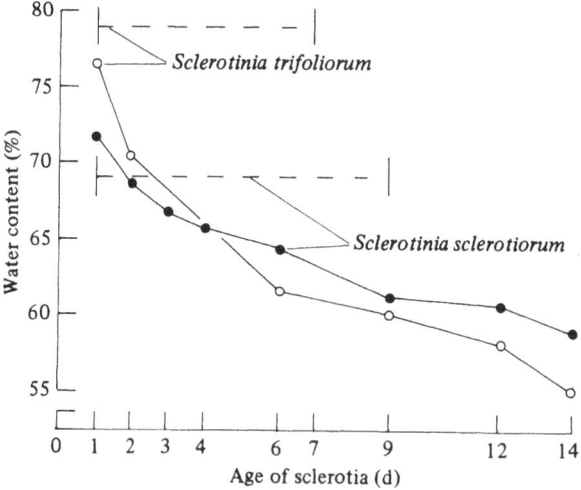

Fig. 3.4. *Sclerotinia sclerotiorum* (●) and *Sclerotinia trifoliorum* (○). Changes in water content (percentage of oven dry weight) of sclerotia during development on autoclaved carrot tissue. Broken lines indicate the period of exudation. From Cooke (1969).

Table 3.2. *Optimum and minimum water potentials (MPa) for hyphal growth and sclerotium formation (based on data from Grogan & Abawi, 1975; Shokes, Lyda & Jordan, 1977; Imolehin, Grogan & Duniway, 1980; Cook & Duniway, 1981)*

	Hyphal growth		Sclerotium formation	
	Optimum	Minimum	Optimum	Minimum
Macrophomina phaseolina	−1.7	−3.8		
Rhizoctonia solani		−4.5		
Sclerotinia minor	−1.2	−7.3	−2.4	−4.3
Sclerotinia sclerotiorum	−1.4	−10.0	−2.4	−6.4
Typhula idahoensis	−0.2	−3.0		
Typhula incarnata	−0.2	−3.0		
Verticillium dahliae	−2.0	−10.0	−2.0 to −3.2	−7.0 to −8.0

very limited data on the effects of ψ on sclerotium-forming species indicate that their requirements differ little from those of fungi at large and that, as might be expected, sclerotia are produced over a narrower range of ψ than will support hyphal extension (Table 3.2). However, details of the manner in which water availability may affect the size and number, and hence disease-inciting capacity, of sclerotia remains to be determined.

Survival

The influence of water availability on longevity has been viewed almost exclusively in the context of sclerotia lying within soil, even though this may often be inappropriate with respect of species where sclerotia persist on unburied litter. In addition most data relate to behaviour under imprecisely defined conditions, which makes it difficult either to draw valid comparisons or to reach firm conclusions. However, it would appear that sclerotia are generally extremely resistant to desiccation, a notable exception being found in *Phymatotrichum omnivorum* the sclerotia of which, when air-dried, can survive for only a few days (King & Eaton, 1934; Taubenhaus & Ezekiel, 1936). By contrast with their durability when dried, and with the possible exception of *P. omnivorum*, viability of sclerotia declines in moist soil, the fall being particularly rapid where moisture conditions fluctuate widely. This behaviour is related to the inability of sclerotia to regulate their water content and to the absence, in the majority, of a state of profound dormancy.

Agar-grown sclerotia of *Sclerotinia sclerotiorum, Sclerotinia trifoliorum, Sclerotium delphinii* and *Sclerotium rolfsii* have at maturity a water content of 60–70% (oven dry weight basis). When incubated at zero relative humidity intact sclerotia lose water steadily through the rind at approximately the same rate as sclerotia which have been bisected to expose their tissues (Fig. 3.5). Rate of water loss is inversely proportional to sclerotium size and, with the exception of *Sclerotium rolfsii*, dehydration continues until water content has been reduced to 1–3%; in *Sclerotium rolfsii* rate of loss gradually decreases to cease when water content is still 6–14%. Desiccated sclerotia rehydrate rapidly when placed in water, regaining their original water content, but not exceeding it, in a fraction of the time taken for dehydration. Oven-dried sclerotia incubated above a free water surface attain a water content of 15–21% within 24 h (Trevethick & Cooke, 1973).

Here there are clear indications that the rind does not restrict water movement either in or out of the sclerotium and that the water content of tissues is a reflection of water availability in the external *milieu*. However, in terms of water movement, response to reduced external ψ is much slower than to increased external ψ and, furthermore, partial rehydration can occur from a saturated atmosphere in the absence of direct contact with free water. The residual water remaining after prolonged incubation at zero relative humidity is presumably bound, 'absorbed' water *sensu* Griffin (1981), being contained within cellular or extracellular macromolecular components. Why the amount of this

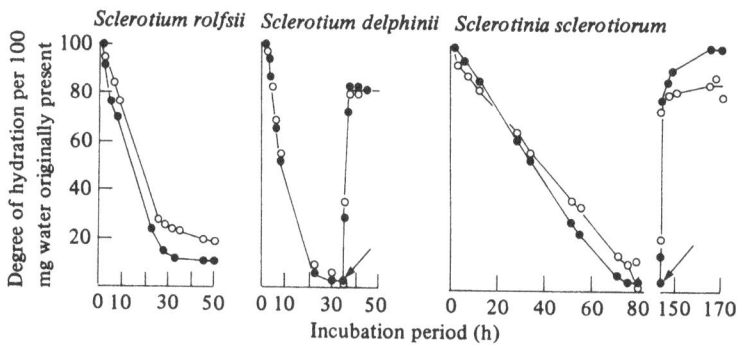

Fig. 3.5. Changes in water content of mature, intact (●) or bisected (○) sclerotia during desiccation at zero rh and rehydration in the presence of free water. Data expressed as mg water present in tissues per 100 mg of water originally present; arrows indicate points at which rehydration commenced. From Trevethick & Cooke (1973).

should be greater in *Sclerotium rolfsii* is not clear. Whilst these observations suggest a general behaviour pattern for sclerotia, they throw no light on mechanisms of water loss or uptake. A major problem is that nothing is known as to how water is compartmented within sclerotium tissues and in what sequence it might be removed from and returned to compartments during conditions of fluctuating water availability. In particular, the possible role of the extracellular polysaccharide matrix in water uptake and retention might repay examination.

When comparing the relative slowness with which sclerotia dehydrate with their ability for rapid uptake of water, it might be supposed that even quite large fluctuations in water availability would not seriously affect durability. Since sclerotia are essentially resting structures, viability of their constituent cells might be expected to be relatively resistant to changes of the magnitude normally experienced in nature. However, there is ample evidence that fluctuating conditions, and indeed the state of hydration itself, can affect cell viability and hence durability of the whole structure. This is because, with a few outstanding exceptions – for instance ergots of *Claviceps* species – sclerotia are exogenously rather than constitutively dormant during their quiescent phase (Coley-Smith & Cooke, 1971). This lack of profound dormancy suggests that metabolic rates will be relatively high during periods when water availability is potentially suitable for germination even though, for a number of other reasons, germination may not take place. A steady state of hydration will therefore lead to diminution of nutrient reserves whilst a fluctuating state will produce wasteful metabolic shifts, both conditions resulting eventually in loss of viability. Constitutively dormant sclerotia, in which germination is activated only by specific physical environmental stimuli, can be presumed to have relatively low metabolic rates and so escape these deleterious effects. A probably more serious consequence of fluctuating water availability has been found with sclerotia of *Sclerotium cepivorum*, *Sclerotium delphinii* and *Sclerotium rolfsii* in which wetting–drying cycles cause severe leakage of metabolites. This not only depletes endogenous reserves but also stimulates activity of microbial antagonists around the sclerotium (Smith, 1972*a*,*b*; Coley-Smith, Ghaffar & Javed, 1974).

Germination

Information on water relations during germination is even more fragmentary than that pertaining to morphogenesis and survival; very few species have been studied and with little uniformity of technique.

Table 3.3. *Germination behaviour in relation to water potential*

	Medium employed	Germination range or optimum (MPa)	No germination (MPa)
Myceliogenic			
Sclerotinia minor	soil	−0.03 to −15	
(Imolehin, Grogan & Duniway, 1980)			
Sclerotium rolfsii	soil, coarse loam	zero to −0.1	−1.0
(Punja & Jenkins,	soil, fine loam	−1.5	
1984)	agar	−0.25 to −3.0	−6.0
Phymatotrichum omnivorum	water agar	zero to −1.0	−3.0
(Stapper, Lyda & Jordan,	PDA	zero to −0.25	−4.0
1984)			
Carpogenic			
Sclerotinia sclerotiorum	agar	zero	−0.6
(Grogan & Abawi, 1975)			−0.05
(Duniway, Abawi &	soil	−0.008 to	
Steadman, 1977)		−0.038	
(Morrall, 1977)	soil	zero to −0.75	−1.07

A summary of published data on conditions for germination is given in Table 3.3, but it should be noted that the ranges of ψ used have commonly been either so narrow, or the steps used so great, that only approximations of optima and minima have usually been obtained. Although the ecological value of such observations is severely limited, two tentative conclusions can be drawn. First, germination takes place over a narrower range of matric potential (ψ_m) than ψ_π, ability to germinate at lower ψ_π being, perhaps, attributable to solute uptake by the sclerotium bringing about a reduction in its internal ψ_π and so allowing maintenance of germination processes. Second, requirements for carpogenic germination are more exacting than those for myceliogenic germination. This is to be expected since the former demands coordinated organogenesis whilst the latter involves independent generation of individual hyphae.

Patterns of water uptake during germination, as opposed to requirements for germination, have also been studied only a little. It seems reasonable to assume that there is a rising demand for water during this phase, the increase being coincident with mobilisation of reserves and enhanced metabolism. Structural studies on *Sclerotinia minor* indicate that utilisation of the polysaccharide matrix occurs during germination and it is possible that, as well as rapidly providing sugars to sclerotium tissues, the generation of osmotically active compounds brings about the increased hydration necessary to support germination (Bullock, Ashford & Willetts, 1980; Bullock, Willetts & Ashford, 1980). This occurs in ergots of *Claviceps purpurea* where mobilisation of lipids, and subsequent synthesis of soluble carbohydrates, results in a rapid increase in the ability of tissues to take up water which is paralleled by increased respiration rates (Fig. 3.6). An additional facet of water relations is shown by this species. In common with some other constitutively dormant sclerotia, ergots require low-temperature activation. However, this is ineffective unless sclerotium tissues are hydrated (Mitchell & Cooke, 1968a; Cunfer & Marshall, 1977). This is also the case in *Typhula erythropus* where high-temperature activation is necessary (Koske, 1975).

Finally, it should be pointed out that, as well as determining survival, pregermination conditions can influence the manner in which a sclerotium germinates when favourable conditions finally obtain. For example, when dried sclerotia of *Sclerotium rolfsii* are rehydrated they tend to germinate eruptively with all resources being diverted to producing plugs

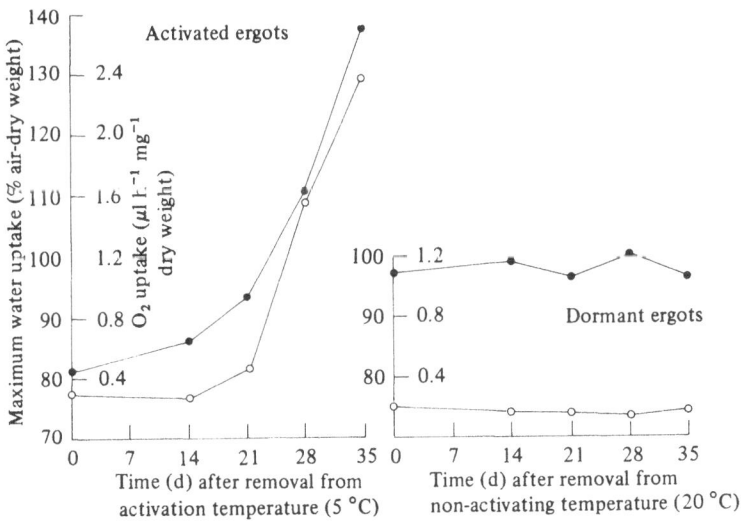

Fig. 3.6. *Claviceps purpurea.* Ability of activated and dormant ergots to take up water (●) in relation of O_2 uptake (○). (Based on data from Mitchell & Cooke, 1968b.)

or strands of aggregated hyphae (Punja & Grogan, 1981). Sclerotia germinating in this way can do so only once and failure to infect results in either elimination or the production of smaller, secondary sclerotia on the erumpent mycelium. Drying cycles in soil can thus have a far-reaching effect on the infective status of sclerotium populations quite apart from influencing survival during the quiescent phase.

Other multicellular structures

A number of other multicellular structures have roles analogous to those of sclerotia. These include acervuli, pycnidia and perithecia which may persist from season to season on dead host litter before dispersing their spores under suitable conditions, and mycelial cords and rhizomorphs which may similarly either persist or migrate to new hosts.

Pycnidial contents have been shown to have a good ability to survive desiccation, spore viability being maintained by the as yet undefined protective influence of their extracellular, polysaccharide matrix; the matrix may also facilitate rapid water absorption on the return of favourable conditions (Louis & Cooke, 1983, 1985). Dehiscence of over-wintered cleistothecia of *Sphaerotheca mors-uvae* depends on water uptake by and rapid swelling of their contained asci; failure to rupture the fruit-body wall results in degeneration of its contents (Jackson & Wheeler, 1974). Observations of this kind, whilst indicating the importance of water relations, serve to emphasise the lack of understanding of the details involved.

Little more is known concerning cords and rhizomorphs. Cords of *Phymatotrichum omnivorum* can survive at below -1.52 MPa for 3–9 months, survival rate over this period being not much higher at -0.01 or -0.03 MPa (Wheeler & Hine, 1972). This contrasts with the 7–9 year survival period of its sclerotia (Table 3.1). With respect to growth of cords and rhizomorphs through soil, although it is acknowledged that soil moisture is a crucial factor in determining distribution it has not, as yet, been considered separately from soil aeration (see Morrison, 1976).

Eco-physiological considerations

The ubiquity of sclerotial fungi attests to the success of sclerotia as perennating and infective structures which can form, survive and germinate over a remarkably wide range of conditions. For the most part, deleterious effects on their development can be demonstrated only by the imposition of severe, and ecologically unrealistic, laboratory regimes.

In the light of this it is difficult to envisage what further progress might be made through a continuation of general studies on the influence of water availability on behaviour. In this regard, and also with respect to their importance in pathogenicity, it is relevant to point out that sclerotia can form, and myceliogenic types can germinate, at ψ lower than those at which both seed germination and root development are curtailed (-1.4 to -2.0 MPa), a feature which, interestingly, they share with spores of vesicular–arbuscular mycorrhizal fungi (Tommerup, 1984). It is, perhaps, the fundamental bases for behaviour which demand attention; in particular the neglected problems as to how water is compartmented in tissues, how this is affected by ecologically realistic fluctuations in water availability and how, in turn, this influences metabolism and subsequent gross behaviour.

References

Brownlee, C. & Jennings, D. H. (1981). Further observations on tear or drop formation by mycelium of *Serpula lacrimans*. *Transactions of the British Mycological Society*, **77**, 33–40.

Bullock, S., Ashford, A. E. & Willetts, H. J. (1980). The structure and histochemistry of sclerotia of *Sclerotinia minor* Jagger. II. Histochemistry of extracellular substances and cytoplasmic reserves. *Protoplasma*, **104**, 333–51.

Bullock, S., Willetts, H. J. & Ashford, A. E. (1980). The structure and histochemistry of sclerotia of *Sclerotinia minor* Jagger. I. Light and electron microscope studies on sclerotial development. *Protoplasma*, **104**, 315–31.

Chet, I, & Henis, Y. (1975). Sclerotial morphogenesis in fungi. *Annual Review of Phytopathology*, **13**, 169–92.

Christias, C. (1980). Nature of the sclerotial exudate of *Sclerotium rolfsii* Sacc. *Soil Biology and Biochemistry*, **12**, 199–210.

Coggins, C. R., Jennings, D. H. & Clarke, R. W. (1980). Tear or drop formation by mycelium of *Serpula lacrimans*. *Transactions of the British Mycological Society*, **75**, 63–7.

Coley-Smith, J. R. & Cooke, R. C. (1971). Survival and germination of fungal sclerotia. *Annual Review of Phytopathology*, **9**, 65–92.

Coley-Smith, J. R., Ghaffar, A. & Javed, Z. U. R. (1974). The effect of dry conditions on subsequent leakage and rotting of fungal sclerotia. *Soil Biology and Biochemistry*, **6**, 307–12.

Colotelo, N. (1973). Physiological and biochemical properties of the exudate associated with developing sclerotia of *Sclerotinia sclerotiorum* (Lib.) de Bary. *Canadian Journal of Microbiology*, **19**, 73–9.

Colotelo, N. (1978). Fungal exudation. *Canadian Journal of Microbiology*, **24**, 1173–81.

Cook, R. J. & Duniway, J. M. (1981). Water relations in life-cycles of soilborne plant pathogens. In *Water Potential Relations in Soil Microbiology*, ed. J. F. Parr., W. R. Gardner & L. F. Elliott, pp. 119–40. Madison, Wisconsin: Soil Science Society of America Special Publication, **9**.

Cooke, R. C. (1969). Changes in soluble carbohydrates during sclerotium formation by *Sclerotinia sclerotiorum* and *S. trifoliorum*. *Transactions of the British Mycological Society*, **53**, 77–86.

Cooke, R. C. (1970). Physiological aspects of sclerotium growth in *Sclerotinia sclerotiorum*. *Transactions of the British Mycological Society*, **54**, 361–5.

Cooke, R. C. (1971*a*). Physiology of sclerotia of *Sclerotinia sclerotiorum* during growth and maturation. *Transactions of the British Mycological Society*, **56**, 51–9.

Cooke, R. C. (1971*b*). Uptake of [^{14}C]glucose and loss of water by sclerotia of *Sclerotinia* during development. *Transactions of the British Mycological Society*, **57**, 379–84.

Cooke, R. C. (1983). Morphogenesis of sclerotia. In *Fungal Differentiation*, ed. J. E. Smith, pp. 397–418. New York: Marcel Dekker.

Cunfer, B. M. & Marshall, D. (1977). Germination requirements of *Claviceps paspali* sclerotia. *Mycologia*, **69**, 1137–41.

Duniway, J. M., Abawi, G. S. & Steadman, J. R. (1977). Influence of soil moisture on the production of apothecia by sclerotia of *Whetzelinia sclerotiorum*. *Proceedings of the American Phytopathological Society*, **4**, 115 (*abstr.*).

Griffin, D. M. (1981). Water and microbial stress. In *Advances in Microbial Ecology*, vol. **5**, ed. M. Alexander, pp. 91–136. New York: Plenum.

Grogan, R. G. & Abawi, G. S. (1975). Influence of water potential on growth and survival of *Whetzelinia sclerotiorum*. *Phytopathology*, **65**, 122–38.

Imolehin, E. D., Grogan, R. G. & Duniway, J. M. (1980). Effect of temperature and moisture tension on growth, sclerotial production, germination and infection by *Sclerotinia minor*. *Phytopathology*, **70**, 1153–7.

Jackson, G. V. H. & Wheeler, B. E. J. (1974). Perennation of *Sphaerotheca mors-uvae* as cleistocarps. *Transactions of the British Mycological Society*, **62**, 73–87.

Jennings, D. H., Thornton, J. D., Galpin, M. F. J. & Coggins, C. R. (1974). Translocation in fungi. In *Transport at the Cellular Level, 28th Symposium of the Society for Experimental Biology*, ed. M. A. Sleigh & D. J. Jennings, pp. 139–56. Cambridge: Cambridge University Press.

King, C. J. & Eaton, E. D. (1934). Influence of soil moisture on the longevity of cotton root-rot sclerotia. *Journal of Agricultural Research*, **49**, 793–8.

Koske, R. E. (1975). *Typhula erythropus*: II. Sclerotial germination and basidiocarp production. *Mycologia*, **67**, 128–46.

Louis, I. & Cooke, R. C. (1983). Influence of the conidial matrix of *Sphaerellopsis filum* (*Darluca filum*) on spore germination. *Transactions of the British Mycological Society*, **81**, 667–70.

Louis, I. & Cooke, R. C. (1985). Conidial matrix and spore germination in some plant pathogens. *Transactions of the British Mycological Society*, **84** (in press).

Mitchell, D. T. & Cooke, R. C. (1968*a*). Some effects of temperature on germination and longevity of sclerotia in *Claviceps purpurea*. *Transactions of the British Mycological Society*, **51**, 721–9.

Mitchell, D. T. & Cooke, R. C. (1968*b*). Water uptake, respiration pattern and lipid utilization in sclerotia of *Claviceps purpurea* during germination. *Transactions of the British Mycological Society*, **51**, 731–6.

Morrall, R. A. A. (1977). A preliminary study of the influence of water potential on sclerotium germination in *Sclerotinia sclerotiorum*. *Canadian Journal of Botany*, **55**, 8–11.

Morrison, D. J. (1976). Vertical distribution of *Armillaria mellea* rhizomorphs in soil. *Transactions of the British Mycological Society*, **66**, 693–9.

Mower, R. L. & Hancock, J. G. (1975). Mechanism of honeydew formation by *Claviceps* species. *Canadian Journal of Botany*, **53**, 2826–34.

Punja, Z. K. & Grogan, R. G. (1981). Eruptive germination of sclerotia of *Sclerotium rolfsii. Phytopathology*, **71**, 1092–9.

Punja, Z. K. & Jenkins, S. F. (1984). Influence of temperature, moisture, modified gaseous atmosphere, and depth in soil on eruptive sclerotial germination of *Sclerotium rolfsii. Phytopathology*, **74**, 749–54.

Shokes, F. M., Lyda, S. D. & Jordan, W. R. (1977). Effect of water potential on the growth and survival of *Macrophomina phaseolina. Phytopathology*, **67**, 239–41.

Smith, A. M. (1972*a*). Drying and wetting sclerotia promotes biological control of *Sclerotium rolfsii* Sacc. *Soil Biology and Biochemistry*, **4**, 119–23.

Smith, A. M. (1972*b*). Biological control of fungal sclerotia in soil. *Soil Biology and Biochemistry*, **4**, 131–4.

Stapper, M. F., Lyda, S. D. & Jordan, W. R. (1984). Temperature × water potential interactions on growth and sclerotial germination of *Phymatotrichum omnivorum. Phytopathology*, **74**, 509–13.

Taubenhaus, J. J. & Ezekiel, W. N. (1936). Longevity of sclerotia of *Phymatotrichum omnivorum* in moist soil in the laboratory. *American Journal of Botany*, **23**, 10–12.

Tommerup, I. C. (1984). Effect of soil water potential on spore germination by vesicular–arbuscular mycorrhizal fungi. *Transactions of the British Mycological Society*, **83**, 193–202.

Trevethick, J. & Cooke, R. C. (1973). Water relations in sclerotia of some *Sclerotinia* and *Sclerotium* species. *Transactions of the British Mycological Society*, **60**, 555–8.

Wheeler, J. E. & Hine, R. B. (1972). Influence of soil temperature and moisture on survival and growth of strands of *Phymatotrichum omnivorum. Phytopathology*, **62**, 828–32.

Willetts, H. J. (1972). The morphogenesis and possible evolutionary origins of fungal sclerotia. *Biological Reviews*, **47**, 515–36.

Willetts, H. J. (1978). Sclerotium formation. In *The Filamentous Fungi, Developmental Mycology*, vol. **3**, ed. J. E. Smith & D. R. Berry, pp. 197–213. London: Edward Arnold.

4

Water availability and fungal reproduction: patterns of spore production, liberation and dispersal

J. LACEY

Rothamsted Experimental Station, Harpenden, Hertfordshire AL5 2JQ, UK

Water is essential to all stages of fungal reproduction but species differ both in their requirements for water at different stages of the process and in their tolerance of dry conditions. Water availability, besides determining whether sporulation can occur and how and when spores are liberated, may also affect the growth and morphology of the sporophore, possibly affecting the ability of spores to be dispersed.

Despite its importance, our knowledge of the effects of water availability on the different stages of reproduction of the majority of fungi, including some well-known pathogens, is limited. For a better appreciation of factors affecting disease development it is essential to know how water availability affects the different stages of reproduction and at what levels it is limiting.

Water availability and spore production

Anamorph production

Soil fungi. Of the soil fungi tested (Table 4.1), *Fusarium* species were the most tolerant of dry conditions with a minimum ψ for sporulation of -6 to -8 MPa. However, the optimum differed between isolates, with those from a drier region tolerating greater water stress than those from a wetter area. Some *Phytophthora* spp. can only produce sporangia at soil water potentials (ψ_{soil}) above -1.5 MPa, while others require wetting of the mycelial mat and yet others require flooding, providing oxygenation is adequate (Zentmyer & Erwin, 1970). Oxygen supply is probably also a factor inhibiting sporangial formation on buried inoculum of *P. cinnamomi*, where maximum production in soil occurred at -16 kPa with none at -1 kPa, while inoculum on the soil surface produced most sporangia when the soil was saturated or flooded (Gisi, Zent-

Table 4.1. *Minimum and optimum water requirements for spore production of some phytopathogenic fungi*

	Control method	Minimum MPa	Optimum MPa	Reference
Soil fungi				
Phytophthora cinnamoni	$\{\psi_m$	−0.25	−0.016	Gisi et al. (1980)
	ψ_m	−0.015	−0.0015 to −0.0075	Benson (1984)
P. cryptogea	ψ_m	−0.4	ND	Cook & Duniway (1980)
P. palmivora	ψ_m	−1.5	−0.001	Gisi et al. (1980)
Fusarium graminearum				Sung & Cook (1981)
F. culmorum	ψ_π	−5 to −8	−1.5 (WA)	
F. avenaceum			−0.14 (PA)	
F. culmorum (ex grain RES)	ψ_π	−11.5	ND	Magan & Lacey (1984)
Stem and Leaf pathogens				
Phytophthora infestans	rh	−7.1	0	Crosier (1934)
Peronospora destructor	rh	−14.5	0	Yarwood (1943)
P. parasitica	rh	−14.5	ND	Hartmann et al. (1982)
P. tabacina	rh	−7.1	−4.2 to 0	Cruickshank (1958)
Bremia lactucae	rh	−2.8	0	Ogilvie (1944)
Sphaerotheca fuliginea	rh	ND	−10.0 to −5.6	Hashioka (1938)
Venturia inaequalis	rh	−22.4	ND	James & Sutton (1982)
Botrytis cinerea	rh	−14.5	ND	Jarvis (1962)
Cladosporum carpophilum	rh	−30.7	−2.8 to 0	Lawrence & Zehr (1982)
Pyricularia grisea	rh	−14.5	0	Meredith (1962b)

ND, not determined; ψ_m, matric potential; ψ_π, osmotic potential; rh, relative humidity; WA, isolate from Washington State (drier climate); PA, isolates from Pennsylvania (wetter climate); RES, isolates from Rothamsted Experimental Station.

myer & Klure, 1980). Zoospore production always requires soil close to or at saturation, or free water (see Chs 6, 7).

Leaf pathogens. Most leaf-infecting fungi require high rh for sporulation (Table 4.1). Lower limits range from 90 to 98%, with optima from 97 to 100%. However, *Sphaerotheca fuliginea* has an optimum of only 93 to 96%, although other powdery mildews, e.g. *Erysiphe graminis* on barley, are favoured by high rh. Some powdery mildews, e.g. *E. cichoracearum*, are indifferent to rh (Rotem, Cohen & Bashi, 1978).

The water relations of fungi sporulating on plant surfaces may differ in several important ways from those of the same species growing in constant conditions in culture. The plant provides a source of water that is renewable and, therefore, more readily available than that in the culture medium, especially if the latter is in equilibrium with low rh. The conidiophore may be growing close to the leaf surface in a higher rh than is general in the surrounding atmosphere but it may also be subject to transient drying in air currents of low rh. Also water may form on the leaf surface from rain or from dew resulting from radiant cooling, even when the rh of the ambient air is as low as 95%. Finally, the morphology or physiology of the plant may change in response to changes (see also Dowding, Ch. 17) in water availability in soil or atmosphere.

Thus the minimum for sporulation of *Phytophthora infestans* on leaves is 95% rh but in culture is -13.0 MPa (91% rh) (Crosier, 1934). In *Peronospora tabacina*, sporulation is controlled both by the rh of the air and by the ψ of the plant tissue (Cruickshank, 1958). Little sporulation occurs at 98% rh when ψ is below -1.0 MPa and none when ψ is below -2.06 MPa. Similarly, when ψ is maintained at -0.26 MPa, sporulation decreases significantly between 97 and 96% rh and little occurs below 94% rh. Most sporulation occurs with ψ above -0.26 MPa or with rh exceeding 97%. With *Drechslera turcica*, more sporulation occurs when conidiophore-bearing leaves are placed briefly in low rh than when they are held at constant high rh (Rotem *et al.*, 1978) while sporulation even at high rh may be inhibited by air movement (Leach, 1985).

The effect of surface water varies with species and may depend in part on size of droplets or depth of water, period of cover and whether these interfere with the oxygen supply to the fungus. Thus free moisture inhibits sporulation of *Peronospora tabacina* and probably most powdery mildews, while 2–3 h of wetness promotes sporulation of *Phytophthora colocasiae*, *Mycosphaerella musicola* and *Sclerospora sorghi*; 7 h are necessary for *P. infestans*, 6–9 h for *Pseudoperonospora cubensis* and 12 h for *Drechslera maydis*. With increases of the wetting period above these minima, the numbers of spores produced increase sigmoidally. Morphological and physiological maturity do not, however, necessarily coincide. For example, sporangia of *Peronospora trifoliorum* morphologically mature after 10 h of wetness but require an additional 2 h of wetness for physiological maturity. Sometimes a wet period longer than those possible in one night is needed for sporulation and, thus, the fungus may need to utilise several separate short wet periods to bring pycnidia to maturity, e.g. *Phyllosticta maydis*, *Ascochyta pisi*, or to produce conidia, e.g. *Alternaria porri* f.sp. *solani*, *Stemphylium botryosum* f.sp. *lycopersici*, *Rhynchosporium secalis*, *Drechslera maydis* and *D. turcica* (Rotem *et al.*, 1978).

Saprotrophic fungi. Saprotrophic fungi are more varied in their water requirements for sporulation. Those from aerial plant surfaces (field fungi), e.g. *Alternaria alternata*, *Cladosporium* spp., *Fusarium culmorum*, *Verticillium lecanii*, have minima for sporulation in the range −11.5 to −14.5 MPa (90 to 92% rh), which are similar to those of leaf and stem pathogens. *Penicillium* spp. from stored products require −16.0 to −29.0 MPa while *Aspergillus* spp. require −7.1 to −34.2 MPa (Mislivec & Tuite, 1970; Pitt, 1975; Magan & Lacey, 1984). *Aspergillus fumigatus* has the greatest water requirement for sporulation and the *Aspergillus glaucus* group (anamorphs of *Eurotium* spp.) the lowest. Curran (1971) found that when ψ_π (osmotic potential) was decreased to −15 MPa conidial production was enhanced in *A. chevalieri* and *A. intermedius* but not in *A. amstelodami*, *A. repens*, *A. mangini* or *A. ruber*. However, the xerophilic *Chrysosporium fastidium* and *Monascus bisporus* can both sporulate at a much lower ψ although up to 90 d are necessary to detect this (Pitt, 1975). Time taken for sporulation often increases with decreasing water availability (Table 4.2) and differences in the limits for sporulation found by different authors may be due at least partly to differences in incubation periods. Thus Magan & Lacey (1984) recorded a minimum for sporulation of *A. glaucus* of −34.2 MPa in 40 d while Pitt (1975) recorded −39.6 MPa in 60 to 90 d.

Teleomorph formation

Water requirements for teleomorph production often differ from those for anamorph production. For example, *Phytophthora* spp. can form oospores in conditions too dry for sporangia or zoospores

Table 4.2. *Time taken (d) by different* Aspergillus *species to produce spores at different water potentials*

Species	MPa				
	0	−2.8	−7.1	−14.5	−22.4
A. aurantiobrunneus	ND	4	6	ND	ND
A. flavus	3	3	4	6	13
A. repens	3	3	4–6	4	8
A. tamarii	3	3	3	6	—
A. versicolor	4	4	4	8	—
A. wentii	4	3–4	4–5	6	13

ND, not determined.

(Sneh & McIntosh, 1974; Reeves, 1975), and flooding of agar plates is necessary for sporangial formation in *P. cactorum* but not for production of oospores (Hawker, 1957). Oospore production by *Peronospora parasitica* and *Sclerospora sorghi* may be stimulated by drying (Rotem *et al.*, 1978). By contrast, zygospores of Mucorales, such as *Sporodinia grandis*, have a higher water requirement for formation than sporangia, although the opposite may be true of *Rhizopus sexualis* (Hawker, 1957).

Formation of apothecia of such ascomycetes as *Ascophanus carneus* requires high rh. By contrast, cleistocarp production by *Magnusia* spp. may occur over the same rh range as vegetative growth, with conidial production restricted to a narrower range (Barnett & Lilley, 1958). It should be remembered that matric and osmotic effects may differ; in *Sclerotinia* apothecial production was prevented by $-0.05\,\text{MPa}\ \psi_m$ or $-0.6\,\text{MPa}\ \psi_\pi$ and was optimal at -8 to $-38\,\text{MPa}\ \psi_m$ (Grogan & Abawi, 1975; Duniway, Abawi & Steadman, 1977).

Isolates of *Emericella aurantiobrunneus* (anamorph *Aspergillus aurantiobrunneus*) from Iranian grain on malt salt and malt extract agars were quite unlike. On malt agar, colonies were restricted and buff, composed almost entirely of hulle cells and ascocarps with only a few conidiophores: those on malt salt agar were spreading, orange and totally conidial. Water relations of this species were compared with those of some other *Aspergillus* spp. and their teleomorphs, controlling water activity with glycerol (Table 4.3). Colonies of *E. aurantiobrunneus* remained essentially the same in appearance down to $-1.4\,\text{MPa}$, with ascocarps predominant and few conidiophores. At $-2.1\,\text{MPa}$ colonies were browner although ascocarps still formed. Hulle cells but not asco-

Table 4.3. *Comparison of minimum water requirements for anamorph and teleomorph production in some fungi*

Species	Anamorph, MPa	Teleomorph, MPa
Emericella aurantiobrunneus	−22.4	−2.1
E. nidulans	−22.4	−7.1
Eurotium spp.	−39.6	−35.9[a] to −20.7
Fennellia flavipes	−14.5	−7.1
Fusarium graminearum	−11.5	−5.0
Monascus bisporus	−57.1	(−55.1[b])
		(−43.3[c])

[a], immature after 120 days incubation; [b], immature after 116 days incubation; [c], mature after 90 days incubation. (Data from: Panasenko, 1967; Pitt, 1975; Sung & Cook, 1981; Lacey, unpubl.)

spores were formed at −2.8 MPa, and colonies were bright orange from conidial production. Between −2.8 and −7.1 MPa, hulle cells were increasingly replaced by large thin-walled cells which decreased in number down to −13.4 to −8.5 MPa depending on the isolate.

Similar relationships have been found with *E. nidulans* although changes in hulle cell appearance have not been found. Similar limits for cleistothecium production in *Eurotium* species were found by Curran (1971) and Pitt (1975). Whether glucose, sucrose, KCl or NaCl was used to control the osmotic potential of the medium, cleistothecia were inhibited by −45.0 MPa although conidia were still produced. However, Curran (1971) did not examine cleistothecia for mature ascospores which Pitt (1975) could not find at −35.9 MPa even after 120 d incubation.

Fennellia flavipes is the teleomorph of *Aspergillus flavipes*. Colonies on media with ψ close to 0 are characterised by bright yellow ascocarps with sausage-shaped hulle cells and sparse conidiophores. Ascocarps dominated colony appearance above −2.8 MPa, mixed with conidiophores at −7.1 MPa and failed to grow at −14.5 MPa. Conidiophores increased in number down to −14.5 MPa but there was no growth of colonies at −22.4 MPa.

No anamorph is known for *Emericella desertorum* but colonies at ψ above −2.8 MPa are characterised by mauve hulle cells and perithecia. However, at −7.1 and −14.5 MPa they became white and sterile.

A water supply from the substratum may be at least as important as the relative humidity of the air with some Basidiomycotina, especially those producing woody or leathery brackets, and with Ascomycotina

Table 4.4. *Conidiophore length of* Aspergillus aurantiobrunneus *at different water potentials (mean of 5 isolates ± standard error of mean)*

MPa	Range (μm)	Mean (μm)
0	49.6–230.0	118.9 ± 12.13
−2.8	40.6–324.7	138.9 ± 15.23
−7.1	76.7–464.5	240.3 ± 29.61
−14.5	99.2–509.6	258.0 ± 14.62

like *Daldinia concentrica* and *Xylaria* spp. Thus *Ganoderma applanatum* produces normal sporocarps from a moist substratum with good aeration at 73% rh but abnormal ones at 53 and 100% rh. The water supply for the rapid extension of the sporophore of *Phallus impudicus* may come from gelatinous tissue in the middle peridium within the subterranean 'egg' from which the stipe emerges (Ingold, 1971).

Effects of water availability on sporophore and spore size and morphology

The effect of water on size and morphology of spores and sporophores varies considerably between species and between spore types. Sporangiophores of *Mucor* elongate only in a moist atmosphere, but those of *Phycomyces blakesleeanus* develop normally in dry atmospheres provided the water supply from the substratum is adequate (Ingold, 1971). *Choanephora cucurbitarum* forms sporangia only at rh close to saturation but forms conidia at lower rh (Barnett & Lilly, 1955).

Conidiophores of *Botrytis cinerea* and *Penicillium* spp. are long and much branched with few conidia in moist atmospheres but short with many conidia in dry atmospheres (Hawker, 1957). However, the reverse would appear to be true in *Aspergillus aurantiobrunneus* in which conidiophore length increased with decreasing ψ of agar (Table 4.4; Lacey, unpubl.).

A marked effect of humidity on sporophore morphology occurs in *Cristulariella pyramidalis*, a pathogen of various plants including pecan in North America. At high rh and on wet leaves this species forms pyramidal heads of conidia. However, at 96% rh, fewer conidiophores were produced and these were often branched pinnately to form several joined pyramidal heads (Latham, 1974).

Rate of water loss may affect the morphology of some basidiomycete fruit bodies. For example, in *Polyporus brumalis*, the ratio of stipe length to cap diameter is smallest when the drying rate is high (Plunkett, 1956). In *Coprinus ephemerus* tremelloid outgrowths are produced on the pileus in a dry air stream (Keyworth, 1942).

There are few reports of any effects of water availability on spore morphology. However, Gregory (1939) found that *Ramularia vallisumbrosae* produced shorter, broader conidia with fewer septa when wet while the converse was true of other species. It is likely that drying is a major factor in the coiling of long filiform ascospores of *Lophodermium* species, *Rhytisma acerinum* and *Colpoma quercinum* (Allitt, 1979).

Spore liberation and dispersal
Spore liberation mechanisms

Before becoming airborne and dispersed a spore must separate from the sporophore, cross the laminar boundary layer surrounding the substratum and enter the turbulent boundary layer above (Gregory, 1973). Fungi have evolved a variety of means of achieving this (Table 4.5).

Dry passive methods are independent of water availability and are important in the absence of rain, often when humidity is also low. Moisture-requiring passive methods may require raindrops to fall onto or into the fruiting structure (rainsplash (see Ch. 5), drip splash, splash cup and bellows mechanisms), or result from vibration when a raindrop strikes dry vegetation, and the puff of air that precedes the resulting fast-spreading water film (tap and puff; Hirst & Stedman, 1963). Mist droplets may collect spores of some fungi with which they come into contact, while spores of aquatic fungi may become airborne in droplets liberated from splashes and bubbles.

Active mechanisms are all greatly affected by water availability because they are activated by drying or require turgid cells for operation. These are discussed individually below. Alternative mechanisms, possibly electrostatic, have been proposed for some fungi previously considered to release spores through hygroscopic movements or water rupture, e.g. *Peronospora destructor*, *Drechslera turcica* (Leach, 1980c, 1982).

Hygroscopic movements. This method of discharge is most characteristically seen in Mastigomycotina. Violent twisting movements of the conidiophores, in response to decreasing or rapidly fluctuating rh, may liberate sporangia of *Phytophthora infestans* and *Peronospora tabacina*,

Table 4.5. *Liberation mechanisms of fungus spores with examples*

Liberation mechanism	Dry		Moisture-requiring	
Passive	Shedding under gravity	Uncertain importance	Mist pick-up	*Cladosporium* spp.
	Shedding in convection currents	Uncertain importance	Rain tap and puff	*Puccinia recondita*
	Deflation	*Dictydium* sp.	Bellows mechanism	*Geastrum* spp.
			Rain splash	*Colletotrichum lindemuthianum*
	Mechanical disturbance	*Erysiphe graminis*	Drip splash	Various
	Insect transmission	Anamorph of *Claviceps purpurea*	Splash cup	*Crucibulum vulgare*
			Bubble scavenging	Aquatic fungi
			Droplet launching	Aquatic fungi
Active	Hygroscopic movements	*Phytophthora infestans*	Squirt gun mechanisms	Ascomycotina
	Water rupture	*Deightoniella torulosa*	Squirting mechanisms	*Pilobolus kleinii*
		Corynespora cassiicola	Rounding of turgid cells	*Entomophthora coronata*
		Alternaria spp.	Ballistospore discharge	Basidiomycotina
				Sporobolomyces spp.

and may also loosen conidia of some hyphomycetes, e.g. *Botrytis cinerea*, *Cladosporium* spp., *Helminthosporium* spp., from organic connection with the conidiophores before subsequent mechanical liberation or deflation (Jarvis, 1962; Meredith, 1965). Many sporangia of *P. destructor* were released as rh decreased below saturation and especially below 59%. Some were also released when rh increased and when leaves were vibrated. No sporangia were released in darkness in a saturated atmosphere, but release was stimulated by brief exposure to infrared irradiation, especially with low rh (Leach, 1982; Leach, Hildebrand & Sutton, 1982).

Water rupture. Under sufficient tension, cohesion between water molecules or their adhesion to the cell wall fails, and a gas bubble forms suddenly and enlarges rapidly. Differential thickening of walls may then result in the thinner wall being drawn in as water is lost by evaporation. The wall suddenly returns to its original position when the bubble forms and conidia are catapulted off. In *Deightoniella torulosa*, the end wall of the final compartment of the conidiophore is the thinner but in *Zygosporium oscheoides*, the concave wall of the large curved cell (falx) which bears the two sporogenous cells is thinner than the convex wall. Its curvature increases with drying and it suddenly springs back flinging off the spores (Meredith, 1961, 1962c). *Cordana musae* closely resembles *Deightoniella* except that a group of spores is liberated and a gas phase in the spores may also contribute to discharge (Meredith, 1962c). In some fungi water rupture may occur simultaneously with hygroscopic movements and in *Cladosporium caryigenum*, release first occurs as rh is decreased from 42 to 26% (Gottwald, 1982).

Spore release in *D. turcica* was triggered by decreasing rh with greater release occurring between 100 and 45% than between 100 and 85% rh. Liberation also occurred with increasing rh but was greatest when conidia were exposed to infrared irradiation at low rh (Leach, 1980a).

Squirt gun mechanisms. Squirt gun mechanisms, typical of many Ascomycotina, are characterised by the bursting of a turgid cell expelling the spores which it contained. Water is necessary to maintain turgidity of the cells, although this does not always come from the atmosphere. The force of discharge may shoot small spores a distance of 1 to 2 cm and larger spores, e.g. *Dasyobolus immersus*, to as much as 60 cm. Asci may discharge singly from perithecia, but almost simultaneously from apothecia.

Dew provides sufficient moisture for ascospore discharge in many species, e.g. *Mycosphaerella pinodes*, *Didymella exitialis* (Carter, 1963; Harries *et al.*, 1985). In others, wetting by rain is necessary to initiate release, while those activated by dew may release larger numbers of spores after rain. The ascocarps of *Venturia inaequalis* on fallen apple leaves, *M. pinodes* on pea straw and *Gaeumannomyces graminis* on wheat debris require 0.18 to 0.25 mm of rain to release ascospores while the stromatial *Eutypa armeniacae* requires at least 1.2 mm (Gregory & Stedman, 1958; Carter & Moller, 1961; Hirst & Stedman, 1962; Moller & Carter, 1965). This may be because rainfall of 0.2 mm can form a continuous film of water on apple leaves while less than 0.05 mm gives only discrete droplets.

Although wetting by rain may be necessary to stimulate ascospore discharge, spores may not be released for the whole period that the leaves are wet. Ascospore release by *G. graminis* is slow until 15 min after wetting, reaches a peak after 30 to 45 min, and is again slow after 2 h (Gregory & Stedman, 1958). Further release may not occur until the leaves have been dried and rewetted. Similarly spore liberation by *D. exitialis* from a barley crop reached a peak 4–8 h after rain but was not reactivated by further rain a few hours later (Fig. 4.1; Harries *et al.*, 1985).

Ascomycetes may draw on large or gelatinous fruit bodies for water when an external supply is lacking. *Daldinia concentrica* can discharge ascospores for up to 30 d after being detached from its substratum and placed in low rh; during this time the density of the stroma may decrease from slightly more than 1.0 to 0.3. Undisturbed, spore discharge continues for about four times as long. *Bulgaria inquinans*, with an apothecium

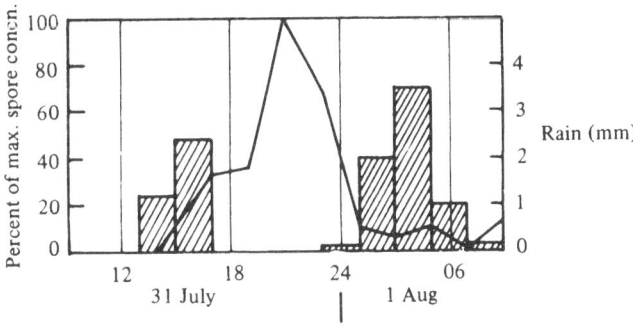

Fig. 4.1. Ascospore release by *Didymella exitialis* following rain (after Harries *et al.*, 1985).

containing much aqueous jelly, also discharges ascospores for several days in the dry, ceasing only after loss of 75% of its original weight. *Epichloe typhina* utilises water from its host. It forms a stroma in close association with shoot tissue of the grass on which it forms, and ascospore discharge ceases rapidly if the stem is cut (Ingold, 1971).

Slight drying may stimulate discharge of *Nectria galligena* and *Sordia fimicola* ascospores by causing contraction of the perithecium, placing pressure on the asci within (Ingold, 1971).

Squirting mechanisms. Squirting mechanisms operated by the bursting of a turgid cell have been described for the liberation of *Pilobolus* spp., *Basidiobolus ranarum*, *Entomophthora muscae*, *Pyricularia oryzae* and *Nigrospora sphaerica* (Webster, 1952; Ingold, 1964, 1971). In *Pilobolus*, the turgid sporangiophore ruptures along a line of weakness below the columella and contraction of the sub-sporangial bulb expels its cytoplasm in a jet forcing the sporangium away to a distance of up to 2 m, initially at $10.8\,\mathrm{m\,s^{-1}}$. Discharge in *B. ranarum* is similar but rupture occurs at the base of the bulb and the conidium is forced away 2–3 cm, by the jet from the attached cell. Discharge in *E. muscae* occurs through the bursting of the conidiophore below the conidium, while the ampulliform basal cell in *N. sphaerica* discharges a jet of cytoplasm through a small channel below the conidium, projecting it several centimetres. The whole cell below the conidium may disintegrate in *P. oryzae*. Leach (1980*b*) found spore liberation to be increased with increasing rh, especially at saturation. Decreasing rh had little effect.

Rounding of turgid cells. The sudden rounding of turgid cells has been used to explain violent spore discharge in several fungi. Commonly spore and sporophore share a common outer layer or sheath while each compartment possesses a separate inner wall. Initially the outer wall ruptures, then the inner walls round off, under the pressure of the cytoplasm, forcing the spore away. In *Entomophthora (Conidiobolus) coronata* the cross-wall bulges into the spore like a columella and it is the inner spore wall that suddenly everts causing the spore to travel up to 4 cm. Discharge ceases rapidly when rh decreases from 100 to 40% (Ingold, 1971). By contrast, spore release in *Epicoccum nigrum* and *Arthrinium cuspidatum* was stimulated by decreases in rh (Webster, 1966). Scanning electron microscopy confirms the rounding mechanism of release of *Epicoccum* conidia (Traquair & Kokko, 1981). Similar types of discharge have been reported in *Sclerospora philippinensis* and in the anamorph of *Xylo-*

sphaera furcata and proposed for discharge of rust aecidiospores (Buller, 1924). In *Uromyces* and *Puccinia* the peridium is not hygroscopic but in *Cronartium ribicola* moisture is also necessary for the aecidial blisters to open (Van Arsdel, Riker & Patton, 1956).

Ballistospore discharge. Ballistospore discharge, characteristic of most Basidiomycotina and of the Sporobolomycetaceae, remains an enigma (see Gregory, 1979) and the controversy surrounding it will not be reopened here. High rh is necessary and discharge declines rapidly when rh decreases from near 100% to less than 56% (Ingold, 1983).

There is little information on the water requirements of basidiospore release but Pearson, Seem & Meyer (1980) reported that basidiospores of *Gymnosporangium juniperi-virginiae* started to discharge a few hours after the onset of rain and ceased when rh fell below 85%.

Diurnal periodicity

In the absence of rain, the abundance of different spore types in the air show various characteristic diurnal changes. Broadly, spores may be divided between those released by day and those by night but subdivision of each category is possible into groups of spores with similar patterns of release. The distribution of spore types between groups is shown in Table 4.6 and the periodicities of selected examples shown in Fig. 4.2. The timings shown can be regarded only as general guides since the actual time of maximum release will be determined by meteorological conditions prevailing at the point of liberation.

Effects of rain on spore liberation

Rain has several successive effects on the air spora. The first drops of rain cause an initial rise in the numbers of airborne spores, particularly of *Cladosporium, Alternaria, Erysiphe* and *Puccinia*, due to spore liberation by the tap and puff mechanism. However, once the surfaces are fully wetted, splash dispersal comes into effect (see Ch. 5). Wetting will initiate ascospore liberation either during or after rain (Fig. 4.1). Thus in a banana plantation in Jamaica ascospore concentration increased from 510 before rain to 2.4×10^5 spores m^{-3} during rain (Meredith, 1962a). Continuous rain eventually washes all spores from the air.

Seasonal periodicity

Seasonal occurrence of fungus spores in the air is determined not only by climatic factors but also by the availability of substratum

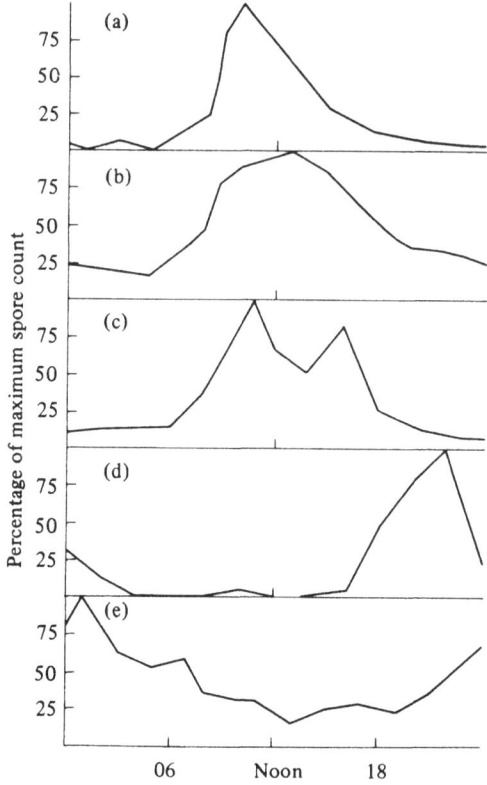

Fig. 4.2. Diurnal periodicity in spore release: (a) early morning pattern, *Nigrospora* spp.; (b) mid-day pattern, *Cladosporium* spp.; (c) double-peak pattern, *Tetraploa* spp.; (d) postdusk pattern, tetraspore type; (e) night pattern, *Pyricularia grisea*. (After Hirst, 1953; Meredith 1962*a,b*; Sreeramulu & Ramalingam, 1966; Shenoi & Ramalingam, 1975.)

or, for plant pathogens, of susceptible host tissue. In general, spores of saprotrophs occur in air over a longer season than those of pathogens since climate is usually the chief limiting factor.

In the UK conditions most favour fungal growth and sporulation in summer: peak numbers of airborne spores often occur late in the season when ripening crops provide abundant substratum. The air spora on dry days is dominated by *Cladosporium* and at night by *Sporobolomyces*. Numbers in winter are limited chiefly by low temperatures. *Erysiphe*, *Ustilago* and *Phytophthora*, in common with other pathogens, are much more restricted in their occurrence by the availability of susceptible host tissue. *Erysiphe* was favoured by high humidity (Hamilton, 1959) when there was abundant susceptible leaf tissue in growing cereal crops in

Table 4.6. *Diurnal periodicities of airborne spore types*

Postdawn pattern (maximum concentration, 07.00–10.00)
Activated by drying or by rapidly changing rh; hygroscopic movement and water
  rupture mechanism

Cercospora	*Gymnosporangium*	*Phytophthora infestans*
Cordana musae	*juniperi-virginiae*	*Polythrincium trifolii*
Corynespora cassiicola	*Nigrospora*	*Trichoconis padwickii*
Deightoniella torulosa	*Peronospora tabacina*	*Zygophiala jamaicensis*
Epicoccum	*Phaeotrichoconis*	*Zygosporium oscheoides*

Midday pattern (maximum concentration, 10.00–16.00)
Released by mechanical disturbance or deflation

Alternaria	*Drechslera*	*Pithomyces*
Aspergillus	(*Helminthosporium*)	*Pseudocercospora*
Aureobasidium	*Erysiphe*	*Puccinia* (uredospores)
Beltrania	*Memnoniella*	*Spegazzinnia*
Botrytis	*Monilinia laxa*	*Stemphylium*
Cladosporium	*Neovossia*	*Tetraploa*
Curvularia	*Penicillium*	*Torula*
Dendryphiella	*Periconia*	*Ustilago*
Dicoccum asperum	*Periconiella*	

Double peak pattern (maximum concentration usually 08.00–10.00 and
  14.00–18.00)
Includes types which elsewhere and in other seasons give midday pattern
Perhaps midday spore cloud is diluted by intense convection

Alternaria	*Entomophthora*	*Periconia*
Botrytis (13.00 and	(05.00 and 13.00)	*Pithomyces*
19.00)	*Epicoccum*	*Puccinia* (uredospores)
Cladosporium	*Fusidium*	*Tetraploa*
Curvularia	*Memnoniella*	
Drechslera		

Post-dusk pattern (maximum concentration, 20.00–22.00)
Liberation mechanisms unknown, probably activated by high rh

Spegazzinia	Tetraspore-type	*Ustilaginoidea virens*

Night pattern (maximum concentration, 02.00–04.00)
High rh required; active mechanisms involving turgid cells

Ascomycotina	*Puccinia* (aecidiospores)	*Tilletiopsis*
Basidiomycotina	*Pyricularia*	*Uromyces*
Fusarium	*Sporobolomyces*	(aecidiospores)

References: Hirst, 1953; Cammack, 1955; Meredith, 1962*a,b*, 1966*a,b*; Jarvis, 1962; Pady *et al.*, 1967, 1969; Sreeramulu, 1959, 1962, 1970; Sreeramulu & Ramalingam, 1962, 1966; Sreeramulu & Vittal, 1966*a,b*; Shenoi & Ramalingam, 1975; Sreeramulu *et al.*, 1971.

early summer. The timing of potato blight epidemics and the occurrence in the air of *P. infestans* sporangia is determined by climatic conditions, provided there is a source of inoculum and crop growth is sufficiently advanced. These have been defined as a 48-h period with temperature exceeding 10 °C and rh exceeding 75% (or 11 h, rh exceeding 90%) once plants meet across the rows to provide a suitable microclimate (Beaumont, 1947; Smith, 1956). Basidiomycotina require high rh for sporophore growth and spore liberation and are thus most common in autumn as temperatures decline. Lignicolous pyrenomycetes are also most frequent in autumn but *D. concentrica* is unusual in releasing spores in summer, perhaps aided by its large water reserve (Ingold, 1971). The occurrence of airborne spores of *Alternaria*, *Botrytis*, *Cladosporium*, *Drechslera* (*Helminthosporium*), *Polythrincium* and *Aureobasidium* on dry days was favoured by high dew points leading to heavy dews overnight, and *Botrytis* by high rh also. *Alternaria* required higher temperatures for maximum numbers in the air spora than other spore types except for *Botrytis* and *Epicoccum* (Hamilton, 1959).

    In tropical climates, temperatures are always high enough to permit fungal growth and the seasonal occurrence of rain and the related patterns of crop growth determine the seasonal occurrence of fungus spores in the air (Fig. 4.3; see also Waller, Ch. 10). *Cladosporium* is the most numerous spore type over the whole year and is dominant in the first part of the cool season; rust uredospores are most numerous in the hot season; basidiospores and *Aspergillus* spp. dominate the wet-season air spora. *Nigrospora* and *Pyricularia oryzae* spores show two peaks associated with rice crops, while *Curvularia* and *Alternaria* tend to occur throughout the year although all are more numerous in the wet season (Sreeramulu & Ramalingam, 1966).

Dispersal and deposition

    Airborne spores are dispersed downwind. Groups or clouds of spores from a single source are diluted and spread vertically and horizontally by wind eddies, obeying the laws of diffusion (Gregory, 1973). Only 5–10% may be carried away from the ground and travel long distances. Deposition of such spores is most likely to be produced by rain-washing. Spores may be impacted by raindrops, and be captured by cloud droplets or even form their nuclei (see Ch. 17). Efficiency of collection depends on spore size so that while 99% of 30-μm diameter spores would be removed from air by rain falling at $2\,\mathrm{mm}\,\mathrm{h}^{-1}$ for 2 h, 72%

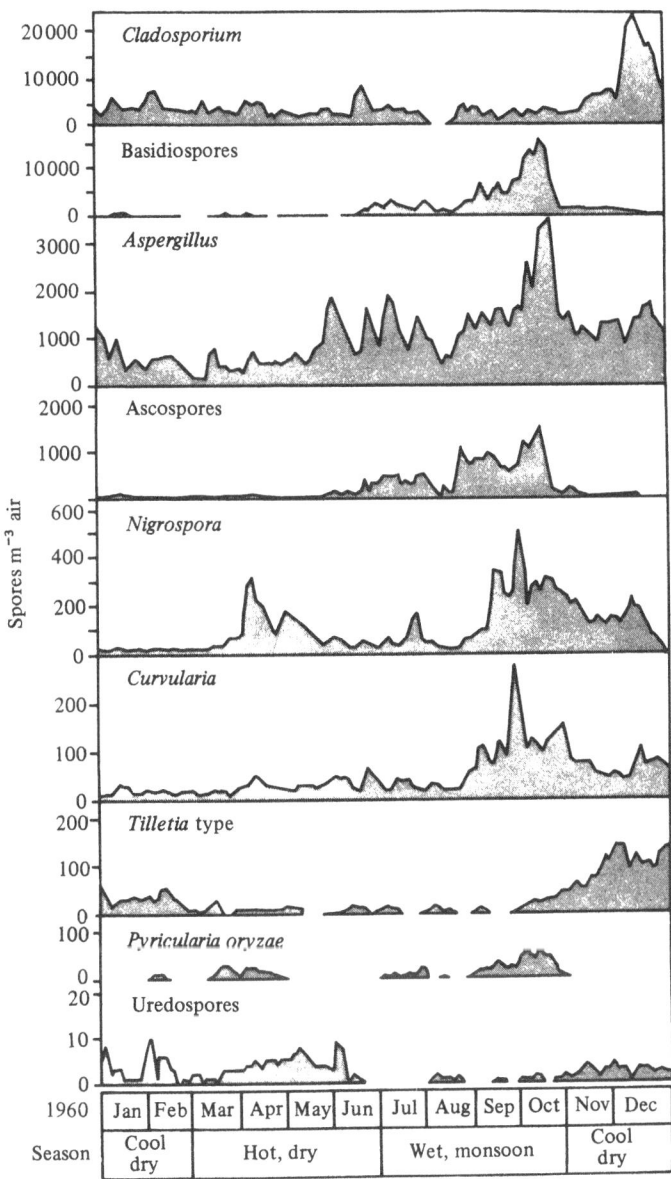

Fig. 4.3. Seasonal periodicity of some airborne fungus spores at Visakhapatnam, India (after Sreeramulu & Ramalingam, 1966). Rice crops are sown in January and July, heading respectively in March–April and October–November and harvested in April–May and November–December.

of 4-μm spores would still be airborne at the end of this period (Chamberlain, 1967).

Dispersal mechanisms are important in determining how a disease spreads, how it is affected by agronomic practices and how it may best be controlled. For instance, with pleomorphic fungi, different methods of dispersal of different spore types can lead to different patterns of disease development. Thus, dispersal of conidia in run-off from dew or rain can lead to streaking of lower leaves, e.g. the heart leaves of banana by *Mycosphaerella musicola*, while ascospore discharge allows upward dispersal, more diffuse speckling and, with *M. musicola*, tip scorch of upper leaves (Meredith, 1973). Ascospores and conidia may be produced at different times of year and may differ in their effectiveness in spreading disease according to the availability of susceptible tissue.

References

Allitt, U. (1979). Coiled ascospores in the Hypodermataceae. *Transactions of the British Mycological Society*, **72**, 147–51.
Barnett, H. L. & Lilly, V. G. (1955). The effect of humidity, temperature and carbon dioxide on the sporulation of *Choanephora cucurbitarum*. *Mycologia*, **47**, 26–9.
Beaumont, A. (1947). The dependence on the weather of the dates of outbreak of potato blight epidemics. *Transactions of the British Mycological Society*, **31**, 45–53.
Benson, D. M. (1984). Influence of pine bark, matric potential and pH on sporangium production by *Phytophthora cinnamomi*. *Phytopathology*, **74**, 1359–1363.
Buller, A. H. R. (1924). *Researches on Fungi*, vol. 3. New York: Longmans Green.
Cammack, R. H. (1955). Seasonal changes in three common constituents of the air spora of southern Nigeria. *Nature*, **176**, 1270–3.
Carter, M. V. (1963). *Mycosphaerella pinodes*. II. The phenology of ascospore release. *Australian Journal of Biological Sciences*, **16**, 800–17.
Carter, M. V. & Moller, W. J. (1961). Factors affecting the survival and dissemination of *Mycosphaerella pinodes* (Berk. & Blox.) Vestergr. in South Australian irrigated pea fields. *Australian Journal of Agricultural Research*, **12**, 878–88.
Chamberlain, A. C. (1967). Deposition of particles to natural surfaces. In *Airborne Microbes*, ed. P. H. Gregory & J. L. Monteith, pp. 138–64. Cambridge: Cambridge University Press.
Cook, R. J. & Duniway, J. M., ed. (1980). Water relations in the life cycles of soilborne plant pathogens. In *Water Potential Relations in Soil Microbiology*, Soil Science Society of America Special Publication, No. **9**, 119–39.
Crosier, W. (1934). Studies in the biology of *Phytophthora infestans*. Memoirs of Cornell Agricultural Experimental Station, No. **155**.
Cruickshank, I. A. M. (1958). Environment and sporulation in phytopathogenic fungi. I. Moisture in relation to the production and discharge of conidia of *Peronospora tabacina* Adam. *Australian Journal of Biological Sciences*, **11**, 162–70).
Curran, P. M. T. (1971). Sporulation in some members of the *Aspergillus glaucus* group

in response to osmotic pressure, illumination and temperature. *Transactions of the British Mycological Society*, **57**, 201–11.

Duniway, J. M., Abawi, G. S. & Steadman, J. R. (1977). Influence of soil moisture on the production of apothecia by sclerotia of *Whetzelinia sclerotiorum*. *Proceedings of the American Phytopathological Society*, **4**, 115.

Gisi, U., Zentmyer, G. A. & Klure, L. J. (1980). Production of sporangia by *Phytophthora cinnamomi* and *P. palmivora* in soils at different matric potentials. *Phytopathology*, **70**, 301–6.

Gottwald, T. R. (1982). Spore discharge by the pecan scab pathogen, *Cladosporium caryigenum*. *Phytopathology*, **72**, 1193–7.

Gregory, P. H. (1939). The life history of *Ramularia vallisumbrosae* Cav. on *Narcissus*. *Transactions of the British Mycological Society*, **23**, 24–54.

Gregory, P. H. (1973). *Microbiology of the Atmosphere*, 2nd edn. Aylesbury: Leonard Hill.

Gregory, P. H. (1976). *Outdoor Aerobiology*. Oxford Biology Readers No. 62. Oxford: The University Press.

Gregory, P. H. (1979). Speculations on basidiospore liberation. *Bulletin of the British Mycological Society*, **13**, 128–31.

Gregory, P. H. & Hirst, J. M. (1957). The summer air-spora at Rothamsted in 1952. *Journal of General Microbiology*, **13**, 135–52.

Gregory, P. H. & Stedman, O. J. (1958). Spore dispersal in *Ophiobolus graminis* and other fungi of cereal foot rots. *Transactions of the British Mycological Society*, **41**, 449–56.

Grogan, R. G. & Abawi, G. S. (1975). Influence of water potential on growth and survival of *Whetzelinia sclerotiorum*. *Phytopathology*, **65**, 122–8.

Hamilton, E. D. (1959). Studies on the air spora. *Acta Allergologica*, **13**, 143–75.

Harries, M. G., Lacey, J., Tee, R. D., Cayley, G. R. & Newman Taylor, A. J. (1985). *Didymella exitialis* and late summer asthma. *Lancet* (in press).

Hartmann, H., Sutton, J. C. & Thurtell, G. W. (1982). An apparatus for accurate control of atmospheric water potentials in studies of foliar plant pathogens. *Phytopathology*, **72**, 914–16.

Hashioka, Y. (1938). Relation of temperature and humidity to *Sphaerotheca fuliginea* (Schecht.) Poll. with special reference to germination, viability and infection. *Transactions of the Natural History Society of Formosa*, **27**, 129–45.

Hawker, L. (1957). *The Physiology of Reproduction in Fungi*. Cambridge: Cambridge University Press.

Hirst, J. M. (1953). Changes in atmospheric spore content: diurnal periodicity and the effect of weather. *Transactions of the British Mycological Society*, **36**, 375–93.

Hirst, J. M. & Stedman, O. J. (1962). The epidemiology of apple scab *Venturia inaequalis* (Cke) Wint. II. Observations on the liberation of ascospores. *Annals of Applied Biology*, **50**, 525–50.

Hirst, J. M. & Stedman, O. J. (1963). Dry liberation of fungus spores by raindrops. *Journal of General Microbiology*, **33**, 335–44.

Ingold, C. T. (1964). Possible spore discharge mechanism in *Pyricularia*. *Transactions of the British Mycological Society*, **47**, 573–5.

Ingold, C. T. (1971). *Fungal Spores: Their Liberation and Dispersal*. London & New York: Oxford University Press (Clarendon).

Ingold, C. T. (1983). A view of the basidium. *Bulletin of the British Mycological Society*, **17**, 82–94.

James, J. R. & Sutton, T. B. (1982). Environmental factors influencing pseudothecial development and ascospore maturation of *Venturia inaequalis*. *Phytopathology*, **72**, 1073–80.

Jarvis, W. R. (1962). The dispersal of spores of *Botrytis cinerea* Fr. in a raspberry plantation. *Transactions of the British Mycological Society*, **45**, 549–59.

Keyworth, W. G. (1942). The occurrence in artifical culture of tremelloid outgrowths on the pilei of *Coprinus ephemerus*. *Transactions of the British Mycological Society*, **25**, 307–10.

Latham, A. J. (1974). Effect of moisture on conidiophore morphology of *Cristulariella pyramidalis*. *Phytopathology*, **64**, 1255–7.

Lawrence, E. G. & Zehr, E. I. (1982). Environmental effects on the development and dissemination of *Cladosporium carpophilum* on peach. *Phytopathology*, **72**, 773–6.

Leach, C. M. (1982*a*). Vibrational release of conidia by *Drechslera maydis* and *D. turcica* related to humidity and red-infra red radiation. *Phytopathology*, **70**, 196–200.

Leach, C. M. (1980*a*). Vibrational release of conidia by *Drechslera maydis* and *D. turcica* spore discharge by *Pyricularia oryzae*. *Phytopathology*, **70**, 201–5.

Leach, C. M. (1980*c*). Evidence for an electrostatic mechanism in spore discharge by *Drechslera turcica*. *Phytopathology*, **70**, 206–13.

Leach, C. M. (1982). Active sporangium discharge by *Peronospora destructor*. *Phytopathology*, **72**, 881–5.

Leach, C. M. (1985). Effect of still and moving moisture-saturated air on sporulation of *Drechslera* and *Peronospora*. *Transactions of the British Mycological Society*, **84**, 179–83.

Leach, C. M., Hildebrand, P. D. & Sutton, J. C. (1982). Sporangium discharge by *Peronospora destructor*: influence of humidity, red-infra red radiation and vibration. *Phytopathology*, **72**, 1052–6.

Magan, N. & Lacey, J. (1984). Effect of temperature and pH on water relations of field and storage fungi. *Transactions of the British Mycological Society*, **82**, 71–81.

Meredith, D. S. (1961). Spore discharge in *Deightoniella torulosa* (Syd.) Ellis. *Annals of Botany (London)* (N.S.), **25**, 271–8.

Meredith, D. S. (1962*a*). Some components of the air spora in Jamaica banana plantations. *Annals of Applied Biology*, **56**, 577–94.

Meredith, D. S. (1962*b*). Spore dispersal in *Pyricularia grisea* (Cooke) Sacc. *Nature*, **195**, 92–3.

Meredith, D. S. (1962*c*). Spore discharge in *Cordana musae* (Zimm.) Hohnel and *Zygosporium oscheoides* Mont. *Annals of Botany (London)* (N.S.), **26**, 233–41.

Meredith, D. S. (1965). Violent spore release in *Helminthosporium turcicum*. *Phytopathology*, **55**, 1099–102.

Meredith, D. S. (1966*a*). Spore dispersal in *Alternaria porri* (Ellis) Neerg. on onions in Nebraska. *Annals of Applied Biology*, **57**, 67–73.

Meredith, D. S. (1966*b*). Diurnal periodicity and violent liberation of conidia in *Epicoccum*. *Phytopathology*, **56**, 988.

Meredith, D. S. (1973). Significance of spore release and dispersal mechanisms in plant disease epidemiology. *Annual Review of Phytopathology*, **11**, 313–42.

Mislivec, P. B. & Tuite, J. (1970) Temperature and relative humidity requirements of species of *Penicillium* isolated from yellow dent corn kernels. *Mycologia*, **62**, 75–88.

Moller, W. J. & Carter, M. V. (1965). Production and dispersal of ascospores in *Eutypa armeniacae*. *Australian Journal of Biological Sciences*, **18**, 67–80.

Ogilvie, L. (1944). Downy mildew of lettuce. *Report of the Agricultural and Horticultural Research Station, Bristol for 1943*, 90–4.

Pady, S. M., Kramer, C. L. & Clary, R. (1967). Diurnal periodicity in airborne fungi in an orchard. *Journal of Allergy*, **39**, 302–10.

Pady, S. M., Kramer, C. L. & Clary, R. (1969). Periodicity in spore release in *Cladosporium. Mycologia*, **61**, 87–98.

Panasenko, V. T. (1967). Ecology of microfungi. *Botanical Review*, **33**, 189–215.

Pearson, R. C., Seem, R. C. & Mayer, F. W. (1980). Environmental factors influencing the discharge of basidiospores of *Gymnosporangium juniperi-virginiae. Phytopathology*, **70**, 262–6.

Pitt, J. J. (1975). Xerophilic fungi and the spoilage of foods of plant origin. In *Water Relations of Food*, ed. R. B. Duckworth, pp. 273–307. New York: Academic Press.

Plunkett, B. E. (1956). The influence of factors of the aeration complex and light upon fruit body form in pure cultures of an agaric and a polypore. *Annals of Botany, London* (N.S.), **20**, 563–86.

Reeves, R. J. (1975). Behaviour of *Phytophthora cinnamoni* Rands in different soils and water regimes. *Soil Biology and Biochemistry*, **7**, 19–24.

Rotem, J., Cohen, Y. & Bashi, E. (1978). Host and environmental influences on sporulation *in vivo. Annual Review of Phytopathology*, **16**, 83–101.

Shenoi, M. M., Ramalingam, A. (1975). Circadian periodicities of some spore components of air at Mysore. *Arogya – Journal of Health Sciences*, **1**, 154–6.

Smith, L. P. (1956). Potato blight forecasting by 90 per cent humidity criteria. *Plant Pathology*, **5**, 83–7.

Sneh, B. & McIntosh, D. L. (1974). Studies on behaviour and survival of *Phytophthora cactorum* in soil. *Canadian Journal of Botany*, **52**, 795–802.

Sreeramulu, T. (1962). Some observations on the *Deightoniella* fruit and leaf-spot disease of the banana. *Current Science*, **31**, 258–9.

Sreeamulu, T. (1962). Some observations on the *Deightoniella* fruit and leaf-spot disease of the banana. *Current Science*, **31**, 258–9.

Sreeramulu, T. (1970). Conidial dispersal in two species of *Cercospora* causing tikka leaf-spots on the groundnut (*Arachis hypogea* L.). *Indian Journal of Agricultural Science*, **40**, 173–8.

Sreeramulu, T. & Ramalingam, A. (1962). Notes on airborne *Tetraploa* spores. *Current Science*, **31**, 121–2.

Sreeramulu, T. & Ramalingam, A. (1965). A two-year study of the air-spora of a paddy field near Visakhapatnam. *Indian Journal of Agricultural Science*, **36**, 111–32.

Sreeramulu, T. & Vittal, B. P. R. (1966a). Periodicity in the air-borne spores of the rice false smut fungus, *Ustilaginoidea virens. Transactions of the British Mycological Society*, **49**, 443–9.

Sreeramulu, T. & Vittal, B. P. R. (1966b). Some aerobiological observations on the rice stackburn fungus, *Trichoconis padwickii. Indian Phytopathology*, **19**, 215–21.

Sreeramulu, T., Vittal, B. P. R. & Ramakrishna, V. (1971). Aeriobiology of *Cercospora koepkei* Kruger causing the yellow spot disease of sugar cane. *Indian Journal of Agricultural Science*, **41**, 655–62.

Sung, J. M. & Cook, R. J. (1981). Effect of water potential on reproduction and spore germination by *Fusarium roseum*, 'Graminearum', 'Culmorum' and 'Avenaceum'. *Phytopathology*, **71**, 499–504.

Traquair, J. A. & Kokko, E. G. (1981). Spore discharge in *Epicoccum nigrum. Canadian Journal of Botany*, **59**, 59–62.

Van Arsdel, E. P., Riker, A. J. & Patton, R. F. (1956). The effects of temperature and moisture on the spread of white pine blister rust. *Phytopathology*, **46**, 307–18.

Webster, J. (1952). Spore projection in the hyphomycete *Nigrospora sphaerica. New Phytologist*, **51**, 229–35.

Webster, J. (1966). Spore projection in *Epicoccum* and *Arthrinium*. *Transactions of the British Mycological Society*, **49**, 339–43.

Westerlund, F. V., Campbell, R. M., Grogan, R. G. & Duniway, J. M. (1978). Soil factors affecting the reproduction and survival of *Olpidium brassicae* and its transmission of big vein agent to lettuce. *Phytopathology*, **68**, 927–35.

Yarwood, C. E. (1943). Onion downy mildew. *Hilgardia*, **14**, 595–691.

Zentmyer, G. A. & Erwin, D. C. (1970). Development and reproduction of *Phytophthora*. *Phytopathology*, **60**, 1120–7.

5
Spore dispersal in splash droplets

B. D. L. FITT and H. A. McCARTNEY

Rothamsted Experimental Station, Harpenden, Hertfordshire, AL5 2JQ, UK

Introduction

Rain-splash is the second most important natural agent, after wind, in the dispersal of spores of plant pathogenic fungi, When a raindrop strikes a solid or liquid surface, its kinetic energy is used to break it into smaller splash droplets and to project them away from the point of impact (Gregory, 1973). If the surface is an infected plant or piece of plant material bearing spores, the splash droplets may pick up and disperse spores. If a large raindrop (e.g. 5 mm diameter) strikes a thin layer (e.g. 0.1 mm deep) of spore suspension, thousands of spore-carrying droplets may be dispersed in a single splash. At a site with an average annual rainfall of 100 cm, each m^2 of ground receives annually about 10^9 raindrops large enough to produce a splash; in a day with 10 mm of rain there would be 10^7 rain-splashes per m^2. Rain-splash may, therefore, disperse millions of spores daily.

Although rain is the most important source of spore-carrying splash droplets, irrigation water, and drops dripping from crop canopies may also disperse spores in splash droplets. In areas with little rainfall, spores of pathogens may be dispersed in splash droplets produced by the overhead irrigation of crops. When crop canopies become saturated by rain, mist, fog or dew, large drips are produced. Consequently, under crop canopies, and especially under trees, drip-splash may be more important than rain-splash. Whereas raindrops are unlikely to contain more than a few spores picked up from the air, drips may contain large numbers of spores washed off spore-bearing leaves or twigs.

The initial splash which occurs when an incident raindrop strikes a surface is termed primary splash (Ingold, 1971). Spores dispersed by a primary splash may be re-dispersed by further raindrops (secondary splash). We need to distinguish between incident drops and splash droplets (Gregory, 1973). Raindrops which reach the ground range from 0.2 to 5 mm in diameter. Drops smaller than 0.2 mm generally evaporate before reaching the ground and drops larger than 5 mm break up when falling at speeds approaching their terminal velocity. Although small raindrops are most numerous, they have little kinetic energy and are less efficient than large raindrops in producing spore-carrying splash droplets. Drips from leaves may be larger than 5 mm because they do not fall far enough to reach terminal velocity but they may have sufficient kinetic energy to produce splash droplets and drip-splash can be an important means of spore dispersal.

Spore-carrying splash droplets range in size from large ballistic droplets with diameter >1000 μm to small airborne droplets with diameter <100 μm. Ballistic droplets follow a trajectory that is little affected by wind, whereas the more numerous airborne droplets may be carried considerable distances by wind and can evaporate to leave airborne spores. Droplets of intermediate size are affected to a greater or lesser extent by the wind, depending on their size, windspeed and turbulence. This chapter discusses the characteristics of splash-dispersed spores, the mechanisms of spore dispersal in splash droplets and their application to epidemiology.

Characteristics of splash-dispersed spores

The classic work on spore dispersal in splash droplets was that of Stepanov (1935), who allowed water drops to fall onto sporulating cultures of fungi and collected the splash droplets on glass slides at distances of 20–25 cm from the point of impact. He collected large numbers of spores from those fungi with spores borne in mucilage (e.g. *Colletotrichum linicola* and *Fusarium culmorum*) but fewer spores from those fungi with spores borne on aerial hyphae (e.g. *Phytophthora infestans*). He concluded that pathogen spores borne in mucilage could be detached from their host by raindrops and dispersed in splash droplets.

There are many plant pathogens with spores that are borne in mucilage and these species are characteristically splash-dispersed (Gregory, 1973). They include fungi with spores borne in pycnidia (e.g. *Septoria nodorum* and *Phoma exigua*, Fig. 5.1a), in acervuli (e.g. *Pyrenopeziza brassicae*, Fig. 5.1b), in sporodochia (e.g. *F. culmorum*) or on stromata (e.g. *Pseudocercosporella herpotrichoides*, Fig. 5.1c). By contrast, spores of pathogens that are characteristically wind-dispersed are often borne away from the mycelium on aerial hyphae (e.g. sporangia of *P. infestans*) or in chains

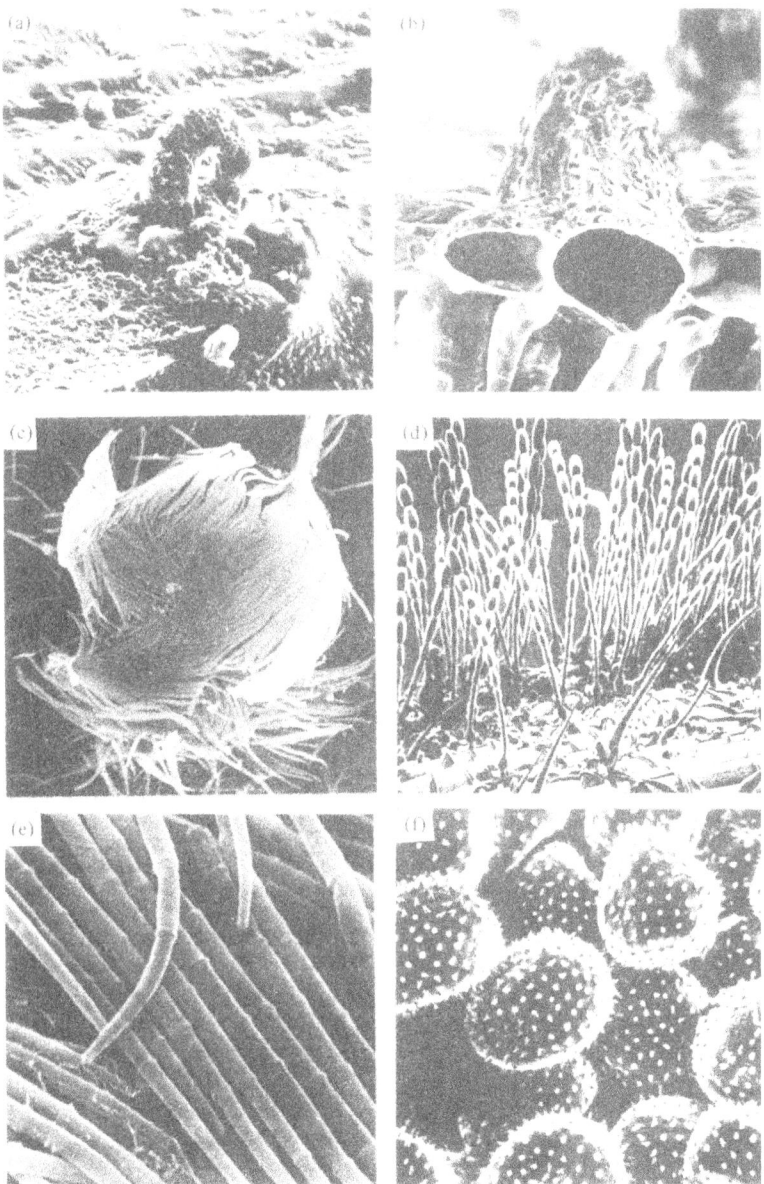

Fig. 5.1. (a) *Phoma exigua* pycnidiospores exuded through pycnidial ostiole (× 300); (b) *Pyrenopeziza brassicae* conidia in acervulus (× 500); (c) *Pseudocercosporella herpotrichoides* conidia on stroma (× 230); (d) *Erysiphe graminis* conidia in aerial chains (× 150); (e) smooth-walled *P. herpotrichoides* conidia (× 1500); (f) rough-walled *Puccinia coronata* uredospores (× 800). (Figs 5.1a, 5.1d, 5.1f, Rothamsted Experimental Station electron microscopy unit; Fig. 5.1b, C. J. Rawlinson; Figs 5.1c, 5.1e, P. T Atkey.)

(e.g. conidia of *Erysiphe graminis*, Fig. 5.1d); some are actively discharged away from the mycelium (e.g. ascospores of *Venturia inaequalis*).

The mucilage, which is composed of proteins and glucides (Fournet, 1969), prevents dispersal of the spores by wind alone. It absorbs water and swells during periods of high relative humidity. In fungi with pycnidia, this causes a cirrhus of spores to be extruded. On wetting, the mucilage dissolves to leave a suspension of spores available for splash dispersal. The mucilage probably protects spores from desiccation and loss of viability, which might occur if they were dispersed in dry weather, and it confines dispersal to periods of rainfall, when free water is available for germination on host surfaces.

Spore germination may be inhibited by substances contained in the mucilage; thus the presence of cirrhus extract inhibits germination of pycnidiospores of *S. nodorum* (Griffiths & Peverett, 1980), and conidia of *P. herpotrichoides* do not germinate when in concentrated suspensions (Glynne, 1953). Consequently, germination may be delayed until spore suspensions are diluted by rain-splash. This decreases the spread of infection on source plants but enhances disease spread to other plants. Components of the mucilage may also affect the surface tension of spore suspensions; the surface tension of water containing suspensions of 2×10^6 *P. brassicae* conidia cm^{-3} was more than 20% less than that of pure water (Rawlinson, 1979). Large decreases in surface tension greatly affect spore dispersal in splash droplets; adding surfactant to suspensions of *S. nodorum* pycnidiospores to cause a 60% decrease in surface tension increased the number, and decreased the average size, of spore-carrying splash droplets (Brennan, Fitt, Taylor & Colhoun, 1985a). However, the surface tension of suspensions containing 6×10^5 spores cm^{-3} was only 3% less than that of water and it seems unlikely that changes in surface tension influence dispersal of *S. nodorum* spores under natural conditions.

Spores borne in mucilage are wettable and are generally carried within splash droplets, although conidia of *Rhynchosporium secalis* can be carried on the droplet surfaces (Stedman, 1980a), like non-wettable, characteristically dry-dispersed spores (Ramalingam & Rati, 1979). There is some evidence that wettable, splash-dispersed spores may have some adhesiveness which enables them to stick to surfaces in the presence of water. Spores of *P. herpotrichoides* in aqueous suspension adhere to plastic centrifuge tubes, but are released by addition of a nonionic surfactant (Fitt & Bainbridge, 1983a). When aqueous suspensions of these spores are shaken after adding paraffin oil, the spores collect at the interface between the oil and aqueous layers, suggesting that they have both hydrophilic and hydrophobic areas on their surfaces. It is not clear if these areas are involved in the adhesiveness, or whether the adhesiveness has any significance *in vivo*.

Characteristically splash-dispersed spores, such as conidia of *P. herpotrichoides* (Figs. 5.1c,e), *S. nodorum* and *P. brassicae* (Fig. 5.1b), are often borne in mucilage, are wettable and have smooth surfaces, thin, hyaline walls and an elongate shape. By contrast, characteristically dry-dispersed spores, such as conidia of *Alternaria*, *Cladosporium* and *Penicillium* spp. and uredospores of *Puccinia coronata* (Fig. 5.1f), are often borne dry, are non-wettable and have rough surfaces, thick coloured walls and are round. There are exceptions, especially of shape. Pycnidiospores of *P. exigua*, which are splash-dispersed (Carnegie, 1980), are ovoid in shape and the dry-dispersed conidia of *Alternaria* spp. (Hirst & Stedman, 1963) and *E. graminis* (Bainbridge & Stedman, 1979) are elongate. Sporangia of *P. infestans*, which are borne dry and are ovoid in shape, but are wettable and have smooth, thin, hyaline walls (D. H. Lapwood, personal communication), are intermediate between the two groups and are dispersed both dry and in splash droplets.

We have described spores as 'characteristically' splash-dispersed or 'characteristically' dry-dispersed, since these two modes of dispersal are not mutually exclusive. However, wind may be important in the dispersal of some spores which are splash-borne. Most of the spores released from their host by rain-splash are carried in large ballistic droplets which travel short distances, but some are carried in small droplets, which become airborne (Fitt & Bainbridge, 1983b) and may be carried considerable distances by wind. Airborne spores of *S. nodorum* (Faulkner & Colhoun, 1977; Wale & Colhoun, 1979), *P. exigua* (Carnegie, 1980) and *P. herpotrichoides* (Fitt & Bainbridge, 1983b) have been collected at heights of up to 2 m above infected crops or stubble, well above the splash zone. On the other hand, rain may be important in dispersal of characteristically dry-dispersed spores. Non-wettable spores, normally released from their host by wind, may be picked up by splash droplets (Ramalingan & Rati, 1979). A small number of the characteristically dry-dispersed conidia of *Botrytis fabae* are dispersed in rain-splash droplets during epidemics of chocolate spot disease in field beans (Fitt, Creighton & Bainbridge, 1985), although these spores are unlikely to affect the development of epidemics, which are favoured by wet weather. However, the dispersal of a few spores in splash droplets may be important epidemiologically in cases when spores that are dispersed dry rapidly lose viability because the free water required for their germination is not present on the host surface.

Rain may also be important in the release of spores which are subsequently dry-dispersed (Hirst & Stedman, 1963). There are often large increases in airborne concentrations of some dry-dispersed spores when rain starts. Hirst & Stedman (1963) allowed drops, 3 mm in diameter, to fall 8 m onto infected sporulating plant material and sampled the air with suction samplers before, during and after the period when drops were falling. They collected the greatest numbers of spores of thirteen pathogens, including *Puccinia graminis, P. infestans*, and *Alternaria* and *Cladosporium* spp., when drops were falling and suggested that the spores were released by 'tap' and 'puff' mechanisms. Impacting water drops may mechanically shake (tap) spores off plant material. As a drop spreads on a surface, radial air movements (puffs) that can exceed $40 \, \mathrm{m \, s^{-1}}$ occur briefly and these too may release spores.

Mechanisms of spore dispersal in splash droplets
The splash process

The remarkable series of photographs taken by Worthington & Cole (1897), using accurately timed spark discharges, have provided much information about the splash process. It consists of a regular sequence of events, under the combined action of the momentum of the falling drop, friction between the drop and the solid or liquid surface on which it impacts, and surface tension. Of particular relevance to spore dispersal in splash droplets is the splash produced by a drop falling onto a thin layer of liquid (Gregory, 1973). The drop flattens on impact and pushes the surface layer of liquid outwards. The resulting radially moving mass of liquid is formed into a crater or 'corona' by the action of surface tension. The photographs by Worthington & Cole (1897) show that liquid from the drop lines the inner surface of the crater and it seems likely that some of the target liquid forms the outer wall. This double wall breaks up into rays of splash droplets. The whole process, which takes less than a hundredth of a second, may produce thousands of splash droplets, some initially travelling at velocities up to $20 \, \mathrm{m \, s^{-1}}$. These splash droplets are an intimate mixture of liquid from the incident drop and from the target liquid; when one dye is incorporated into incident drops and another into target liquids, the resulting splash droplets contain mixtures of both dyes (Gregory, Guthrie & Bunce, 1959; Fitt & Lysandrou, 1984).

Experiments with drops incorporating a fluorescent dye allowed to fall onto targets placed over mown grass, show that the splash process is affected by incident drop size, and the rigidity, angle of inclination and surface texture of the target (Stedman, 1979). As the incident drop size increased more liquid was splashed, and as the angle of inclination of the target increased less liquid was splashed. More liquid was splashed from rigid than from non-rigid tobacco and brussel sprout leaves, and less liquid was splashed from waxed than from unwaxed Perspex. When the target was at a height of 1 m, splash droplets were deposited up to 16 m downwind. Wind affects the distribution of splash droplets greatly; in still-air experiments splash droplets were symmetrically distributed about the target, but in wind of $2 \, m \, s^{-1}$ about 80% of the drops were deposited downwind (Brennan, Fitt, Taylor & Colhoun, 1985b).

How splash droplets pick up fungal spores, which sink to the bottom of a layer of target liquid (Gregory *et al.*, 1959), is unknown. However, a clear spore-free circle, about 1 cm diam., is commonly seen at the point of impact (Gregory, 1973), suggesting that the spreading drop dislodges the spores. They are presumably picked up as the crater forms and transferred to the splash droplets.

Methods for studying spore dispersal in splash droplets

The first experiments on mechanisms of spore dispersal in splash droplets were those of Gregory *et al.* (1959), who collected spore-carrying splash droplets on microscope slides coated with a layer of gelatin containing naphthol green B dye. On touching the gelatin layer, droplets spread out to leave permanent, circular traces about three times their original diameter. However, Gregory's method is not easy to use, since the thickness of the gelatin layer, which affects droplet spread after impact and therefore the ratio of droplet diameter to trace diameter, needs to be constant from slide to slide. This uniformity is difficult to achieve. An improved method for sampling spore-carrying splash droplets uses fixed photographic film (Fitt, Lysandrou & Turner, 1982). In still air, the droplets leave clear, circular, permanent traces within which spores are easily visible (Fig. 5.2). The droplet-spread factor is constant because the manufactured gelatin layer is of uniform thickness. Estimates of numbers of splash droplets, spore-carrying droplets and spores dispersed by a 5-mm drop falling onto a spore suspension (0.5 mm deep) are similar to those obtained by Gregory *et al.* (1959). However, in moving air the droplets leave elliptical traces, which are difficult to interpret (O. C. Macdonald, personal communication).

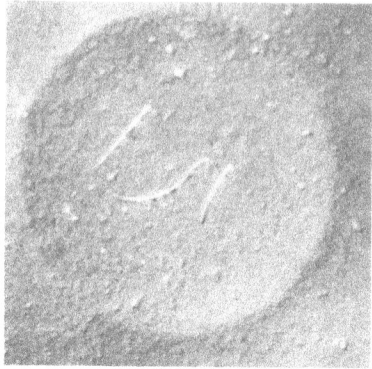

Fig. 5.2. Trace of a splash droplet, containing *Pseudocercosporella herpotrichoides* conidia, on photographic film (× 270).

Factors affecting spore dispersal in splash droplets

One splash may disperse many thousands of spore-carrying droplets from a thin layer of spore suspension. Gregory *et al.* (1959) estimated that the splash from a drop of water, 5 mm diam., falling from a height of 7 m onto an aqueous suspension of *Fusarium solani* spores (0.1 mm deep, 1.6×10^5 spores cm^{-2}), dispersed over 5000 splash droplets, of which more than 50% carried spores. Similar numbers of spore-carrying droplets were dispersed by a single splash when drops 4 or 5 mm diam. fell 13 m down a raintower (Fitt & McCartney, 1986) onto spore suspensions, 0.5 mm deep, of *P. herpotrichoides* (Fitt & Lysandrou, 1984), *P. brassicae* (Fatemi & Fitt, 1983) or *S. nodorum* (Brennan *et al.*, 1985a) in still-air experiments. In these experiments, up to 90 000 spores were dispersed by a single splash from suspensions containing 1.2×10^5 spores cm^{-3}. In general, fewer spores are dispersed from infected plant material than from artificial spore suspensions. When simulated raindrops fell onto straw infected by *S. nodorum* (11×10^6 spores g^{-1} dry weight: Brennan *et al.*, 1985a) or *P. herpotrichoides* (10^4 spores per straw: Fitt & Nijman, 1983), *ca.* 25 spores were dispersed per splash.

In experiments with spore suspensions, the number of spore-carrying droplets increases as the incident drop diameter and velocity (related to the height of fall) increases and as the depth of the target spore suspension decreases (Table 5.1). Increasing the drop diameter or the height of fall increases the kinetic energy of the drop at the moment of impact and hence increases the number of spore-carrying splash drop-

Table 5.1. *The effects of height of fall, incident drop diameter and depth of target spore suspension* (1.6×10^5 *spores* cm^{-2}) *on numbers of splash droplets carrying macroconidia of* Fusarium solani *dispersed from a single splash*

Height of fall (m)	Incident drop diameter (mm)	Estimated kinetic energy (mJ)	Number of spore-carrying splash droplets		
			depth of spore suspension (mm)		
			0.1	0.5	1.0
2.9	2	0.07	82	26	24
2.9	3	0.20	348	109	82
2.9	4	0.59	567	211	127
2.9	5	1.43	1560	829	522
7.4	3	0.42	559	283	470
7.4	4	1.18	2010	603	391
7.4	5	2.48	2133	1690	518

Gregory, Guthrie & Bunce (1959).

lets produced. Whether spores are incorporated into incident drops or into target liquids, the resulting splash droplets contain spores (Gregory *et al.*, 1959; Fitt & Lysandrou, 1984), which is further evidence that splash droplets come from both drop and target liquids. Gregory (1973) suggested that the size and shape of spores may be related to the ease with which they are picked up in splash droplets. However, when 5-mm drops fell onto spore suspensions containing 1.2×10^5 spores cm^{-3} of *P. brassicae* (cylindrical, $12 \times 3 \mu m$) or *P. herpotrichoides* (needle-shaped, $52 \times 2 \mu m$), the distributions of spores with splash-droplet size and with distance from the target were surprisingly similar for the two fungi (Fatemi & Fitt, 1983).

Sizes of spore-carrying droplets

Spore-carrying splash droplets range from 50 to 2500 μm in diameter (Fitt *et al.*, 1982). The smallest splash droplets are most numerous but few carry spores, whereas most of the large splash droplets carry spores. The majority of spore-carrying droplets are of intermediate diameter, as shown in an experiment with spore suspensions of *P. herpotrichoides* (Fig. 5.3). Similar distributions have been obtained with single drops falling onto spore suspensions of *F. solani* (Gregory *et al.*, 1959), *S. nodorum* (Brennan *et al.*, 1985a) and *P. brassicae* (Fatemi & Fitt,

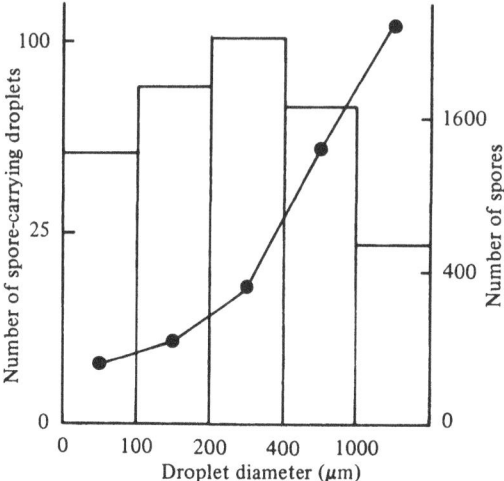

Fig. 5.3. Numbers (square-root transformed) of spore-carrying droplets (histogram) and spores (●) in different droplet-size categories collected up to 100 cm from conidial suspensions of *Pseudocercosporella herpotrichoides* (3.2×10^5 spores cm^{-3}, 0.5 mm deep) on to which 20 drops (5 mm diameter) fell in still air. Sizes marked on *x*-axis are boundaries of categories. (Fitt & Lysandrou, 1984.)

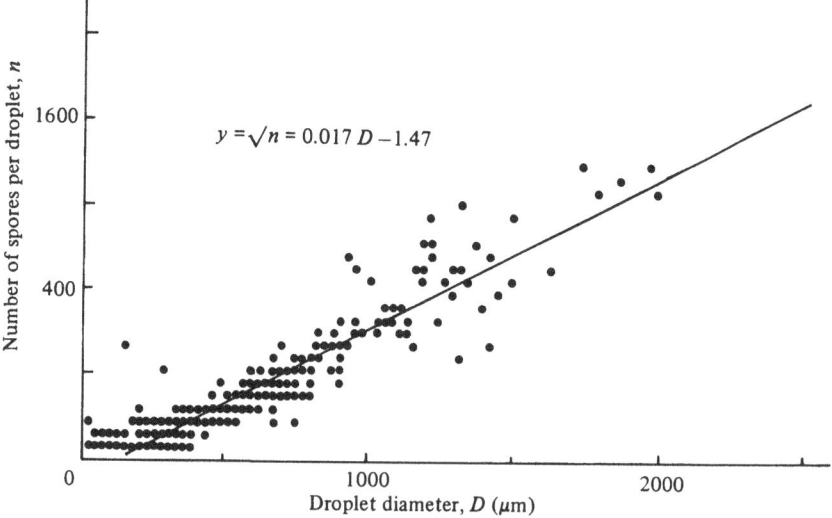

Fig. 5.4. Numbers (square-root transformed) of spores in splash droplets in relation to diameter of droplets collected up to 100 cm from conidial suspensions of *Pyrenopeziza brassicae* (1.2×10^5 spores cm^{-3}, 0.5 mm deep) on to which 20 drops (5 mm diameter) fell in still air. (Fatemi & Fitt, 1983.)

1983) or with simulated rain (rate 13.8 dm^3 h^{-1} m^{-2}; volume mean diameter 2.9 mm) falling onto infected straw bearing sporulating *P. herpotrichoides* (Fitt & Nijman, 1983). Over 60% of the spores were carried in droplets >1000 μm in diameter and less than 2% in droplets <100 μm in diamter. In general, the number of spores per droplet was proportional to the square of the droplet diameter (Fig. 5.4). The constant of proportionality, estimated as the slope of a regression line, increased with increasing spore concentration in the target suspension and with increasing incident drop diameter.

Distance and height of spore dispersal

In still air, few spore-carrying splash droplets travel beyond 1 m from the point of impact and the number of spore-carrying droplets and spores deposited on horizontal surfaces decreases with increasing distance from the target. When 3-mm drops fell onto suspensions of *S. nodorum* pycnidiospores, more than 60% of the spore-carrying droplets and more than 70% of the spores were deposited within 25 cm of their origin (Fig. 5.5). Similar results have been obtained with spore suspensions of *F. solani*, *P. brassicae* and *P. herpotrichoides*.

The decrease in the number of spores deposited (y) with distance (x) is known as a spore deposition gradient. Such gradients may be described by empirical models, such as the power law model of Gregory

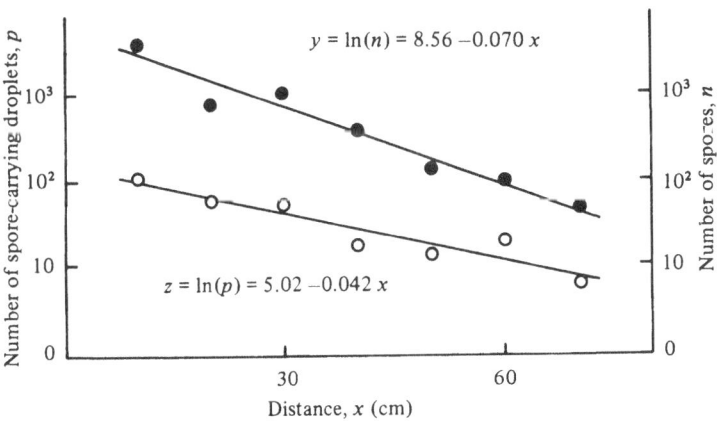

Fig. 5.5. Estimated numbers (ln-transformed) of spore-carrying droplets (p) and spores (n) dispersed to different distances by one drop from suspensions of *Septoria nodorum* pycnidiospores (6.5×10^5 spores cm^{-3}, 0.5 mm deep) on to which 25 drops (3 mm diameter) fell in still air. Regressions of ln(p) and ln(n) on distance, x. (Brennan *et al.*, 1985a.)

(1968) or the exponential model of Kiyosawa & Shiyomi (1972). The power law model (Eq. (1)) assumes that the number of spores deposited is inversely proportional to some power of the distance and the exponential model (Eq. (2)) assumes that the number of spores deposited decreases exponentially with distance:

$$y = ax^{-b} \tag{1}$$

$$y = c.\exp(-dx) \tag{2}$$

Where a, b, c and d are constants. The exponents b and d both represent the rate of decrease in y with distance; b is dimensionless and d has the dimension of length^{-1}. In the exponential model, the value of y decreases by half in a fixed distance, the half-distance (α), whatever the starting point. The half-distance is related to the coefficient d, such that $\alpha = 0.693/d$.

Deposition gradients of splash-dispersed spores are usually steep and are generally fitted better by the exponential model than by the power law model (McCartney & Fitt, 1985a,b). When splash-dispersed and wind-dispersed clubmoss spores were released simultaneously, the splash-dispersed spores had a much steeper deposition gradient (Stedman, 1979). Half-distances of spore deposition for *S. nodorum, P. brassicae* and *P. herpotrichoides* in the still-air experiments described were less than 20 cm. Vertical deposition gradients of splash-dispersed spores are equally steep. In still air, droplets carrying pycnidiospores of *S. nodorum* have been collected at heights up to 50 cm, but most were collected below 20 cm and the numbers collected decreased with increasing height (Brennan *et al.*, 1985a). Similar vertical profiles have been observed in numbers of conidia of *P. herpotrichoides* collected at 0.75 and 1 m downwind of infected straw onto which simulated rain fell in a wind of 2 m s^{-1} (Fitt & Nijman, 1983).

Horizontal spore deposition gradients are flattened by wind. When simulated raindrops fell onto suspensions of *S. nodorum* pycnidiospores, the downwind half-distance was greater in a wind of 3 m s^{-1} (Fig. 5.6) than in still air (Fig. 5.5). Some spores, presumably carried in ballistic droplets, are deposited upwind of sources, but most are deposited downwind (Fitt & Nijman, 1983). Whereas in still air the spore-carrying droplets which travel furthest and highest are the large ballistic droplets (Gregory *et al.*, 1959), in wind it is the small spore-carrying droplets that travel furthest (Brennan *et al.*, 1985b). Although downwind spore deposition gradients are still steep, with most spores deposited within 2 m of sources, small numbers of spores travel further. Thus spores of

S. nodorum (Brennan *et al.*, 1985*b*) and *P. herpotrichoides* (Fitt & Nijman, 1983) have been deposited up to 4 m downwind of sources in winds of 2 to 3 m s^{-1}. In wind, some small droplets become airborne and a few spores of *S. nodorum* (Brennan *et al.*, 1985*b*) and *P. herpotrichoides* (Fitt & Nijman, 1983) have been collected by suction samplers at a height of 40 cm as far as 10 m downwind of targets.

Spore dispersal in splash droplets: epidemiological implications
Samplers for splash-dispersed spores
Knowledge about the sizes of splash droplets carrying spores in raintower and wind-tunnel experiments can guide the choice and positioning of samplers for splash-dispersed spores in the field. The size range of spore-carrying splash droplets produced by rain falling onto naturally infected plant material is similar to that observed in experiments with drops falling onto spore suspensions; most spore-carrying droplets are between 200 and 600 µm in diameter, some are <100 µm and others are >1000 µm (Fitt & Bainbridge, 1983*b*). This enormous size range suggests that no one sampler is appropriate for collecting all sizes of spore-carrying droplets. Thus, the choice of sampler for spores borne in splash droplets depends not so much on the size of the spores

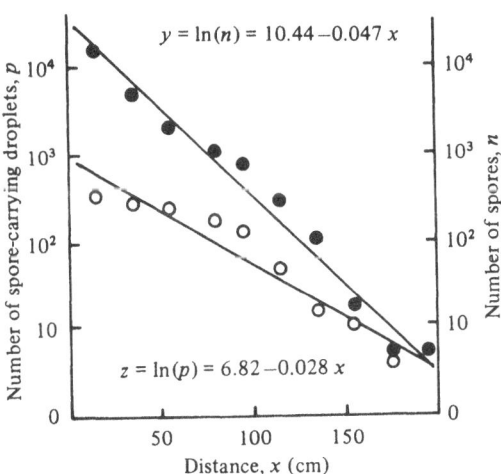

Fig. 5.6. Numbers (ln-transformed) of spore-carrying droplets (*p*) and spores (*n*) collected at different distances downwind (windspeed 3 m s^{-1}) of suspensions of *Septoria nodorum* pycnidiospores (7.8 × 10^5 spores cm^{-3}, 0.5 mm deep) on to which 25 drops (4.3 mm diameter) fell. Regressions of ln(*p*) and ln(*n*) on distance, *x*. (Brennan *et al.*, 1985*b*.)

themselves as on the size of the spore-carrying droplets which are to be collected.

Evidence that the larger ballistic droplets are little affected by wind (Brennan *et al.*, 1985*b*) suggests that they would not be collected effectively by volumetric samplers facing into the wind, which are often used for collecting airborne spores (Fitt & McCartney, 1986). Samplers with a horizontal collecting surface are more suitable. Thus horizontal slides (under rainshields to prevent wash-off), funnels (draining into beakers) or impingers collected more spores of *P. herpotrichoides* than did cylinders or a Burkard sampler placed in an area of infected straw (Fitt & Bainbridge, 1983*b*). Slides are simple and easily replicated, which is important with a patchily distributed disease, but lose some spores through run-off. Funnels, which present a large horizontal collecting surface, collect large numbers of spores, but samples may require concentration and it is sometimes necessary to separate spores from soil particles (Fitt & Bainbridge, 1983*a*). Impingers collect more spores per cm^2 of sampling surface than funnels, probably because they draw in some airborne droplets as well as ballistic droplets (Fitt & Bainbridge, 1983*b*). However, they require a source of power for the suction pump and are less easily replicated than funnels or slides, although they can be connected to a rain-activated switch to confine sampling to periods of rainfall (Fitt, Rawlinson & Smith, 1982).

The steep horizontal and vertical spore deposition gradients observed in splash-dispersal experiments, which have been confirmed in field conditions (Rowe & Powelson, 1973; Fitt & Bainbridge, 1983*b*), suggest that samplers for ballistic spore-carrying droplets should be placed near to sources of inoculum. Thus spores of *S. nodorum* have been collected in funnels at ground level and at a height of 40 cm, but not at a height of 80 cm, and at a distance of 50 cm, but not at 100 cm, from sources (Griffiths & Ao, 1976). Similarly, most spores of *R. secalis* are collected at ground level in a short winter barley crop; however, as the season progresses and sporulating lesions spread up the plants, spores may be collected in small numbers at heights of up to 1 m (Stedman, 1980*b*).

If the small airborne spore-carrying droplets, generally present in only low concentrations, are to be collected, rather than the ballistic droplets, a different type of sampler is required. Airborne droplets are not collected efficiently by samplers with horizontal surfaces near to ground level and should be sampled with a volumetric sampler, facing into the wind. Thus, airborne spores of *S. nodorum* have been collected successfully with a pre-impinger (11 dm^3 min^{-1}; Faulkner & Colhoun,

1977) while high volume samplers, such as a cyclone separator (720 dm³ min⁻¹; Fitt & Bainbridge, 1983*b*) and a Casella high volume bacterial sampler (350 dm³ min⁻¹; Carnegie, 1980), have been used to collect airborne spores of *P. herpotrichoides* and *P. exigua* respectively.

Timing and distance of disease spread

The increase in the number of spores dispersed from suspensions as incident drop size increases (Gregory *et al.*, 1959; Fitt & Lysandrou, 1984) suggests that heavy rainfall may be more effective than other precipitation in spreading splash-dispersed pathogens, particularly when inoculum is at ground level on open ground or in a short crop. This is the case for *P. herpotrichoides*; spores are dispersed from infected straw in large numbers during heavy rain but are not dispersed during dry weather or light rain (Fitt & Bainbridge, 1983*b*). Similarly, dispersal of pycnidiospores of *S. nodorum* from crop debris requires at least 10 mm of rain, reaching an intensity of 2 mm h⁻¹, in a 48-h period (Jordan & Overthrow, 1980). However, under crop canopies, where drip-splash may be more important in spore dispersal than direct rain-splash, light rain, mist or dew may spread disease, e.g. *Gleosporium* spores are spread between apple trees in this way. Nevertheless, D. J. Royle (personal communication) attributes sudden spread of *S. nodorum* from lower to upper leaves of wheat crops to the occurrence of heavy rainstorms in the summer, although these occasions are sometimes difficult to detect in periods of wet weather when disease is gradually spreading up plants.

Unfortunately, data from epidemics have rarely been accompanied by the accurate measurements of rainfall intensity needed to identify precisely the conditions under which pathogen spores are spread by rain-splash. It has been difficult to obtain accurate rainfall intensity data with standard rain gauges or chart records from natural siphon rainfall recorders (Fitt *et al.*, 1982). However, accurate measurements can now be obtained with electronic rainfall recorders connected to data loggers.

The short distances travelled by spores in experiments with spore suspensions suggest that splash-borne pathogens will have steep spore-deposition gradients, and consequently steep disease gradients, in crops. Steep disease gradients, with an effective dispersal range of only *ca.* 1 m, have been observed in field experiments with eyespot (*P. herpotrichoides*) (Rowe & Powelson, 1973). Although gradients in the incidence of leaves infected by *S. nodorum* in spring wheat plots may be steep 2–3 weeks after inoculation, disease gradients in winter wheat have often been flattened by secondary disease spread 2–4 months after inoculation

(Jeger, Jones & Griffiths, 1983). Nevertheless, disease spread by rain-splash is generally far more limited than disease spread by wind because most spores are carried only a short distance in ballistic splash droplets.

Spores carried in small airborne splash droplets, which have been detected in some wind-tunnel experiments (Fitt & Nijman, 1983; Brennan *et al.*, 1985*b*) have, however, also been observed in the field. Small numbers of airborne spores of *P. herpotrichoides* (Fitt & Bainbridge, 1983*b*) and *S. nodorum* (Faulkner & Colhoun, 1977; Wale & Colhoun, 1979) have been collected up to 2 m above infected stubble. Large numbers of airborne *P. exigua* spores have been detected up to 600 m downwind of infected potato crops during heavy rain (Carnegie, 1980, 1984). The significance of these airborne spores in disease spread is not clear. Droplets small enough to become airborne generally contain only one or two spores and for many pathogens it is not known how long the spores remain viable or how many spores are required to initiate infections. Although eyespot lesions can be initiated by a few spores of *P. herpotrichoides*, they develop much more rapidly when initiated by several hundred spores (Higgins & Fitt, 1984). However, there is evidence that airborne droplets may be important in the spread of black pod (*Phytophthora* spp.) from ground level inoculum up to heights of 5 m in the canopy of cocoa trees (Gregory, Griffin, Maddison & Ward, 1984).

References

Bainbridge, A. & Stedman, O. J. (1979). Dispersal of *Erysiphe graminis* and *Lycopodium clavatum* spores near to the source in a barley crop. *Annals of Applied Biology,* **91,** 187–98.

Brennan, R. M., Fitt, B. D. L., Taylor, G. S. & Colhoun, J. (1985*a*). Dispersal of *Septoria nodorum* pycnidiospores by simulated raindrops in still air. *Phytopathologische Zeitschrift,* 281–90.

Brennan, R. M., Fitt, B. D. L., Taylor, G. S. & Colhoun, J. (1985*b*). Dispersal of *Septoria nodorum* pycnidiospores by simulated rain and wind. *Phytopathologische Zeitschrift,* 291–97.

Carnegie, S. F. (1980). Aerial dispersal of potato gangrene, *Phoma exigua* var. *foveata. Annals of Applied Biology,* **94,** 165–73.

Carnegie, S. F. (1984). Seasonal occurrence of the potato gangrene pathogen, *Phoma exigua* var. *foveata,* in the open air. *Annals of Applied Biology,* **104,** 443–9.

Corke, A. T. K. (1966). The role of rainwater in the movement of *Gleosporium* spores on apple trees. In *The Fungus Spore,* ed. M. F. Madelin, pp. 143–9. London: Butterworths.

Fatemi, F. & Fitt, B. D. L. (1983). Dispersal of *Pseudocercosporella herpotrichoides* and *Pyrenopeziza brassicae* spores in splash droplets. *Plant Pathology,* **32,** 401–4.

Faulkner, M. J. & Colhoun, J. (1977). An automatic spore trap for collecting pycnidiospores of *Leptosphaeria nodorum* and other fungi from air during rain and maintaining them in a viable condition. *Phytopathologische Zeitschrift*, **89**, 50–9.

Fitt, B. D. L. & Bainbridge, A. (1983a). Recovery of *Pseudocercosporella herpotrichoides* spores from rain-splash samples. *Phytopathologische Zeitschrift*, **106**, 177–82.

Fitt, B. D. L. & Bainbridge, A. (1983b). Dispersal of *Pseudocercosporella herpotrichoides* spores from infected wheat straw. *Phytopathologische Zeitschrift*, **106**, 214–25.

Fitt, B. D. L., Creighton, N. F. & Bainbridge, A. (1985). The role of wind and rain in the dispersal of *Botrytis fabae* conidia. *Transactions of the British Mycological Society*, 307–12.

Fitt, B. D. L. & Lysandrou, M. (1984). Studies on mechanisms of splash dispersal of spores, using *Pseudocercosporella herpotrichoides* spores. *Phytopathologische Zeitschrift*, **111**, 323–31.

Fitt, B. D. L., Lysandrou, M. & Turner, R. H. (1982). Measurement of spore-carrying splash droplets using photographic film and an image-analysing computer. *Plant Pathology*, **31**, 19–24.

Fitt, B. D. L. & McCartney, H. A. (1986). Spore dispersal in relation to epidemic models. In *Plant Disease Epidemiology*, vol. 1, ed. K. J. Leonard & W. E. Fry. New York: Macmillan Publishing Company (in press).

Fitt, B. D. L. & Nijman, D. J. (1983). Quantitative studies on dispersal of *Pseudocercosporella herpotrichoides* spores from infected wheat straw by simulated rain. *Netherlands Journal of Plant Pathology*, **89**, 198–202.

Fitt, B. D. L., Rawlinson, C. J. & Smith, C. B. (1982). A comparison of two rain-activated switches used with samplers for spores dispersed by rain. *Phytopathologische Zeitschrift*, **105**, 39–44.

Fournet, J. (1969). Propriétés et rôle du cirrhe du *Septoria nodorum* Berk. *Annales de Phytopathologie*, **1**, 87–94.

Glynne, M. D. (1953). Production of spores by *Cercosporella herpotrichoides*. *Transactions of the British Mycological Society*, **36**, 46–51.

Gregory, P. H. (1968). Interpreting plant disease dispersal gradients. *Annual Review of Phytopathology*, **6**, 189–212.

Gregory, P. H. (1973). *The Microbiology of the Atmosphere*, 2nd edn. London: Leonard Hill.

Gregory, P. H., Griffin, M. J., Maddison, A. C. & Ward, M. R. (1984). Cocoa black pod: a reinterpretation. *Cocoa Grower's Bulletin*, **35**, 5–22.

Gregory, P. H., Guthrie, E. J. & Bunce, M. E. (1959). Experiments on splash dispersal of fungus spores. *Journal of General Microbiology*, **20**, 328–54.

Griffiths, E. & Ao, H. C. (1976). Dispersal of *Septoria nodorum* spores and spread of glume blotch of wheat in the field. *Transactions of the British Mycological Society*, **67**, 413–18.

Griffiths, E. & Peverett, H. (1980). Effects of humidity and cirrhus extract on survival of *Septoria nodorum* spores. *Transactions of the British Mycological Society*, **75**, 147–50.

Higgins, S. & Fitt, B. D. L. (1984). Production and pathogenicity to wheat of *Pseudocercosporella herpotrichoides* conidia. *Phytopathologische Zeitschrift*, **111**, 222–31.

Hirst, J. M. & Stedman, O. J. (1963). Dry liberation of fungus spores by raindrops. *Journal of General Microbiology*, **33**, 335–44.

Ingold, C. T. (1971). *Fungal Spores: Their Liberation and Dispersal*. Oxford: Clarendon Press.

Jeger, M. J., Jones, D. G. & Griffiths, E. (1983). Disease spread of non-specialised fungal pathogens from inoculated point sources in intraspecific mixed stands of cereal cultivars. *Annals of Aplied Biology*, **102**, 237–44.

Jordan, V. W. L. & Overthrow, R. B. (1980). Epidemiology and control of splash-dispersed and other cereal diseases. *Report of Long Ashton Research Station for 1979*, pp. 110–11.

Kiyosawa, S. & Shiyomi, M. (1972). A theoretical evaluation of the effect of mixing resistant variety with susceptible variety for controlling plant diseases. *Annals of the Phytopathological Society of Japan*, **38**, 41–51.

McCartney, H. A. & Fitt, B. D. L. (1985a). Construction of dispersal models. In *Mathematical Modelling of Crop Disease*, ed. C. A. Gilligan. *Advances in Plant Pathology*, **3**, 107–43.

McCartney, H. A. & Fitt, B. D. L. (1985b). Spore dispersal gradients and disease development. In *Populations of Plant Pathogens: Their Dynamics and Genetics*, ed. M. S. Wolfe & C. E. Caten. Oxford: Blackwell Scientific Publications (in press).

Ramalingam, A. & Rati, E. (1979). Role of water in the dispersal of nonwettable spores. *Indian Journal of Botany*, **2**, 8–11.

Rawlinson, C. J. (1979). Light leaf spot of oilseed rape: an appraisal with comments on strategies for control. In *Proceedings of the 1979 British Crop Protection Conference – Pests and Diseases*, pp. 137–43. London: British Crop Protection Council.

Rowe, R. C. & Powelson, R. L. (1973). Epidemiology of Cercosporella foot rot of wheat: disease spread. *Phytopathology*, **63**, 984–8.

Stedman, O. J. (1979). Patterns of unobstructed splash dispersal. *Annals of Applied Biology*, **91**, 271–85.

Stedman, O. J. (1980a). Dispersal of spores of *Rhynchosporium secalis* from a dry surface by water drops. *Transactions of the British Mycological Society*, **75**, 339–40.

Stedman, O. J. (1980b). Observations on the production and dispersal of spores, and infection by *Rhynchosporium secalis*. *Annals of Applied Biology*, **95**, 163–75.

Stepanov, K. M. (1935). Dissemination of infective diseases of plants by air currents (Ru.). *Bulletin of Plant Protection, Leningrad, Series 2, Phytopathology*, **8**, 1–68.

Wale, S. J. & Colhoun, J. (1979). Further studies on aerial dispersal of *Leptosphaeria nodorum*. *Phytopathologische Zeitschrift*, **94**, 185–9.

Worthington, A. M. & Cole, R. S. (1897). Impact with a liquid surface, studied by the aid of instantaneous photography. *Philosophical Transactions of the Royal Society A*, **189**, 137–48.

6
The zoospore and its problems

MICHAEL J. CARLILE

Department of Pure and Applied Biology, Imperial College at Silwood Park, Ascot, Berkshire, SL5 7PY, UK

'These unique infective units provide tremendous and explosive disease potential'

–George A. Zentmyer on *Phytophthora* zoospores (Zentmyer, 1983)

The zoospore can be regarded as a vehicle with a single role – the transport of the fungus genome from one suitable environment to another. Many of the questions that we can ask about a zoospore are analogous to those that we would ask about a man-made vehicle such as an aeroplane. What does it look like, and what sort of engine does it have? What is its speed, fuel consumption and range? What guidance systems does it possess? In addition to such details, we also want to know about a typical zoospore's trip – take-off, the 'flight', approach to destination and 'landing'. Some of the above questions can be answered for zoospores of the important Oomycete plant pathogen *Phytophthora* (Carlile, 1983) and to a lesser extent for the saprotrophic Chytridiomycete *Blastocladiella emersonii* (Cantino & Mills, 1976), and it is upon these fungi that this chapter concentrates. The diversity of zoospore structure, form and behaviour is discussed by Fuller (1977), Webster (1980) and Lange & Olson (1983).

Form and structure

The zoospores of Oomycetes are biflagellate (Fig. 6.1a). The anterior 'tinsel' flagellum carries on each side a row of stiff hairs (mastigonemes). These are lacking in the apparently smooth posterior flagellum although fine hairs have been detected. The soma (body) tapers towards the front and ranges in length in the genus *Phytophthora* from 9 to 15 μm. In the aquatic Oomycete *Saprolegnia* there are two types of zoo-

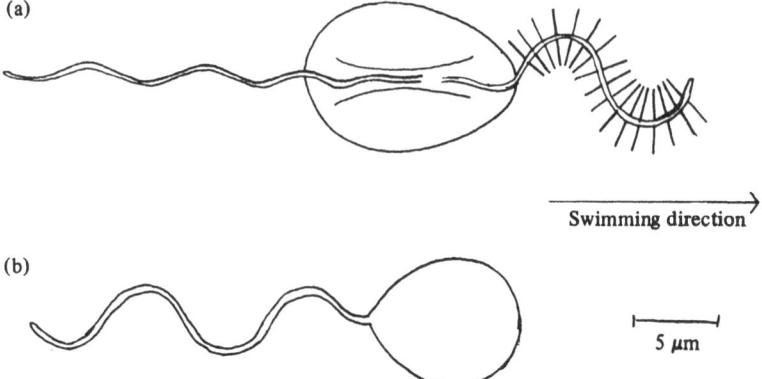

Fig. 6.1. Diagram of zoospores, approximately to scale and swimming towards the right. (a) A zoospore of *Phytophthora palmivora* viewed from the ventral surface showing the deep groove from which the flagella emerge. The length of the soma which tapers towards the front is *ca.* 13 μm, the width 8 μm and the dorsal–ventral distance rather less. The anterior flagellum which bears mastigonemes 2 μm long undulates vigorously and provides about 90% of the thrust. The generally less-active posterior flagellum can undulate in a plane perpendicular to that of the soma and anterior flagellum and also acts as a rudder.
(b) A zoospore of *Blastocladiella emersonii*. The soma is *ca.* 9 μm long and 7 μm wide and the single posterior flagellum undulates in a plane with waves of amplitude *ca.* 2.5 μm and wavelength 18 μm.

spores. differing in form and internal structure (Holloway & Heath, 1977). The *primary zoospore* swims sluggishly and soon after emerging from the sporangium encysts to give a primary cyst. This germinates after a few hours to give a *secondary zoospore* which swims vigorously and has the typical Oomycete zoospore form. In other members of the Saprolegniales the primary zoospore may be absent, may encyst within the sporangium or immediately it emerges, or may be vigorous and take the place of the secondary zoospore (Webster, 1980). The zoospores of Chytridiomycetes have a single smooth posterior flagellum (Fig. 6.1b). The soma is generally smaller than that of Oomycetes, the length being *ca.* 9 μm in *Blastocladiella* and 3 μm in *Synchytrium endobioticum*. The flagellum in both uniflagellate and biflagellate zoospores is *ca.* 20 μm long.

   Zoospores of Oomycetes and Chytridiomycetes also differ in their internal structure – for detail see Hemmes (1983) for *Phytophthora* and Cantino & Mills (1976) for *Blastocladiella*. Zoospores lack a rigid wall, although a fluffy coat *ca.* 25 nm thick has been detected outside the plasmalemma in *Phytophthora* (Bartnicki-Garcia & Wang, 1983). This nearly naked condition presents problems. In solutions hypotonic with respect to protoplasm, such as rain, ponds and rivers, water will enter the cell by osmosis and create a positive turgor potential (ψ_p). In other fungal cells rupture is prevented by the mechanical strength of the cell wall. In the zoospore, however, excessive ψ_p must be avoided by the expulsion of water. The Oomycete zoospore has a water expulsion apparatus consisting of a contractile vacuole and an associated vesicular zone near the ventral groove. Such a striking, localised apparatus is lacking in the Chytridiomycetes but some less obvious structures presumably fulfil a similar role (Lange & Olson, 1983). The absence of a cell wall may also render zoospores more susceptible than other fungal cells to damage in solutions with an inappropriate ion content or traces of toxic materials.

Swimming, osmoregulation and the maintenance of the correct internal ion concentrations require energy, and the necessary fuel stores are partly in the form of lipid droplets. Zoospores of *Phytophthora palmivora* also contain the $(1 \rightarrow 3)$-β-D-glucan mycolaminaran (Bartnicki-Garcia & Wang, 1983) and zoospores of *Blastocladiella* contain glycogen. A few zoospore-producing fungi, such as the Oomycete *Aqualinderella fermentans* (Held, 1970) and the Chytridiomycete *Blastocladia* (Emerson & Robertson, 1974), are exclusively fermentative in their metabolism and energy is provided by substrate-level phosphorylation. Most fungi however depend on oxidative phosphorylation as their source of energy and zoospores of Oomycetes contain many small mitochondria. *Blastocladiella* has a single compact mitochondrion adjacent to the base of the flagellum and such an arrangement is common in small uniflagellate zoospores. A remarkable feature of the zoospores of *Blastocladiella* and other Blastocladiales is the nuclear cap, a package of about a million ribosomes (Cantino, 1975) adjacent to the nucleus. Small uniflagellate zoospores tend to show greater ultrastructural organization than do the large zoospores of Oomycetes. They can be thought of as having evolved towards the lower limit of size for a flagellate eukaryotic cell. This would permit the production of a greater number of zoospores per unit mass of parental protoplasm. Miniaturisation implies, however, the need for an efficient packaging of organelles and so a variety of fascinating layouts are observed in small zoospores (Lange & Olson, 1983).

Metabolism and performance

Zoospores are capable of swimming in nutrient-free solutions for many hours. This implies that swimming, osmoregulation and the maintenance of an appropriate ionic balance can be supported by endogenous metabolism. Suberkropp & Cantino (1973) found that zoospores of *Blastocladiella* with an average weight of 47 pg initially contained *ca.* 26 pg of protein, 4 pg of lipid and a glycogen content varying from 1.4 to 2.3 pg. After 5 h, *ca.* 3 pg of protein and 1 pg of lipid had disappeared. The fall in the amount of glycogen depended on the initial concentration and varied from 0.5 to 1 pg, being greatest when the initial level was high. A respiratory quotient of 0.92 was obtained, consistent with the oxidative metabolism of the amounts of protein, lipid and glycogen that had disappeared. Suberkropp & Cantino (1972) found that about 15% less oxygen was consumed by zoospores that had been deprived of their flagella, suggesting that this is the proportion of the energy consumption that is devoted to motility. Cantino, Truesdell & Shaw (1968) found that at low oxygen tensions swimming could continue for 6 h and that lactic acid was produced, suggesting that motility and other functions were being maintained by fermentation, as indeed must occur in zoospores of obligately fermentative species. The zoospores of *P. palmivora*, however, require oxygen for survival (Cameron, 1979). Bimpong (1975) observed that about half the lipid in *P. palmivora* zoospores disappears in 6 h.

Zoospores of *Blastocladiella* are propelled at about $100 \, \mu m \, s^{-1}$ by waves that travel from the base to the tip of the flagellum (Miles & Holwill, 1969). They undergo frequent, apparently random changes of direction including periods of cessation of movement of *ca.* 0.1 s. Waves cease to be propagated along the flagellum, which then bends near its base causing a reorientation of the soma. Wave propagation is then resumed and the zoospore moves off in the new direction.

The swimming of *Phytophthora* zoospores is reviewed by Carlile (1983). In *P. palmivora* waves are propagated from base to tip in both the anterior and posterior flagella, but because of the presence of mastigonemes on the former, both produce thrust in the same direction (Holwill, 1982). Cine film of waveform and hydrodynamic analysis shows that the anterior flagellum contributes ten times the thrust of the posterior. Directional changes occur through a brief (*ca.* 0.1 s) cessation of wave propagation by the posterior flagellum which bends at the base to produce an orientation perpendicular to that of the soma. The anterior flagellum continues to beat so the posterior flagellum acts as a rudder and a direction change occurs with little reduction in velocity.

While smooth swimming with both flagella bending in the same plane is observed in a shallow layer of suspending medium, zoospores of *P.*

palmivora and other species commonly rotate when not restricted by shallow media, with the posterior flagellum undulating in a plane perpendicular to that of the soma and the anterior flagellum. Rotation converts a tendency to yaw into a helical path enabling a constant direction to be maintained. It is probable that both modes of progress are natural, the smooth path being normal in thin films or close to obstacles and the helical path where restrictions on movement are lacking. The swimming speeds (i.e. displacement along the axis of their helical path) for *Phytophthora* zoospores range, depending on species and temperature, from 90 to 160 μm s^{-1}.

Zoospores of *Blastocladiella* swim for 5 h without appreciable encystment (Suberkropp & Cantino, 1973) and in several *Phytophthora* spp. a substantial proportion of a zoospore population is still swimming after 6 to 10 h (Carlile, 1983). There are a few reports of motility over longer periods but often these apply to a small part of a population and may be based on zoospores that have had a period in an encysted state. It seems reasonable to regard the potential period of activity of a zoospore as at least 6–10 h but usually less than 24 h.

Assuming that a zoospore remains motile for 10 h, how far can it swim? Taking 160 μm s^{-1} as approximately the greatest swimming speed, the zoospores could swim 576 mm in 1 h and 5.8 m in 10 h. However the distance reached is limited by frequent spontaneous direction changes, which for *P. cinnamomi* average *ca.* 0.14 rad s^{-1} or 8° s^{-1}. Newhook, Young, Allen & Allen (1981) studied the swimming of *P. cinnamomi* zoospores through coarse wet sand from a reservoir. They found that zoospores reached 4 cm in 3 h, 5 cm in 6 h and 6 cm in 10 h, although the numbers at the latter distance were only 0.1–0.2% of those remaining in the reservoir. Thus, although zoospores may be able to swim *ca.* 6 m in 10 h, they reach at most 6 cm from their starting point by swimming. Experiments with soils (see p. 111) also show that zoospores are unable to spread far by swimming which, with its numerous directional changes, is a means of thoroughly exploring a limited volume rather than travelling long distances.

Take-off: zoospore discharge and departure

The conditions under which zoosporangia are formed and zoospores produced and discharged have been most intensively studied in *Phytophthora* (Duniway, 1983; Gisi, 1983; see also Duniway & Gordon, Ch. 7; Holderness & Pegg, Ch. 11). Species which form sporangia on the aerial parts of plants require high relative humidities and those that form them in the soil need high ambient water potentials (ψ), i.e. soil solutions low in solutes. The sporangia of essentially root-infecting species of *Phytophthora* remain attached to the mycelium but species which attack the aerial parts of plants produce sporangia which are detachable and can be dispersed by rain and in a few species by wind.

Sporangia of *Phytophthora* can germinate directly by means of a germ tube, thus acting as a conidium, or indirectly by the discharge of zoospores. Discharge of zoospores is associated with high ψ (Gisi, 1983) and, in species which attack the aerial parts of plants, also with low temperatures. It is at low temperatures that dew or rain will persist longest on leaves and permit zoospore motility. High temperatures or low ψ favour production of germ tubes, less dependent on water than are zoospores for their survival. Heavy infection by *Phytophthora* tends to occur under conditions favouring indirect rather than direct germination, testifying to the effectiveness of zoospores in spreading infection (Duniway, 1983). Zoosporangium formation and discharge by other Oomycetes and by Chytridiomycetes also appears to be favoured by high ψ since in many species both processes are promoted by repeated washing of mycelia with distilled water or dilute salt solutions.

As discussed earlier, randomly swimming zoospores can travel only a short distance from their starting point. Zoosporangia of aquatic fungi may well be produced in a layer of static water adjacent to their substratum and those of soil inhabitants in wet conditions below the soil surface. In both instances negative geotaxis is likely to take zoospores to locations where passive dispersal is more likely to occur; it has been demonstrated in *Phytophthora cactorum*, *Phytophthora nicotianae* and *P. palmivora* (Cameron & Carlile, 1977). Bean (1984) has discussed the ways in which geotaxis of motile unicellular organisms may occur. One theory is that an organism that tapers towards the front will tend to point upwards due to a more rapid sedimentation of the bulky posterior. Another theory requires that cell rotation is combined with a centre of propulsive effect anterior to the centre of gravity. Zoospores of *Phytophthora* have a bulky posterior, flagella inserted well forward and cell rotation, satisfying the requirements of both theories. Since zoospores of other Oomycetes, including aquatic species, are similar in form to those of *Phytophthora*, it could well be that negative geotaxis is a universal feature of Oomycete zoospores. The zoospores of Chytridiomycetes, however, have forms which, from physical theories of geotaxis, seem inappropriate for negative geotaxis.

Positive phototaxis has been demonstrated in the zoospores of several aquatic Chytridiomycetes (Kazama, 1972; Robertson, 1972). This would be a satisfactory process for bringing about upward swimming in zoospores of aquatic but not of soil fungi. A third way in which upward swimming could occur is by positive aerotaxis, swimming up an oxygen gradient. This has been excluded in several *Phytophthora* spp. (Cameron & Carlile, 1977) but perhaps should be sought in Chytridiomycetes not known to show phototaxis.

Flight: movement of zoospores over long distances

It is possible that rain-splash followed by transport of droplets by wind may disperse *Phytophthora* zoospores over tens of metres (Gregory, 1983). 'Flight', however, must in other instances be construed metaphorically rather than literally.

It is probable that water currents have a role in the dispersal of zoospores of aquatic fungi. Zoospores of terrestrial fungi such as *Phytophthora* may also be transported by water flowing over the surface of soil and also through soil (Duniway, 1983, and Ch. 7). The extent to which water flowing through soil will transport zoospores will depend on soil texture and the degree of disturbance by agricultural and other human activities. The transport of *P. cinnamomi* zoospores by water flowing at various rates has been studied with soil models composed of glass beads of various sizes (Young, Newhook & Allen, 1979), with soil–sand mixtures and with cores of natural soil (Newhook *et al.*, 1981). It was found that provided the soil or soil model had pores of a size in excess of zoospore diameter, a large proportion of the zoospores could both be carried through the system and retain motility. Zoospores of some other *Phytophthora* spp. were more liable to encyst during passage (Benjamin & Newhook, 1982). This need not be a disadvantage, because cysts can, depending on conditions (see p. 115), germinate either directly to give a germ-tube or indirectly to give a zoospore.

If water-flow rates are greatly in excess of the swimming speed of a zoospore, then the zoospore will have no control over its direction of movement. If, however, the flow rate is similar to or less than the swimming speed, then a zoospore can be expected to respond in an appropriate way to environmental cues, as has been shown for *Phytophthora* zoospores with chemical attractants (Young *et al.*, 1979; Newhook *et al.*, 1981). Negative geotaxis may help to keep *Phytophthora* zoospores near the surface of soil where most roots occur, as may the positive rheotaxis (upstream swimming) demonstrated in *Phytophthora capsici* by Katsura & Miyata (1971). Negative chemotaxis (repulsion) in response to hydrogen ions and to varying extents to a range of other cations has been demonstrated in *P. cinnamomi* (Allen & Harvey, 1974) and *P. palmivora* (Cameron & Carlile, 1980). Repulsion by hydrogen ions occurs at $150\,\mu m$ (pH 3.8) in *P. palmivora* and $37\,\mu M$ (pH 4.4) in *P. cinnamomi*, and by most other cations at rather higher concentrations. Byrt, Irving & Grant (1982*b*) show that the threshold for damage to zoospores by cations is in general near the thresholds for repulsion demonstrated by Allen & Harvey (1974) and Cameron & Carlile (1980).

Approach to destination: positive chemotaxis

In order to fulfil their roles the zoospores of plant pathogens must reach a host plant, and those of saprotrophs suitable organic material. Contact might ultimately be achieved as a result of random swimming, but will occur sooner if a zoospore can detect its host or food material at a distance and then modify its direction of swimming accordingly. Roots slough off cells in their progress through the soil and also release remarkable quantities of amino acids, sugars, organic acids and macromolecules (Gaworzewska & Carlile, 1982). Microbial attack on sloughed-off cells and macromolecules may increase the amount of soluble, low molecular weight compounds, which will diffuse away from the root. The remains of plants and animals will also contain low molecular weight compounds, and the amounts diffusing from fragments may be increased by autolysis or microbial attack.

Disturbance by currents is minimal close to solid objects, so plant roots or organic fragments will be the centre of reasonably stable diffusion gradients of organic solutes. It is clear that the zoospores of many fungi can respond appropriately by positive chemotaxis. The most extensive information is for zoospores of *Phytophthora*, which have been shown to be attracted (Carlile, 1983) to root tips, both excised and growing in soil, root exudates, amino acids, sugars, organic acids, ethanol (likely to be emitted by roots in waterlogged soils) and a range of higher alcohols, aldehydes and fatty acids known or likely to be emitted by plants. Attraction to roots occurs in other root-infecting Oomycetes such as *Pythium aphanidermatum* (Royle & Hickman, 1964) and attraction to stomata occurs in Downy Mildews that infect leaves (Royle & Thomas, 1973). The zoospores of *Allomyces* are attracted by amino acids (Machlis, 1969), those of the rumen fungus *Neocallimastrix* by sugars (Orpin & Bountiff, 1978) and the secondary zoospores of *Saprolegnia* by a mixture of amino acids and salts (Fischer & Werner, 1958).

Electrotaxis as well as chemotaxis occurs in *Phytophthora* (Khew &

Zentmyer, 1974). This response may be of significance in the final stages of a zoospore's approach to its host since weak electric currents occur in the vicinity of roots.

Landing: zoospore encystment and adhesion

Production of a cyst wall is a necessary preliminary to production of a germ-tube and vegetative growth, so a zoospore will need to encyst on arriving at its destination. It is likely therefore that encystment will be induced by substances emitted by a host or, for saprophytes, by a suitable non-living substrate. It is possible that the same substances that act as attractants as a gradient and at low concentration will induce encystment at higher concentrations. Thus Dill & Fuller (1971) showed that leucine and lysine, attractants for *Allomyces* zoospores with a threshold of 10^{-5} M (Machlis, 1969) will, at concentrations above 3×10^{-3} M, immobilise zoospores in about one minute and cause encystment. Lysine and arginine induce encystment in *Phytophthora cinnamomi* (Byrt, Irving & Grant, 1982*a*). Living hosts will be likely to emit a greater variety of compounds with greater consistency, and thus for parasites there will be the possibility of encysting in response to high molecular weight compounds that are unlikely to diffuse far from the host surface. Thus *P. cinnamomi* has been shown to encyst in response to pectins and polygalacturonic acid, and to carbohydrates washed from roots of axenically grown plants (Byrt *et al.*, 1982*b*; Irving & Grant, 1984).

It has been found that a brief period of violent agitation produces rapid and synchronous encystment of *Phytophthora palmivora* zoospores. This has permitted detailed study of the events that terminate motility (Bartnicki-Garcia & Wang, 1983). The zoospore becomes sluggish, rounds off and sheds or retracts its flagella. Meanwhile peripheral vesicles fuse with the plasmalemma and discharge their contents onto the cell surface to give an amorphous coat which is at least in part an adhesive, since the zoospore becomes sticky for a short period. If during this brief period it comes into contact with a solid object it becomes firmly attached. Whether attachment occurs or not, the production of a cyst wall follows. Protein synthesis is not required for encystment in *P. palmivora*, so the necessary enzymes for wall biosynthesis must already be present but in an inactive state or separated from their substrates. The same is true for *Blastocladiella* (Soll & Sonneborne, 1971) and may well be general for zoospore encystment.

As indicated above, a phase of adhesiveness is one way in which zoospores become attached to a host. There are however other possible tactics, as shown by the secondary cysts of *Saprolegnia* which bear double-headed recurved hooks (Webster, 1980) and readily attach to the skin of fish (Willoughby & Pickering, 1977).

The passengers emerge: germination and tropism

Germination rapidly follows encystment for zoospores that have reached a host or a suitable substratum. In *Blastocladiella* encystment automatically leads to germination. The nuclear cap disintegrates, releasing its ribosomes, protein synthesis commences and a germ-tube is produced. In *Phytophthora* germ-tube production, which as in *Blastocladiella* involves protein synthesis, need not automatically follow encystment. However, germination in the vicinity of roots will swiftly follow encystment since it is stimulated by a variety of factors present in root exudates. Some of these such as lysine and arginine are themselves encystment-inducing factors, others (such as the other amino acids and sugars) are not (Byrt *et al.*, 1982*b*).

The germ-tubes of *Blastocladiella* and the rhizoids that emerge from them show positive chemotropism, growing up concentration gradients of amino acids and inorganic phosphate (Harold & Harold, 1980; Kropf & Harold, 1982). Germ-tubes of *Achlya* are also attracted by amino acids (Musgrave *et al.*, 1977). It is clear that *Phytophthora* germ-tubes are attracted by roots (Carlile, 1983) but the nature of the attractant is uncertain. Ho & Hickman (1967*b*) found that germ-tubes would grow up a H^+ gradient and Duggan & Carlile (unpublished, see Carlile, 1983) found indications that germ-tubes would grow down an O_2 gradient. Claims for attraction by root exudates or amino acid mixtures have generally been based on growth into the mouth of a capillary tube containing the postulated attractant, but such growth could be down an O_2 gradient. Tests with the quantitative and objective assays of Musgrave, Ero, Scheffer & Oehlers (1977) are needed with root exudates; a limited study along these lines with *Phytophthora citricola* (Tew & Carlile, unpubl.) indicated that amino acids are not chemotropically active for germ-tubes of that species.

Coping with emergencies

The likelihood of any single zoospore reaching a suitable destination is slight and probably most journeys end in disaster. Zoospores have however some ability to cope with emergencies, escaping from hostile conditions by evasive action (negative chemotaxis) or encystment. *Phytophthora* zoospores can avoid potentially toxic concentrations of cations (Byrt *et al.*, 1982*a*) by negative chemotaxis (p. 111). Swift escape may however not always be easy, perhaps being hampered by water movements or obstacles. Byrt *et al.* (1982*a*) have shown that many cations induce encystment at concentrations near the threshold for toxicity, and it is probable that encysted zoospores are less susceptible to damage by toxic chemicals than when in a wall-less state. Induction of encystment by cations also occurs in *Blastocladiella* (Soll & Sonneborne, 1972; Van Brunt & Harold, 1980). The most common emergency likely to be encountered by zoospores will be failure to reach a suitable destination after many hours of motility. With fuel supplies running low, spontaneous (i.e. endogenously determined) encystment is an appropriate response. It is likely that in the encysted state, with no need for osmoregulation and motility, reserves are consumed much less rapidly.

Encysted zoospores of *Blastocladiella* are incapable of a resting phase and germinate immediately. It is probable, however, that the germ-tube with its wall is better protected from adverse conditions than is a naked zoospore. The zoospores of *Phytophthora* need not germinate immediately, although there is a lack of study on how long they can survive. Even a brief period of survival may give time for a growing rootlet to reach them. Ho & Hickman (1967*a*) showed that *Phytophthora* cysts usually germinate to give zoospores if in distilled water but germ-tubes if nutrients or root exudates are present.

Conclusions

Successful modern airliners are the end products of a process of intense economic competition. Their speeds are as great as can be achieved without an excessive cost in fuel. Their range, for the different classes of airliners, is governed by typical distances between major cities plus a margin for safety. Zoospores are highly successful devices for the propagation of many fungi and comparable optimisations will have been achieved by intense intraspecific competition.

References

Allen, R. N. & Harvey, J. D. (1974). Negative chemotaxis of zoospores of *Phytophthora cinnamomi*. *Journal of General Microbiology*, **84**, 28–38.
Bartnicki-Garcia, S. & Wang, M. C. (1983). Biophysical aspects of morphogenesis in *Phytophthora*. In *Phytophthora: Its Biology, Taxonomy, Ecology and Pathology*, ed. D. C. Erwin, S. Bartnicki-Garcia & P. H. Tsao, pp. 121–37. St Paul, Minnesota: The American Phytopathological Society.

116   M. J. Carlile

Bean, B. (1984). Microbial geotaxis. In *Membranes and Sensory Transduction*, ed. G. Colombetti & F. Lenci, pp. 163–98. New York: Plenum Publishing Corporation.

Benjamin, M. & Newhook, F. J. (1982). Effect of glass microbeads on *Phytophthora* zoospore motility. *Transactions of the British Mycological Society*, **78**, 43–6.

Bimpong, C. E. (1975). Changes in metabolic reserves and enzyme activities during zoospore motility and cyst germination in *Phytophthora palmivora*. *Canadian Journal of Botany*, **53**, 1411–16.

Byrt, P. N., Irving, H. R. & Grant, B. R. (1982a). The effect of cations on zoospores of the fungus *Phytophthora cinnamomi*. *Journal of General Microbiology*, **128**, 1189–98.

Byrt, P. N., Irving, H. R. & Grant, B. R. (1982b). The effect of organic compounds on the encystment, viability and germination of zoospores of the fungus *Phytophthora cinnamomi*. *Journal of General Microbiology*, **128**, 2343–51.

Cameron, J. N. (1979). *Taxes of the Zoospores of* Phytophthora. Ph.D. Thesis, University of London.

Cameron, J. N. & Carlile, M. J. (1977). Negative geotaxis of zoospores of the fungus *Phytophthora*. *Journal of General Microbiology*, **98**, 599–602.

Cameron, J. N. & Carlile, M. J. (1980). Negative chemotaxis of zoospores of the fungus *Phytophthora palmivora*. *Journal of General Microbiology*, **120**, 347–53.

Cantino, E. C. (1975). Direct counts of the ribosomes per cell in the zoospores of *Blastocladiella emersonii*. *Biochemical and Biophysical Research Communications*, **63**, 343–8.

Cantino, E. C. & Mills, G. L. (1976). Form and function in chytridiomycete spores. In *The Fungal Spore: Form and Function*, ed. D. J. Weber & W. H. Hess, pp. 501–57. New York: Wiley.

Cantino, E. C., Truesdell, L. C. & Shaw, D. S. (1968). Life history of the motile spore of *Blastocladiella emersonii*: a study in cell differentiation. *The Journal of the Elisha Mitchell Scientific Society*, **84**, 125–46.

Carlile, M. J. (1983). Motility, taxis and tropism in *Phytophthora*. In *Phytophthora: Its Biology, Taxonomy, Ecology and Pathology*, ed. D. C. Erwin, S. Bartnicki-Garcia & P. H. Tsao, pp. 95–107. St Paul, Minnesota: The American Phytopathological Society.

Dill, B. C. & Fuller, M. S. (1971). Amino acid immobilization of fungal cells. *Archiv für Mikrobiologie*, **78**, 92–8.

Duniway, J. M. (1983). Role of physical factors in the development of *Phytophthora* diseases. In *Phytophthora: Its Biology, Taxonomy, Ecology and Pathology*, ed. D. C. Erwin, S. Bartnicki-Garcia & P. H. Tsao, pp. 175–87. St Paul, Minnesota: The American Phytopathological Society.

Emerson, R. & Robertson, J. A. (1974). Two new members of the Blastocladiaceae. I. Taxonomy, with evaluation of genera and inter-relationships in the family. *American Journal of Botany*, **61**, 303–17.

Fischer, F. G. & Werner, G. (1958). Die Chemotaxis der Schwärmssporen von Wasserpilzen (Saprolegniaceen). *Hoppe-Seyler's Zeitschrift für Physiologische Chemie*, **310**, 65–91.

Fuller, M. S. (1977). The zoospore, hallmark of aquatic fungi. *Mycologia*, **69**, 1–20.

Gaworzewska, E. T. & Carlile, M. J. (1982). Positive chemotaxis of *Rhizobium leguminosarum* and other bacteria towards root exudates of legumes and other plants. *Journal of General Microbiology*, **128**, 1179–88.

Gisi, U. (1983). Biophysical aspects of the development of *Phytophthora*. In *Phytopthora: Its Biology, Taxonomy, Ecology and Pathology*, ed. D. C. Erwin, S. Bartnicki-Garcia & P. H. Tsao, pp. 109–19. St Paul, Minnesota. The American Phytopathological Society.

Gregory, P. H. (1983). Some major epidemics caused by *Phytophthora*. In *Phytophthora: Its Biology, Taxonomy, Ecology and Pathology*, ed. D. C. Erwin, S. Bartnicki-Garcia & P. H. Tsao, pp. 271–8. St Paul, Minnesota: The American Phytopathological Society.

Harold, R. L. & Harold, F. M. (1980). Oriented growth of *Blastocladiella emersonii* in gradients of ionophores and inhibitors. *Journal of Bacteriology*, **144**, 1159–67.

Held, A. A. (1970). Nutrition and fermentative energy metabolism of the water mold *Aqualinderella fermentans*. *Mycologia*, **62**, 339–58.

Hemmes, D. E. (1983). Cytology of *Phytophthora*. In *Phytophthora: Its Biology, Taxonomy, Ecology and Pathology*, ed. D. C. Erwin, S. Bartnicki-Garcia & P. H. Tsao, pp. 9–40. St Paul, Minnesota: The American Phytopathological Society.

Ho, H. H. & Hickman, C. J. (1967a). Asexual reproduction and behaviour of zoospores of *Phytophthora megasperma* var. *sojae*. *Canadian Journal of Botany*, **45**, 1963–81.

Ho. H. H. & Hickman, C. J. (1967b). Factors governing zoospore responses of *Phytophthora megasperma* var. *sojae* to plant roots. *Canadian Journal of Botany*, **45**, 1983–94.

Holloway, S. A. & Heath, I. B. (1977). An ultrastructural analysis of the changes in organelle arrangement and structure between the various spore types of *Saprolegnia*. *Canadian Journal of Botany*, **55**, 1328–39.

Holwill, M. E. J. (1982). Dynamics of eukaryotic flagellar movement. In *Prokaryotic and Eukaryotic Flagella*, ed. E. B. Amos & J. G. Duckett, pp. 289–312. Cambridge: Cambridge University Press.

Irving, H. R. & Grant, B. R. (1984). The effect of pectin and plant root surface carbohydrates on encystment and development of *Phytophthora cinnamomi* zoospores. *Journal of General Microbiology*, **130**, 1015–18.

Katsura, K. & Miyata, Y. (1971). Swimming behaviour of *Phytophthora capsici* zoospores. In *Morphological and Biochemical Events in Plant–Parasite Interaction*, ed. S. Akai & S. Ouchi, pp. 107–28. Tokyo: The Phytopathological Society of Japan.

Kazama, F. Y. (1972). Ultrastructure and phototaxis of the zoospores of *Phlyctochytrium* sp., an estuarine chytrid. *Journal of General Microbiology*, **71**, 555–66.

Khew, K. L. & Zentmyer, G. A. (1974). Electrostatic response of zoospores of seven species of *Phytophthora*. *Phytopathology*, **64**, 500–7.

Kropf, D. L. & Harold, F. M. (1982). Selective transport of nutrients via the rhizoids of the water mold *Blstocladiella emersonii*. *Journal of Bacteriology*, **151**, 429–37.

Lange, L. & Olson, L. W. (1983). The fungal zoospore: its structural and biological significance. In *Zoosporic Plant Pathogens: a Modern Perspective*, ed. S. T. Buczacki, pp. 1–40. London: Academic Press.

Machlis, L. (1969). Zoospore chemotaxis in the watermold *Allomyces*. *Physiologia Plantarum*, **22**, 126–39.

Miles, C. A. & Holwill, M. E. J. (1969). Asymmetrical flagellar movement in relation to the orientation of the spore of *Blastocladiella emersonii*. *Journal of Experimental Biology*, **50**, 683–7.

Musgrave, A., Ero, L., Scheffer, R. & Oehlers, E. (1977). Chemotropism of *Achlya bisexualis* germ hyphae to casein hydrolysate and amino acids. *Journal of General Microbiology*, **101**, 65–70.

Newhook, F. J., Young, B. R., Allen, S. D. & Allen, R. N. (1981). Zoospore motility of *Phytophthora cinnamomi* in particulate substrates. *Phytopathologische Zeitschrift*, **101**, 202–9.

Orpin, C. G. & Bountiff, L. (1978). Zoospore chemotaxis in the rumen phycomycete *Neocallimastrix frontalis*. *Journal of General Microbiology*, **104**, 113–22.

Robertson, J. A. (1972). Phototaxis in a new *Allomyces*. *Archiv für Mikrobiologie*, **85**, 259–66.

Royle, D. J. & Hickman, C. J. (1964). Analysis of factors governing *in vitro* accumulation of zoospores of *Pythium aphanidermatum* on roots. 1. Behavior of zoospores. *Canadian Journal of Microbiology*, **10**, 151–62.

Royle, D. J. & Thomas, G. G. (1973). Factors affecting zoospore responses towards stomata in hop downy mildew (*Pseudoperonospora humuli*) including some comparisons with grapevine downy mildew (*Plasmopara viticola*). *Physiological Plant Pathology*, **3**, 405–17.

Soll, D. R. & Sonneborne, D. R. (1971). Zoospore germination in the water mold *Blastocladiella emersonii*: cell differentiation without protein synthesis? *Proceedings of the National Academy of Sciences, USA*, **68**, 459–63.

Soll, D. R. & Sonneborne, D. R. (1972). Zoospore germination in *Blastocladiella emersonii*. IV. Ion control over cell differentiation. *Journal of Cell Science*, **10**, 315–33.

Suberkropp, K. F. & Cantino, E. C. (1972). Environmental control of motility and encystment in *Blastocladiella emersonii* zoospores at high population densities. *Transactions of the British Mycological Society*, **59**, 463–75.

Suberkropp, K. F. & Cantino, E. C. (1973). Utilization of endogenous reserves by swimming zoospores of *Blastocladiella emersonii*. *Archiv für Mikrobiologie*, **89**, 205–21.

Van Brunt, J. & Harold, F. M. (1980). Ionic control of germination of *Blastocladiella emersonii* zoospores. *Journal of Bacteriology*, **141**, 735–44.

Webster, J. (1980). *Introduction to Fungi*, 2nd edn. Cambridge: Cambridge University Press.

Willoughby, L. G. & Pickering, A. D. (1977). Viable Saprolegniaceae spores on the epidermis of the salmonid fish *Salmo trutta* and *Salvelinus alpinus*. *Transactions of the British Mycological Society*, **68**, 91–5.

Young, B. R., Newhook, F. J. & Allen, R. N. (1979). Motility and chemotactic response of *Phytophthora cinnamomi* zoospores in 'ideal' soils. *Transactions of the British Mycological Society*, **72**, 395–401.

Zentmyer, G. A. (1983). The world of *Phytophthora*. In *Phytophthora; Its Biology, Taxonomy, Ecology and Pathology*, ed. D. C. Erwin, S. Bartnicki-Garcia & P. H. Tsao, pp. 1–7. St Paul, Minnesota: The American Phytopathological Society.

7
Water relations and pathogen activity in soil

J. M. DUNIWAY and T. R. GORDON

Department of Plant Pathology, University of California, Davis, California 95616, USA

Interactions between plant diseases and soil water potential (ψ_{soil}) take many forms. Some of the interactions between plant pathogens and ψ_{soil} are similar to those between many aerial pathogens and their environment in that the behaviour of the pathogen and, therefore, disease development is determined largely by the pathogen's requirements for water for growth, sporulation, dispersal, germination, or survival. To an extent, Phytophthora root rots, when aggravated by saturated soil conditions, are such examples. In contrast, many important plant pathogenic fungi in soil, such as *Fusarium* spp., are tolerant of a wider range in ψ_{soil} than are their living host plants. In diseases such as Fusarium foot rot of wheat, the effects of ψ_{soil} on disease development are often mediated by a response of the host, or of competitive microorganisms, to changing ψ_{soil}. Mechanisms by which ψ_{soil} affects the development of Phytophthora root rot and Fusarium foot rot will be compared and contrasted. In this brief review, only these two, relatively well studied examples, are discussed in detail and the reader is referred to more inclusive reviews of water potential effects on plant pathogens in soil for other examples (e.g. Cook & Duniway, 1981; Griffin, 1981; Duniway, 1982; and others cited therein). Soil ψ is frequently partitioned into its major components of osmotic (=solute) potential (ψ_{π}) and negative pressure or matric potential (ψ_m).

Phytophthora root rot

There are now a variety of experimental data to show that diseases incited by soilborne members of the genus *Phytophthora* are aggravated by saturated soil conditions (see Zentmyer, 1980; Duniway, 1983). Among the more clear experiments to be done in a field is one on the

effects of irrigation on the development of Phytophthora root rot in
safflower at Davis, California (Duniway, 1978, 1983). No rain fell during
crop growth. When susceptible cultivars were grown on water stored
in the soil, without irrigation, no above-ground symptoms of root rot
developed. In contrast, when the soil surface was irrigated once during
crop growth, about 30% of the susceptible plants developed severe symptoms of root rot within 30 d; four irrigations stimulated disease development even more (Fig. 7.1). Evidently, the development of perceptible
root rot required the saturated conditions caused by irrigation, a conclusion that has been substantiated by a number of experiments on plants
in containers. For example Kenerley, Papke & Bruck (1984) grew seedlings of Fraser fir in pots of soil infested with *P. cinnamomi* and found
that when the pots were flooded once with water for 24 or 48 h, rather
than being allowed to drain freely after watering, the incidence of infection and mortality of seedlings in infested soil increased dramatically.
Flooding uninfested soil had little effect on seedlings. Similar results
were obtained previously with fruit tree seedlings and more precise
experiments have now demonstrated the effects of soil moisture on
Phytophthora root rot of cherry. Wilcox & Mircetich (1985) grew
Mahaleb cherry seedlings in soil on tension plates. When ψ_m was maintained constant at -1.0 or -2.5 kPa, *P. cryptogea*, *P. megasperma*, and
P. drechsleri caused little disease. However, when infested soil normally
maintained at -2.5 kPa was flooded for 48 h once every two weeks,
all three *Phytophthora* spp. caused severe disease. Experiments employing tension plates also show quite clearly that saturated periods, even

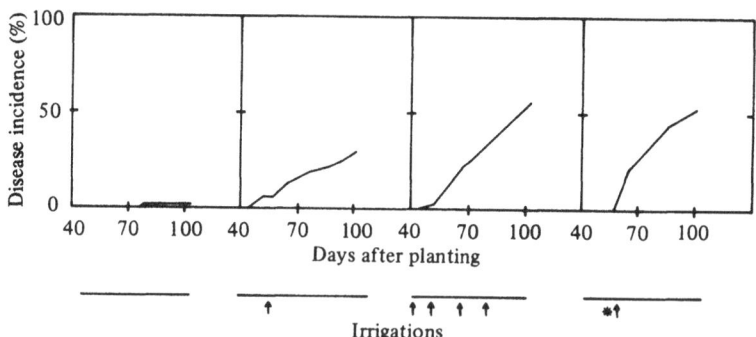

Fig. 7.1. The development of Phytophthora root rot in the susceptible safflower cultivar Nebraska 10 subjected to various irrigation schedules in
the field at Davis, California. Surface irrigations were applied at the times
indicated by arrows and the asterisk denotes the occurrence of water stress
before irrigation. (From Duniway, 1978, 1982, 1983.)

when relatively short, are highly stimulatory to the development of black shank in tobacco seedlings caused by *P. parasitica* (Shew, 1983).

While the preponderance of data show that saturated soil conditions are highly stimulatory or are even required for soilborne *Phytophthora* spp. to infect roots, there are cases where infections have been demonstrated to occur in unsaturated soil, i.e. at ψ_m values considerably less than -1 kPa. Of course, such results may be obtained if ψ_m is not controlled adequately. However, given inoculum levels high enough that large numbers of spores or mycelia are contacted by or reside close to roots, direct infection is possible. Some *Phytophthora* species may be more adapted to infect directly than others. For example, *P. cambivora* has been shown to cause extensive root and crown rot in cherry seedlings maintained in soil at a ψ_m of -2.5 kPa (Wilcox & Mircetich, 1985), perhaps because its propagules tend to germinate directly and it grows extensively in the host.

Although infections by directly germinated propagules are possible, a variety of observations suggest that the indirect germination of sporangia and subsequent infections by zoospores are likely to be of greater significance in the epidemic development of Phytophthora root rot. For example, the more dormant propagules of *Phytophthora* spp. in soil frequently germinate to form sporangia (e.g. Zentmyer, 1980; Duniway, 1983; Ioannou & Grogan, 1985). In addition, under controlled conditions that restrict the production and dispersal of zoospores, high levels of *Phytophthora* inoculum are generally required in soil for root disease to develop extensively in individual plants (Kuan & Erwin, 1982; Shew, 1983). These results suggest that individual infections do not grow rapidly or extensively, so many infections are required for epidemic development. Given the proper conditions, sporangia are produced abundantly on infected roots, and zoospores are the likely means by which large numbers of secondary infections can be initiated. In other words, the zoospore stage makes secondary spread possible in soil and allows Phytophthora root rots to be potentially 'multiple cycle' or 'compound interest' diseases that spread and increase rapidly. As will be discussed below, another piece of correlative evidence for this proposal is that the water requirements for infection that have been demonstrated in relatively precise experiments are the same as those for zoospore release and dispersal (e.g. Duniway, 1976, 1983; Shew, 1983; Wilcox & Mircetich, 1985).

Because the asexual reproductive cycle prevails during the development of *Phytophthora* epidemics, some of the direct influences of water status on these stages are reviewed here. In addition, it should be noted that soil-water status also has large effects on persistence and germination by chlamydospores and oospores of *Phytophthora* spp. in soil (Duniway, 1979, 1983). Furthermore, in many of their relationships to water in soil, *Phytophthora* spp. have aspects in common with a variety of other zoosporic fungi that cause root infections and seedling blights in soil (e.g. Duniway, 1979).

Sporangia

When buried in soil under constant conditions, mycelia of *P. cactorum, P. cinnamomi*, and *P. cryptogea* formed maximum numbers of sporangia at -2 to -20 kPa ψ_m and formed measurable numbers at ψ_m or Ψ values as low as -0.25 to -0.4 MPa (Duniway, 1979, 1983; Gisi, Zentmyer & Klure, 1980; Gisi, 1983). In fully saturated soil, *P. cryptogea* and *P. cinnamomi* formed few sporangia when buried but formed many when placed on the soil surface (Duniway, 1983; Gisi, 1983). The effects of depth, as well as soil texture (Sidebottom & Shew, 1984), suggest that the uppermost ψ_m at which some species form sporangia within soil is determined by a need for gas exchange with the air. However, some *Phytophthora* spp. that form sporangia in drained soil, such as *P. cactorum* and *P. palmivora*, can also form them within saturated soil (Gisi, 1983). In fact, it appears that several *Phytophthora* spp. are capable of forming sporangia over a fairly wide range of ψ_m values. Additional research on a few *Phytophthora* spp. also suggests that sporangium formation is not limited by the salinity levels or the ψ_π component of most agricultural soils (Blaker & MacDonald, 1985) and that sporangia remain viable in soil at moderately low Ψ values (Duniway, 1979, 1983).

An important finding relative to the stimulatory effects of saturated-soil conditions on disease development is that a transition from a drained to a saturated state in soil is highly stimulatory to sporangium formation by some *Phytophthora* spp. For example, while *P. parasitica* can form sporangia in soil at ψ_m values as low as -20 or -30 kPa, it forms large numbers of sporangia rapidly after a drained soil is saturated (Bernhardt & Grogan, 1982). *P. cambivora* has even more precise water requirements for sporangium formation in that few sporangia are formed in drained soil but many are formed within a few hours when soil is saturated (Wilcox & Mircetich, 1985).

The release or discharge of zoospores by sporangia appears to be the developmental stage having the most exacting ψ requirements, especially in terms of its direct sensitivity to differences of one or a few kilopascals ψ_m. Zoospores may exit sporangia singly under their own locomotion, but in many fungi, including *Phytophthora* spp., they appear to be forcibly expelled from sporangia (Gisi, 1983). A variety of fungi have generally similar water requirements for zoospore discharge and possible mechanisms are reviewed elsewhere (Duniway, 1979; Gisi, 1983). The quantitative effects of soil ψ_m on zoospore release were first determined by wetting soil in which *P. cryptogea* and *P. megasperma* had formed sporangia in order to raise its ψ_m (MacDonald & Duniway, 1978). When soil was wetted to 0 (fully saturated) or −0.1 kPa ψ_m, 77–99% of the sporangia released zoospores within 4 h. In contrast, at final ψ_m values of −0.5, −1.0, and −2.5 kPa, the percentages of sporangia releasing zoospores within 6 h were approx. 30, 10, and 0, respectively (Fig. 7.2). While various solutes can retard or even inhibit zoospore release (MacDonald & Duniway, 1978; Duniway, 1979; Gisi, 1983), recent experiments indicate that when the ψ_m of soil is increased sufficiently, zoospore release is stimulated so long as the ψ_π component is not lowered below that at which the sporangia were formed (Blaker & MacDonald, 1985). Therefore, in soils wetted by rain or irrigation, zoospore release is likely to be determined largely by the ψ_m component of soil Ψ.

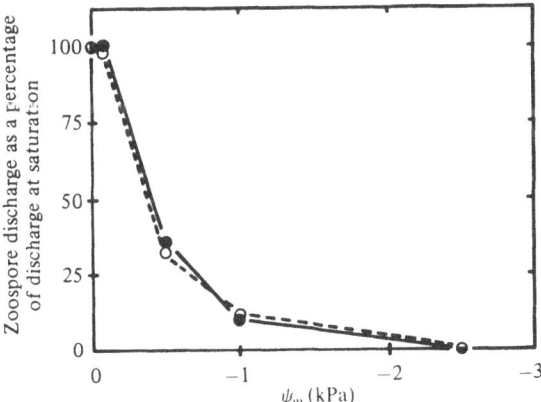

Fig. 7.2. Influence of soil ψ_m on zoospore discharge by *Phytophthora cryptogea* (●——●) and *P. megasperma* (○----○). Sporangia formed in a coarse (>250 μm) fraction of Yolo fine sandy loam at $\psi_m = -15$ kPa were maintained at ψ_m values shown for 6 h. (From MacDonald & Duniway, 1978.)

Zoospores

The active movement and taxis of zoospores offer opportunities for directed dispersal in soil that are unavailable to nonmotile forms. The swimming activities of zoospores in soil are clearly confined to pore spaces that are water filled and, because soil structure and water status determine the size distribution of water-filled pores in soil, these parameters largely govern the extent to which zoospores move in soil. Griffin (1981) described the probable importance of these interactions for all motile microorganisms in soil, and work on *P. cryptogea* (Duniway, 1976) shows that its zoospores can readily swim 25 to 35 mm in standing surface water or through a coarse-textured potting mix at ψ_m values higher than -0.1 kPa (Fig. 7.3). Active movement in the potting mix, however, was greatly reduced at -1 kPa ψ_m and was barely detected at -5 kPa. In finer loam soils active movement was more limited at -0.1 kPa and was confined to a distance of 5 mm or less at -1 kPa ψ_m (Fig. 7.3). Soil pores in the order of 300 μm diameter drain water at -1 kPa ψ_m and it appears that the larger and less tortuous are the water-filled pores in soil, the more suitable they will be for zoospore dispersal. The effects of particle size in systems saturated with water under static conditions, which are reviewed elsewhere (Carlile, 1983; Duniway, 1983), also suggest that the chemotactic movement by zoospores of *P. cinnamomi* requires pore neck diameters larger than 150–180 μm, i.e. pore necks that are only likely to remain water-filled at $\psi_m > -2$ kPa.

Although dispersal of zoospores by their own locomotion, even if confined to a few millimetres distance in soil, may allow new infections of roots, passive movement with water flow is likely to disperse zoospores farther in soil. The swimming activities of zoospores may reduce impaction on solid surfaces and entrapment in small or dead-end channels, and thereby enhance dispersal through soil with water flow (see also Carlile, Ch. 6; Holderness & Pegg, Ch. 11). For example, motile zoospores of *P. megasperma* moved as far as 17 to 44 cm with water that infiltrated dry soil to a distance of 65 cm, while nonmotile zoospores moved about one half as far (Wilkinson, Miller & Millar, 1981). Furthermore, the leading boundary for infestation by motile zoospores occurred at ψ_m values between -1.4 and -1.7 kPa. Although many attributes of zoospore dispersal in soil still need to be researched, it is now apparent that both the passive and active dispersal of flagellated zoospores in soil are largely confined to ψ_m values higher than -2.5 to -1.0 kPa and as soils become more fully saturated the likelihood of effective zoospore dispersal is increased. In fact, it can be concluded that the levels of soil saturation that promote zoospore formation and dispersal are precisely the same as those that are required for many Phytophthora root rots to develop extensively (e.g. Duniway, 1976, 1979, 1983; Kuan & Erwin, 1982; Shew, 1983; Wilcox & Mircetich, 1985). Furthermore, while zoospores of *Phytophthora* spp. are likely to be very sensitive to desiccation in the flagellated, nonwalled state, they can encyst rapidly. Once encysted and walled, zoospores of at least some *Phytophthora* spp. can survive considerable desiccation in soil and can germinate to form another zoospore when wetted (Duniway, 1979, 1983). In other words, a zoospore that fails to reach a host or other substrate because the soil dries may have a second chance for dispersal when soil is re-wetted.

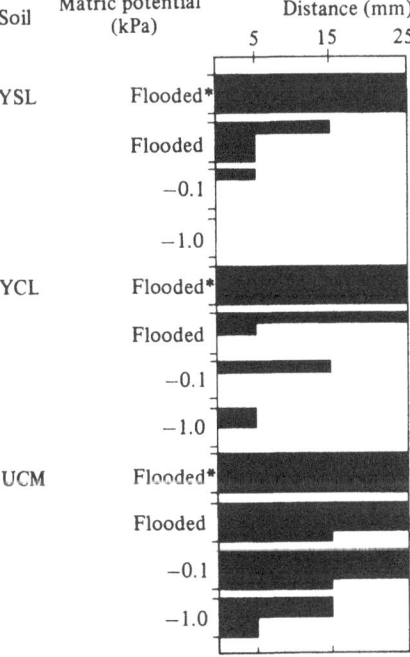

Fig. 7.3. Net effect of ψ_m on the release and movement of *Phytophthora cryptogea* zoospores in three soils. Sporangia formed on mycelial disks at $\psi_m = -30$ kPa before soil ψ_m was raised to the final values shown for 24 h. An asterisk indicates mycelial disks bearing sporangia were on the soil surface rather than 3–8 mm below the surface. Bar width is proportional to the incidence of *P. cryptogea* in safflower seedlings 5, 15, or 25 mm from the nearest sporangia. Legend to soils: YSL = Yolo fine sandy loam; YCL = Yolo clay loam; UCM = a mixture of fine sand and peat. (From Duniway, 1976.)

Infection and colonisation

We know little about the water requirements for *Phytophthora* spp. to penetrate and grow in host roots. Initiation of infection by zoospores, however, does appear to occur rapidly at the soil ψ_m values that allow their release and effective dispersal (Duniway, 1983). A number of studies have examined the effects of Ψ on the growth of mycelium by *Phytophthora* spp. in culture. Even though growth by *Phytophthora* spp. is confined to higher ψ_π values than is growth of many plant-pathogenic Ascomycotina and Fungi Imperfecti, the results suggest that *Phytophthora* spp. grow well at Ψ_{medium} corresponding to those they are most likely to encounter in the roots of growing plants (Duniway, 1979, 1983; Gisi, 1983). In addition, recent research has shown that mycelia of *Phytophthora* spp. do have effective mechanisms of osmotic adjustment to maintain turgor at low Ψ values (Luard, 1982) (see Chapters 1 and 2 for further discussion). Furthermore, a variety of Ψ values to which susceptible safflower plants were allowed to dry in soil after roots were inoculated with *P. cryptogea* were all suitable for the continued development of root rot symptoms (Duniway, 1977*b*). It should be noted, however, that no research has been done to demonstrate the actual effects of tissue ψ on the growth rate of a *Phytophthora* sp. in a host.

Predisposition

Although the physical requirements of a pathogen establish some of the water requirements for disease development, in the final analysis, pathogen behaviour and disease epidemiology represent an integration of pathogen and host physiology. It is becoming more apparent that some of the changes in Ψ_{soil} that occur routinely in agriculture can significantly predispose or otherwise modify the resistance of plants to attack by *Phytophthora* spp. For example, the same very wet soil conditions that enhance zoospore formation and dispersal have been demonstrated to predispose some hosts to infection by *Phytophthora* spp. Predisposition of alfalfa was demonstrated by flooding plants in sterilised soil before inoculating them with zoospores of *P. megasperma* (Kuan & Erwin, 1980). The rhododendron cultivar Caroline, which is normally resistant to Phytophthora root and crown rot, also became severely diseased if it was flooded for 48 h before inoculation with zoospores of *P. cinnamomi* (Blaker & MacDonald, 1981). The mechanisms of flood predisposition to Phytopthora root and crown rot are not clearly known but low oxygen levels may make roots unusually leaky and thus more attractive to zoospores (Kuan & Erwin, 1980). It is also noteworthy that post-inoculation treatments of safflower roots with oxygen concentrations less than 5% greatly increased the rate of pathogenic attack by *P. cryptogea* (Heritage & Duniway, 1985). Such results suggested that low oxygen concentrations may compromise resistance mechanisms that are normally manifested in roots after infection.

Even though saturated soil conditions enhance pathogen activity and may increase host susceptibility, a lack of water has also been shown to predispose some hosts to Phytophthora root rot. An induction of such water stress in safflower plants before they were watered and inoculated with zoospores of *P. cryptogea* increased the severity of root rot that developed subsequently (Duniway, 1977*b*). Significant predisposition occurred in the susceptible cultivar Nebraska 10 if water was withheld to depress ψ_{leaf} from $-0.6\,MPa$ to less than $-1.3\,MPa$, and comparable depressions of ψ_{leaf} predisposed the normally resistant cultivar Biggs to a greater extent. Furthermore, in the field, the severity of Phytophthora root rot which developed in several safflower cultivars after one irrigation was greatest in those plants subjected to the greatest water stress before irrigation (Duniway, 1978). The effects of water stress and irrigation on the development of Phytophthora root rot in a susceptible safflower cultivar are illustrated in Fig. 7.1. Under controlled conditions, an induction of water stress before inoculation also rendered the Caroline cultivar of rhododendron susceptible to extensive attack by *P. cinnamomi* (Blaker & MacDonald, 1981).

Levels of salinity stress that can occur in some irrigated or heavily fertilised soils have been shown to predispose chrysanthemum (Mac-Donald, 1982, 1984), citrus (Blaker & MacDonald, 1984) and tomato (T. J. Swiecki & J. D. MacDonald, unpubl.) to Phytophthora root rot. In citrus and tomato, field and controlled environment experiments demonstrated a relationship between salinity and root rot. In addition, the incidence of Phytophthora root rot of soybean in a field study was positively correlated with levels of fertilisation (Dirks, Anderson & Bolton, 1980). Little is known about the mechanisms by which salinity predisposes plants. In chrysanthemum, a pulsed exposure to salinity before inoculation increased zoospore aggregation on roots, and exposure after inoculation increased the rate at which mycelium colonised roots (MacDonald, 1982, 1984). Although salinity effects on other species need to be researched more in soil, the isolates of *P. parasitica* pathogenic to citrus in a saline soil were able to form sporangia and release zoospores at relatively high levels of salinity (Blaker & MacDonald, 1985).

Phytophthora spp. cause root and crown rots in a large number of diverse plant species, and predisposition to Phytophthora root rot by flooding, drought, or salinity stresses has only been investigated in a few examples. However, the diversity of the known examples suggests that they are not unique. Therefore, even though the water requirements for the pathogen can be very precise and may influence disease development greatly, it appears that the effects of water status on host susceptibility also need consideration when determining the effects of ψ on the behaviour of *Phytophthora* spp. as plant pathogens in soil. In addition to affecting host susceptibility and pathogen behaviour, stresses caused by flooding, drought, or salinity are likely to increase the impacts of Phytophthora root rot on a host. The length of functional roots decreases and internal resistance to water uptake increases as root rot develops (Duniway, 1977*a*, 1983). The effects of such changes on a host are accentuated as the water supply in soil is depleted or under saline conditions. In addition, a lack of regeneration or growth of new roots under drought and saline conditions in soil, as well as chronically oxygen-deficient conditions due to saturation, may seriously impair a plant's capacity to tolerate Phytophthora root rot (Duniway, 1983; Blaker & MacDonald, 1984). It appears that a number of interactions between the host and pathogen can contribute to the net effects of ψ_{soil} on the development of Phytophthora root rot.

Fusarium foot rot of wheat

Fusarium spp. cause diseases of wheat in most wheat-growing regions of the world (Nelson, Toussoun & Cook, 1981). Infections by soilborne Fusaria may manifest themselves as seedling blight, foot and crown rot of mature plants, or infected plants may be symptomless. The correlation between comparatively dry soil and the occurrence of seedling blight caused by *Fusarium* spp. has long been recognised (e.g. Dickson, 1923). Low ψ in soil and infected host tissue has also been shown to increase the incidence and severity of *Fusarium* diseases of mature wheat plants (Papendick & Cook, 1974). Most information concerning the various ways in which water availability influences the pathogenic activities of wheat-infecting Fusaria pertains to *F. culmorum* and *F. graminearum*, the most important *Fusarium* species pathogenic on wheat. The following discussion will therefore focus on these species.

Survival

In the absence of a growing wheat crop, *F. graminearum* survives as an occupant of host residues in or on the soil; chlamydospores are not thought to be important to its survival (Sitton & Cook, 1981). In order to persist, it is therefore necessary for *F. graminearum* to retain possession of at least some portion of the tissue it colonised prior to the death of its host. Its ability to do so is greatly influenced by the moisture regime to which the infected straw is exposed. Conditions which are suitable for microbial activity in general, will lead to decomposition of the straw and elimination of the pathogen. For example, Burgess & Griffin (1968) found that *F. graminearum* could be recovered from only a very small percentage of infected straws incubated at 100% rh and 35 °C for 24 weeks. At lower rh (e.g. 87% at 25 °C), it was displaced by xerophilic fungi such as *Penicillium* spp. Under still drier circumstances (e.g. 75.5% rh at 25 °C), *F. graminearum* was recovered from all the originally infected straw, reflecting the fact that even xerophilic competitors were inactive.

Competing saprotrophs also affect the activity of *F. culmorum* in infected residues. Cook & Bruehl (1968) showed that colonisation of wheat straw by *F. culmorum* was usually restricted to the stem base, which was occupied during the parasitic growth of the fungus. Following the death of the host, *F. culmorum* did not grow into the upper portion of the stem even though the ψ of this tissue was usually well within the range conducive to saprotrophic growth of *F. culmorum* (Cook & Christen, 1976). Inability to colonise the wheat stem further can be attributed to competition from saprotrophic fungi such as *Alternaria* and *Cladosporium* spp. Only when ψ dropped as low as −8 to −9 MPa did *F. culmorum* become an effective coloniser of wheat straw in natural soil (Cook & Bruehl, 1968).

F. culmorum produces chlamydospores which, unlike those of *F. graminearum*, are very effective survival structures. Its chlamydospores persist under a wide range of ψ_{soil} even when subjected to rapid drying (Sitton & Cook, 1981). Although chlamydospore germination is potentially a water-limited process, Sung & Cook (1981) found that chlamydospores of *F. culmorum* had a uniformly high rate of germination between 0 and −2 MPa ψ. The rate of germination declined as ψ dropped below −2 MPa, but this is clearly well beyond the ψ range within which seed germination and subsequent growth of the seedling would take place. Extensive chlamydospore germination in soil too dry for the host is probably prevented, in part, by the absence of solutes which in moist soil would diffuse away from seeds and seedlings and serve to stimulate spore germination (Cook & Duniway, 1981). In contrast to the germination process, growth of chlamydospore germlings may often be limited by soil-water status. In field soils above -1 MPa ψ, germ tubes lysed or formed new chlamydospores within three days following germination, while at lower ψ values germ tubes continued to grow (Fig. 7.4) (Cook & Papendick, 1970). The upper ψ limit for growth of germlings coincided with the lower ψ limit for bacterial activity in this soil. The implication that bacterial antagonism limited the growth of *F. culmorum* in this system was strengthened by the observation that the addition of streptomycin and neomycin promoted germling survival in soil as wet as -0.1 MPa (Cook & Papendick, 1970).

Growth and infection

Vegetative growth of both *F. culmorum* and *F. graminearum*, whether originating from a germinated chlamydospore or mycelium in residues, can occur over a wide range of ψ_{soil} (see for example, Fig. 7.4). Measurable growth of *F. graminearum* has been recorded at Ψ_{soil}

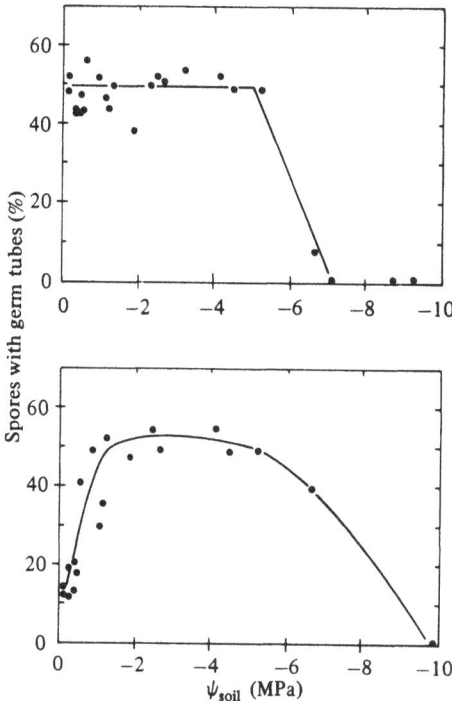

Fig. 7.4. Influence of ψ_{soil} on the percentage of chlamydospores of *Fusarium culmorum* with germ tubes in Ritzville silt loam. Observations were made 24 h (upper graph) and 72 h (lower graph) after glucose and ammonium sulphate were added to stimulate chlamydospore germination in nonsterile soil. (From Cook & Papendick, 1970.)

values as low as -7.7 MPa (Wearing & Burgess, 1979). In sterile soil, growth is maximal at ψ near zero (Cook, Papendick & Griffin, 1972) but, as noted above, the microflora may restrict the growth of these fungi when ψ_{soil} exceeds -1 MPa. Accordingly, it is not surprising that infection of seedlings is influenced by ψ_{soil}. Cassell & Hering (1982) reported that *F. culmorum* caused more injury to wheat seedlings in soil at -0.1 than at -0.01 MPa ψ. A ψ value of -0.1 MPa would not be prohibitive to bacterial activity in general but it may be dry enough to affect one or more species which contribute to the inhibition of *F. culmorum*. Alternatively the greater disease in drier soil may reflect constraints on nonbacterial components of the soil biotica which influence growth and infection by *F. culmorum*, such as soilborne amoebae known to parasitise fungi, which have very exacting moisture requirements (Homma & Cook, 1985). Cassell & Hering considered the possibility that greater aeration in drier soil might be important. They found that for soil–sand mixtures at equal ψ (i.e. maintained at -7 kPa) but with differing ratios of soil and sand, and hence differing water contents and volumes of air-filled pore spaces, disease was most severe at the highest level of aeration. Since disease severity was evaluated only on a visual rating of symptoms, the effects of ψ_{soil} and/or aeration may reflect the influence of these factors on the infection process, or subsequent development of the disease, or both. Studies by Liddell & Burgess (1985) showed a clear effect of soil-water status on the infection of wheat seedlings by *F. graminearum*. They used wax layers to isolate soil containing inoculum at the tops of pots from deeper soil that was maintained at near field capacity. When the top soil was maintained at $\psi = -0.1$ MPa, 17.3% of the seedlings were infected as compared with 37.0% at -0.7 MPa. This is consistent with the concept that elements of the microflora were antagonistic to the pathogen at higher Ψ values. However, at $\psi_{soil} = -1.5$ MPa only 3.5% of the seedlings were infected. This latter result is somewhat surprising since ψ_{soil} of -1.5 MPa should not greatly restrict the growth of *F. graminearum* but should, on the other hand, be too dry to support bacterial growth. Actinomycetes are capable of growing at ψ as low as -1.5 MPa and some species have been implicated as fungal antagonists (Wong, 1975).

It is difficult to establish and maintain ψ_{soil} in the range employed by Liddell & Burges, as these authors note, but as methods for accomplishing this task are refined further, more unexpected results may emerge. Subtle selective effects, not apparent when treatment differences are large, such as wet (e.g. $\psi > -0.1$ MPa) versus dry (e.g.

$\psi < -1\,\text{MPa}$), may be revealed when comparisons are made within a narrower range of ψ_{soil}. It may be that workers in different areas, using equally adequate methods, will find different effects of ψ_{soil} on infection of wheat by *Fusarium* spp., reflecting differences in the microflora associated with local field soils and/or *Fusarium* isolate used (Sung & Cook, 1981). A continuing effort in this area, however, should clarify the basis for such differences and expand our understanding of how wheat-infecting Fusaria interact with other soilborne microorganisms.

Of course, experiments which employ constant ψ values cannot, by themselves, provide a complete understanding of how the pathogen behaves in the field where ψ_{soil} fluctuates greatly, especially near the surface where most inoculum of *F. culmorum* and *F. graminearum* resides. For this reason, studies which monitor the activities of the pathogen in soil subjected to one or more cycles of drying and rewetting can be very informative (e.g. Burgess & Griffin, 1968). Studies of this type have recently been initiated in California field soils to examine the effects of soil moisture on infection of wheat seedlings by *F. graminearum* (T. R. Gordon & R. K. Webster, unpubl.). In a greenhouse experiment using potted plants, inoculum was incorporated into the upper 3 cm of a clay loam soil. Soil in these pots was wetted to field capacity at the time seeds were sown and was then either maintained near field capacity ($\psi > -0.1\,\text{MPa}$) by regular watering, or allowed to dry so that ψ_{soil} was -1.0 to $-1.2\,\text{MPa}$ when the experiment was terminated 24 d after planting. Plants in the drier soil were well supplied with water from the subsoil and showed no visible signs of stress. As expected, a higher level of infection occurred when the soil was allowed to dry (82%), than when soil remained wet (58%). Similar results were obtained when the same soil was sterilised prior to incorporation of the inoculum; 64% of the seedlings were infected in the wet soil as compared to 96% of the seedlings in the soil which was allowed to dry. Statistical analysis confirmed that, in this experiment, the significant effect of drier soil on seedling infection was operative in sterile as well as nonsterile soil. If the primary effect of Ψ_{soil} is mediated through its influence on the activity of microbial antagonists, then sterilising the soil should reduce the effect, but it did not. As previously indicated, some data reported by Cassell & Hering (1982) suggest a possible effect of limited aeration at high ψ_{soil}. Under different experimental circumstances, however, Cook & Papendick (1970) determined that reduced aeration in wet soil was not responsible for the lysis of *F. culmorum* germlings in wet soil, which they attributed to bacterial antagonism. Perhaps the high level of inoculum (2000 macroconidia g^{-1} soil) used in the experiment of Gordon & Webster was sufficient to overcome an inhibitory effect of the soil microflora. It is also worth considering a possible effect of the soil-moisture treatments on the host. While measurements of ψ_{leaf} at the conclusion of the experiment confirmed that there was no significant difference in the water status of plants in the two treatments, it remains possible that a localised difference in tissue ψ at the site of infection near the soil surface was important. More extensive drying of the outer cell layers of the coleoptile and leaf sheaths, as would occur in dry soil, may render the plant more susceptible to infection by the pathogen.

Disease development

Whatever environmental constraints may limit the infection process it is clear that under field conditions a high level of infection commonly occurs where sufficient inoculum is present and the upper layers of the soil are dry (Cook, 1968). However, infected plants which escape the seedling blight phase of the disease may remain infected and yet apparently healthy. Here again the water status of the soil is a critical factor, in this case because of its effect on the water status of the host. When soil moisture becomes limiting for wheat and ψ_{leaf} drops to -3.5 MPa and below, wheat plants infected by *F. culmorum* develop foot rot (Papendick & Cook, 1974). Although it could not be determined with certainty from their results, Papendick & Cook felt that the infection did not cause the measured water stress; they concluded that water stress predisposed infected plants to rapid development of foot rot.

Inasmuch as field-grown wheat must develop very low ψ values to produce the foot-rotting phase of the disease, it is not surprising that foot rot is difficult to reproduce in a greenhouse. Potted plants in a greenhouse will not ordinarily dry to ψ as low as those required for disease development (Cook, 1980). Because of their restricted root systems, plants growing in small pots tend to dry abruptly precluding the possibility of osmotic adjustment. Such adjustment, resulting in lower ψ_{π}, allows water-stressed plants to maintain turgor at progressively lower ψ (Turner & Jones, 1980). When this occurs in the field and ψ drops sufficiently, infected plants develop foot rot. Why must a threshold ψ be reached in order for the parasite to become a pathogen? It is probably not correct to say that a wheat plant in the field at -3.5 MPa is more severely stressed than a greenhouse-grown plant at -2 MPa, since the latter may in fact be nearer to death. Perhaps the important difference is in the rate at which the stress develops. When stress develops slowly physiological alterations occur that are absent in rapidly stressed plants. Susceptibility may be related to the metabolic cost of osmotic adjustment. Alternatively, a stress-induced alteration which may be relevant to *Fusarium*-infected plants is enhanced carbon allocation to the root system. Perhaps a pathogen residing in the cortical tissue of the crown is stimulated by the enhanced flow of carbon to the roots. If so, the ability of a crown-infecting fungus to become a pathogen may be determined by its ability to take advantage of such a stress-induced change in the pattern of translocation.

Concluding remarks

As the diseases reviewed here clearly show, studies of the effects of ψ_{soil} on pathogen behaviour and disease development also need to consider interactions between the pathogen, host, and antagonistic microorganisms. Even though responses to ψ_{soil} may be complex or indirect, and may be modified by a number of other factors, there is frequently much to be gained by knowing more about them. For example, knowledge of predisposing factors may be essential for reproduction of a disease by inoculation and for realistic screening for stable forms of genetic resistance. Of course, the water requirement of the pathogen also needs to be satisfied in attempts to reproduce a disease or screen for genetic resistance and other controls. Also, in agriculture it is possible to manipulate irrigation schedules, volumes applied, and means of delivery, and, in most agriculture, some features of surface and soil drainage can be manipulated to advantage (see also Ch. 10). For example, studies of both irrigation and draining practices in the field strongly suggest that measures that reduce the frequency and duration of saturation periods in soil will reduce a variety of diseases incited by soilborne *Phytophthora* spp. (e.g. Dirks *et al.*, 1980; Zentmyer, 1980; Duniway, 1983; Hoy, Ogawa & Duniway, 1984). It may also be possible to devise irrigation schedules that allow antagonists to suppress disease, as has been done with common scab of potato (Lapwood, Willings & Rosser, 1979). In many instances, irrigation can be scheduled to avoid levels of stress that predispose plants to disease.

References

Bernhardt, E. A. & Grogan, R. G. (1982). Effect of soil matric potential on formation and indirect germination of sporangia of *Phytophthora parasitica*, *P. capsici*, and *P. cryptogea*. *Phytopathology*, **72**, 507–11.

Blaker, N. S. & MacDonald, J. D. (1981). Predisposing effects of soil moisture extremes on the susceptibility of rhododendron to Phytophthora root and crown rot. *Phytopathology*, **71**, 831–4.

Blaker, N. S. & MacDonald, J. D. (1984). Effect of soil salinity on development of Phytophthora root rot of citrus. *Phytopathology*, **74**, 866 (Abstr.).

Blaker, N. S. & MacDonald, J. D. (1985). The effect of soil salinity on the formation of sporangia and zoospores by three isolates of *Phytophthora*. *Phytopathology*, **75**, 270–4.

Burgess, L. W. & Griffin, D. M. (1968). The recovery of *Gibberella zeae* from wheat straws. *Australian Journal of Experimental Agriculture & Animal Husbandry*, **8**, 364–70.

Carlile, M. J. (1983). Motility, taxis, and tropism in *Phytophthora*. In *Phytophthora: Its Biology, Taxonomy, Ecology, and Pathology*, ed. D. C. Erwin, S. Bartnicki-Garcia & P. H. Tsao, pp. 95–107. St Paul: American Phytopathological Society.

Cassell, D. & Hering, T. F. (1982). The effect of water potential on soil-borne diseases of wheat seedlings. *Annals of Applied Biology*, **101**, 367–75.

Cook, R. J. (1968). Fusarium root and foot rot of cereals in the Pacific Northwest. *Phytopathology*, **58**, 127–31.

Cook, R. J. (1980). Fusarium foot rot of wheat and its control in the Pacific Northwest. *Plant Disease*, **64**, 1061–6.

Cook, R. J. & Bruehl, G. W. (1968). Relative significance of parasitism versus saprophytism in colonization of wheat straw by *Fusarium roseum* 'culmorum' in the field. *Phytopathology*, **58**, 306–8.

Cook, R. J. & Christen, A. A. (1976). Growth of cereal root-rot fungi as affected by temperature–water potential interactions. *Phytopathology*, **66**, 193–7.

Cook, R. J. & Duniway, J. M. (1981). Water relations in the life cycles of soilborne plant pathogens. In *Water Potential Relations in Soil Microbiology*, Spec. Pub. 9, ed. J. F. Parr, W. R. Gardner & L. F. Elliott, pp. 119–39. Madison: Soil Science Society of America.

Cook, R. J. & Papendick, R. I. (1970). Soil water potential as a factor in the ecology of *Fusarium roseum* f. sp. *cerealis* 'culmorum.' *Plant and Soil*, **32**, 78–82.

Cook, R. J., Papendick, R. I. & Griffin, D. M. (1972). Growth of two root-rot fungi as affected by osmotic and matric water potentials. *Soil Science Society of America Proceedings*, **36**, 78–82.

Dickson, J. G. (1923). Influence of soil temperature and moisture on the development of seedling blight caused by *Fusarium saubinettii*. *Journal of Agricultural Research*, **23**, 837–70.

Dirks, V. A., Anderson, T. R. & Bolton, E. F. (1980). Effect of fertilizer and drain location on incidence of Phytophthora rot in soybeans. *Canadian Journal of Plant Pathology*, **2**, 179–83.

Duniway, J. M. (1976). Movement of zoospores of *Phytophthora cryptogea* in soils of various textures and matric potentials. *Phytopathology*, **66**, 877–82.

Duniway, J. M. (1977a). Changes in resistance to water transport in safflower during the development of Phytophthora root rot. *Phytopathology*, **67**, 331–7.

Duniway, J. M. (1977b). Predisposing effect of water stress on the severity of Phytophthora root rot in safflower. *Phytopathology*, **67**, 884–9.

Duniway, J. M. (1978). Irrigation, plant water stress and varietal interactions affecting the development of Phytophthora root rot of safflower in the field. *Phytopathology News*, **12**, 148 (Abstr.).

Duniway, J. M. (1979). Water relations of water molds. *Annual Review of Phytopathology*, **17**, 431–60.

Duniway, J. M. (1982). Soil–plant–water relations and disease. In *Role of Biometeorology in Integrated Pest Management*, ed. J. L. Hatfield & I. J. Thomason, pp. 307–25. New York: Academic Press.

Duniway, J. M. (1983). Role of physical factors in the development of *Phytophthora* diseases. In *Phytophthora: Its Biology, Taxonomy, Ecology, and Pathology*, ed. D. C. Erwin, S. Bartnicki-Garcia & P. H. Tsao, pp. 175–87. St Paul: American Phytopathology Society.

Gisi, U. (1983). Biophysical aspects of the development of *Phytophthora*. In *Phytophthora: Its Biology, Taxonomy, Ecology, and Pathology*, ed. D. C. Erwin, S. Bartnicki-Garcia & P. H. Tsao, pp. 109–19. St Paul: American Phytopathology Society.

Gisi, U., Zentmyer, G. A. & Klure, L. J. (1980). Production of sporangia by *Phytophthora cinnamomi* and *P. palmivora* in soils at different matric potentials. *Phytopathology*, **70**, 301–6.

Griffin, D. M. (1981). Water potential as a selective factor in the microbial ecology of soils. In *Water Potential Relations in Soil Microbiology*, Spec. Pub. No. 9, ed. J. F. Parr, W. R. Gardner & L. F. Elliott, pp. 141–51. Madison: Soil Science Society of America.

Heritage, A. D. & Duniway, J. M. (1985). Influence of depleted oxygen supply on Phytophthora root rot of safflower in nutrient solution. In *Ecology and Management of Soil-borne Plant Pathogens*, Proc. Sec. V, 4th Int. Congr. Plant Pathol., ed. C. A. Parker *et al.*, pp. 199–202. St Paul: American Phytopathological Society.

Homma, Y. & Cook, R. J. (1985). Influence of matric and osmotic water potentials and soil pH on the activity of giant vampyrellid amoebae. *Phytopathology*, **75**, 243–6.

Hoy, M. W., Ogawa, J. M. & Duniway, J. M. (1984). Effects of irrigation on buckeye rot of tomato fruit caused by *Phytophthora parasitica*. *Phytopathology*, **74**, 474–8.

Ioannou, N. & Grogan, R. G. (1984). Water requirements for sporangium formation by *Phytophthora parasitica* in relation to bioassay in soil. *Plant Disease*, **68**, 1043–8.

Kenerley, C. M., Papke, K. & Bruck, R. I. (1984). Effect of flooding on development of Phytophthora root rot in Fraser fir seedlings. *Phytopathology*, **74**, 401–4.

Kuan, T.-L. & Erwin, D. C. (1980). Predisposition effect of saturating soil with water on the severity of Phytophthora root rot of alfalfa. *Phytopathology*, **70**, 981–6.

Kuan, T.-L. & Erwin, D. C. (1982). Effect of soil matric potential on Phytophthora root rot of alfalfa. *Phytopathology*, **72**, 543–8.

Lapwood, D. H., Willings, L. W. & Rosser, W. R. (1970). The control of common scab of potatoes by irrigation. *Annals of Applied Biology*, **66**, 397–405.

Liddell, C. M. & Burgess, L. W. (1985). Wax layers for partitioning soil moisture zones to study the infection of wheat seedlings by *Fusarium graminearum*. In *Ecology and Management of Soil-borne Plant Pathogens*, Proc. Sec. V, 4th Int. Congr. Plant Pathol., ed. C. A. Parker *et al.*, pp. 206–8. St Paul: The American Phytopathology Society.

Luard, E. J. (1982). Growth and accumulation of solutes by *Phytophthora cinnamomi* and other lower fungi in response to changes in external osmotic potential. *Journal of General Microbiology*, **128**, 2583–90.

MacDonald, J. D. (1982). Effect of salinity stress on the development of Phytophthora root rot of chrysanthemum. *Phytopathology*, **72**, 214–19.

MacDonald, J. D. (1984). Salinity effects on the susceptibility of chrysanthemum roots to *Phytophthora cryptogea*. *Phytopathology*, **74**, 621–4.

MacDonald, J. D. & Duniway, J. M. (1978). Influence of the matric and osmotic components of water potential on zoospore discharge in *Phytophthora.* *Phytopathology*, **68**, 751–7.

Nelson, P. E., Toussoun, T. A. & Cook, R. J. ed. (1981). *Fusarium: Diseases, Biology, and Taxonomy.* University Park: Pennsylvania State University Press.

Papendick, R. I. & Cook, R. J. (1974). Plant water status and the development of Fusarium foot rot in wheat subjected to different cultural practices. *Phytopathology*, **64**, 358–63.

Shew, H. D. (1983). Effects of soil matric potential on infection of tobacco by *Phytophthora parasitica* var. *nicotianae. Phytopathology*, **73**, 1160–3.

Sidebottom, J. R. & Shew, H. D. (1984). The effects of soil matric potential and soil texture on sporangial formation by *Phytophthora parasitica* var. *nicotianae. Phytopathology*, **74**, 814 (Abstr.).

Sitton, J. W. & Cook, R. J. (1981). Comparative morphology and survival of chlamydospores of *Fusarium roseum* 'culmorum' and 'graminearum.' *Phytopathology*, **71**, 85–90.

Sung, J. M. & Cook, R. J. (1981). Effect of water potential on reproduction and spore germination by *Fusarium roseum* 'graminearum', 'culmorum', and 'avenaceum.' *Phytopathology*, **71**, 499–504.

Turner, N. C. & Jones, M. M. (1980). Turgor maintenance by osmotic adjustment: a review and evaluation. In *Adaptation of Plants to Water and High Temperature Stress*, ed. N. C. Turner & P. J. Kramer, pp. 87–103. New York: J. Wiley & Sons.

Wearing, A. H. & Burgess, L. W. (1979). Water potential and the saprophytic growth of *Fusarium roseum* 'graminearum.' *Soil Biology and Biochemistry*, **11**, 661–7.

Wilcox, W. F. & Mircetich, S. M. (1985). Influence of soil water matric potential on the development of Phytophthora root and crown rots of Mahaleb cherry. *Phytopathology*, **75**, 648–53.

Wilkinson, H. T., Miller, R. D. & Millar, R. L. (1981). Infiltration of fungal and bacterial propagules into soil. *Soil Science Society of America Proceedings*, **45**, 1034–9.

Wong, P. T. W. (1975). Effect of osmotic potential on streptomycete growth and antibiotic production. *Soil Biology and Biochemistry*, **6**, 319–25.

Zentmyer, G. A. (1980). *Phytophthora cinnamomi and the Diseases it Causes.* Monograph No. 10. St Paul: American Phytopathology Society.

8

Epidemiological significance of liquid water in crop canopies and its role in disease forecasting

D. J. ROYLE and D. R. BUTLER

University of Bristol Department of Agricultural Sciences, Long Ashton Research Station, Bristol, BS18 9AF, UK

All fungal pathogens of aerial parts of plants depend in some degree on weather for their development and spread in crops. For the majority, however, disease progress is conditioned by a total reliance of one or more phases of their life-cycle on some form of water. This dominating and potentially limiting role of water has been exploited frequently in attempts to forecast disease development in crops, to enable fungicides to be applied according to need rather than in a routine manner. Whether or not the interest has been in forecasting, there has been much research into relationships between water, epidemics and different phases of life-cycles, especially infection. Accounts are given in many text-books on plant pathology and in reviews (e.g. Colhoun, 1973; Rotem, Cohen & Bashi, 1978; Yarwood, 1978).

A common approach on which forecasting methods have been based has been to investigate the pathogen's response to wet conditions imposed in artificial environments, and then to use the information to interpret the responses in field crops. In this research, emphasis has been on the pathogen, giving little regard to the complex ways in which water behaves in crop canopies. Application of these ways is limited and they have not yet influenced forecasting developments.

Of the natural sources of water, rain is most commonly associated with movement of water in canopies and splash droplets may move within the canopy, or from the canopy to the ground or to the air above the crops. For many pathogens this is important for the transport of resident inoculum and for the redistribution of inoculum already deposited on plant parts. Mist and fog can also transport inoculum but dew is usually only effective to the extent of bringing about local movement on a particular organ. All these sources of water lead to surface wetness on plants, and many pathogens depend on its persistence for infection, sporulation and occasionally colonisation of tissues. The different sources provide wetness with different and complex characteristics which depend on the species and age of the vegetation. The effects of these differences on pathogen behaviour and subsequent disease have been little studied and so it is not surprising that the value of understanding them in developing forecast methods has not been judged.

This paper discusses how environmental variables based on water have been used in forecasting fungal disease; it considers how a better understanding of the mechanisms of wetting, water movement and persistence of surface wetness in crops in relation to important epidemiological parameters might contribute to forecasting.

Water-related variables in forecasting

Variables related to water in crops have been the basis of attempts to develop disease forecasting methods ever since periods of wetness for infection were used in the first prediction systems, for vine downy mildew (*Plasmopara viticola*) in Germany and France in 1912–13 (Populer, 1981). Indeed, most subsequent forecasting methods which have been proposed in different countries have concerned water-dependent pathogens and have used environmental variables exclusively. Exceptions have been those methods for some powdery mildews (Butt, 1978), in which host × inoculum rather than environmental factors primarily determine disease progress, and the methods for forecasting mildew and rusts of winter wheat in the Dutch EPIPRE supervised disease control system (Zadoks, 1981). Here, a series of disease assessments in a crop is required to integrate the effects of physical and biological factors on disease.

The majority of forecasting proposals are for individual diseases and exist as simple rules or guidelines, sometimes called 'models'. They do not use weather forecasts, which are considered to be too unreliable, but interpret weather conditions retrospectively and thus try to warn of possible disease changes within the span of one latent period for a particular pathogen. In contrast, one advisory scheme for the application of fungicide to control carrot leaf blight (*Alternaria dauci*) in Canada, uses 36-h regional weather forecasts (Gillespie & Sutton, 1979). Few of the forecasting methods which have been proposed in the UK have so far been used to help guide spray decisions at the farm or individual field level (Royle, 1985). This is mainly because they have not been developed with due regard to the problems of integrating them into

Table 8.1. *Some forecasting proposals based on daily rainfall*

Pathogen, host (name)	Other incorporated variables	Origin
Phytophthora infestans, potato (BLITECAST)	temp., rh	USA (Krause, Massie & Hyre, 1975)
Phytophthora phaseoli, lima bean	temp.	USA (Hyre, 1959)
Pseudoperonospora humuli, hop	rh	UK (Royle, 1973)
Septoria nodorum, wheat	rh	UK (Tyldesley & Thompson, 1980)
Septoria tritici, wheat	temp.	USA (Shaner & Finney, 1976)
Septoria spp., wheat	rh	UK (J. E. King, unpubl.)

farming practices. One problem lies in the conception and utilisation of the water-related variables. In this section these variables have been grouped into four categories, giving emphasis to their application in UK forecasting methods, although examples from other countries are also included.

Daily rainfall

Daily rainfall has been employed as a key variable in a large number of forecasting proposals, some examples of which are listed in Table 8.1. The majority have been derived relatively crudely by comparing records of disease levels in crops with concurrent weather factors measured synoptically at nearby standard meteorological stations. Daily rainfall has been used for forecasting because it indicates the occurrence of wet weather in which pathogens that rely strongly on water are able to multiply quickly. It is a convenient variable because it avoids having to understand complex ways in which rainfall affects the pathogen, and also because it is the only measure of rainfall routinely available at standard weather stations.

Forecasts which incorporate daily rainfall can be valuable for warning farmers that conditions favourable for a disease to develop are likely to have occurred in a region and that a fungicide should be applied as a precaution. In the UK, the Agricultural Development and Advisory Service (ADAS) issues regional warnings using the forecast rules for *Septoria* spp. and for hop downy mildew (*Pseudoperonospora humuli*). However, because a particular set of rules incorporates so general a relationship between rainfall and disease, it offers no greater forecast precision than the simplest, arbitrary statements of daily rainfall. The proposals in Table 8.1, whose origins and application both depend on regional weather stations, cannot be expected to be applicable to individual crops as rainfall can vary appreciably within a locality. A probable exception is BLITECAST, used in the USA to indicate the timing of the first seasonal spray against potato late blight (*Phytophthora infestans*). It is a synthesis of two earlier empirical methods and is adapted for local operation by using meterological sensors linked to a programmed data-logging device situated on the farm (MacKenzie & Schimmelpfennig, 1978).

Perhaps the most serious criticism of the use of daily rainfall records is that it offers no explanation of the ways in which rainfall influences each phase of the pathogen life cycle. It is an empirical variable and the forecast rules that embody it are empirical statements based only on a restricted set of data that often do not cover a realistic range. Whether applied regionally or locally, it is therefore inevitable that these forecasts will not be reliable.

Rainfall intensity and duration

Although these variables are far more descriptive of pathogen activity than daily rainfall, there have been very few forecast proposals which utilise them. Bourke (1970) cites one, in which rainfall intensity and temperature are used to predict wheat eyespot (*Pseudocercosporella herpotrichoides*) in France (Ponchet, 1958). Rainfall intensity has strong implications for the splash dispersal of inoculum, and duration relates to the persistence of wetness for infection and sporulation. However, it is not known how to interpret these variables in terms of pathogen response. Although the information is available on charts from tilting syphon rain gauges on meteorological sites, its extraction is laborious and, possibly because there has never been a demand, it is not routinely provided. The time scale of measures of intensity and duration is critical. Thus, rates of rainfall in five-minute periods rather than hourly ones are needed to be related to dispersal. Whilst the number of hours of continuous or intermittent rain in a day is a valuable concept, it requires interpretation alongside estimates of leaf wetness duration.

Humidity

Table 8.2 lists a few of the many forecast methods which have been proposed that use humidity as the central environmental variable.

Table 8.2. *Some forecasting proposals based on humidity*

Pathogen, host (name)	Other incorporated variables	Origin
Phytophthora infestans, potato (Beaumont Period)	temp.	UK (Beaumont, 1940)
P. infestans, potato	temp.	UK (Smith, 1956)
P. infestans, potato (PHYTPROG)	temp.	W. Germany (Schrödter & Ullrich, 1965)
P. infestans, potato	temp.	UK (W. R. Sparks, unpubl.)
Pseudocercosporella herpotrichoides, wheat	temp.	W. Germany (Fehrmann & Schrödter, 1972)
Pyrenophora teres, barley	temp.	UK (M. R. Thomas, unpubl.)
Septoria nodorum, wheat	temp.	W. Germany (Mangstl *et al.*, 1982)
Venturia inaequalis, apple	temp.	UK (Preece & Smith, 1961)

Humidity directly influences sporulation in many fungi (see Ch. 4) and also has implications for wetness persistence but not for dispersal. Even so, humidity is most commonly used in an empirical manner in forecasting because it is thought to be a reasonably good indicator of whether plants are wet or dry. Some of the forecasting rules have been derived from empirical associations between disease increases and weather, in a similar way to those based on daily rainfall. The Beaumont and Smith Period rules for potato blight are examples and suffer the same criticisms as any empirical methods. In some forecast methods, attempts have been made to interpret humidity in a more explanatory way, as it affects separate phases of the pathogen's life cycle. In the PHYTPROG model (Schrödter & Ullrich, 1965), for example, data from field and laboratory experiments have been used to develop a relationship between the rate of disease increase and regression functions, describing the effects of humidity and other factors on infection, sporulation and lesion expansion. This is one of few 'negative forecast' systems which have been developed to advise when conditions do not favour disease, and has been in operation in West Germany for several years.

In a computer model (by W. R. Sparks) which predicts the date on which seasonal outbreaks of potato blight occur, the humidity variable is the dew point depression of the air. Practically all other forecast methods however, use relative humidity (rh) either at a particular time of day (1400 h in the W. German rules for *Septoria nodorum*), or more usually as the number of hours when a threshold value (commonly 75 or 90%), is exceeded. In the UK, rh is measured routinely on meteorological sites at 0900 h GMT, and sometimes also at 6-h intervals during the day. Although hygrographs can record rh continuously, extraction of hourly values is inconvenient; inclusion of a critical period of rh in forecast proposals therefore seems impractical. Further problems lie in assigning a cut-off point to measurements of rh and in serious errors which occur because instruments are not usually sufficiently precise. Occasionally, hours of rh greater than a threshold value have been substituted for leaf wetness duration in a forecast method, as in the Smith Period rules for apple scab (*Venturia inaequalis*). This has probably been done because of the convenience of using thermohygrographs which provide simultaneous records of rh and temperature. However, rh will not necessarily correlate with wetness when rain is held on leaves as discrete drops (Brain & Butler, 1985).

Surface wetness

Of water-related variables, the duration of surface wetness has the most direct influence on pathogen activity because for most foliar pathogens surface water is necessary for infection and sometimes also for sporulation. Periods of surface wetness are often regarded as synonymous with infection periods and can account for a large amount of the variability in subsequent disease. Relationships between the minimum length of the wet period required at different temperatures and a specified amount of infection are embodied in many proposals for forecasting methods, some examples of which are given in Table 8.3. These relationships are usually determined in growth-room experiments which assume that the character and persistence of wetness on plants in artificial environments are appropriate to those encountered in field crops. This is often not the case. However, when based on surface wetness, there is a more serious difficulty which has largely prevented forecast rules from being successful in practice. Periods of leaf wetness cannot be measured directly, only estimated with the aid of special instruments. Thus, although surface wetness is causally associated with infection, it can be applied only in an empirical manner. Any estimate of surface wetness duration is likely to be of value for specific crops and cannot be applied regionally, but the relation between wetness on an exposed sensor and that on leaves in crops is unknown. There are few examples in which attempts have been made to give disease forecasts through local operation of wetness-recording instruments. Warnings of apple scab, based on the Mills Period rules, were issued for a time in the 1960s through

Table 8.3. *Some forecasting proposals based on surface wetness ascribed to different sources of water*

Pathogen, host (name)	Other incorporated variables	Origin
Rain wetness		
Coccomyces hiemalis, cherry	temp.	USA (Eisensmith & Jones, 1981)
Pseudoperonospora humuli, hop	temp., daily rainfall	UK (Royle, 1973)
P. humuli, hop	temp., airborne spores	W. Germany (Kremheller & Diercks, 1983)
Venturia inaequalis, apple (Mills Period)	temp.	USA (Mills & LaPlante, 1954)
Dew wetness		
Puccinia hordei, barley	temp.	UK (King & Polley, 1976)
Unspecified wetness source		
Botrytis squamosa, onion (BOTCAST)	temp., rh	Canada (Sutton, James & Rowell, 1984)
Puccinia hordei, barley	temp.	UK (Anon.[a])
Rhynchosporium secalis, barley	temp.	UK (Polley, 1971)
Septoria spp., wheat	temp.	UK (Anon.[a])

[a] Incorporated in Meteorological Office/ADAS Crop Disease Environment Monitor (Sparks & Wass, 1983).

the use of mechanical wetness recorders (Hirst, 1957) in Wisbech apple orchards. Electronic devices to estimate wetness have been developed in W. Germany (Richter & Haussermann, 1975) and in the USA (Jones *et al.*, 1984) specifically to forecast this disease.

Water in crop canopies

The above brief survey of the ways in which key water-related variables have been selected and utilised in forecasting proposals indicates some reasons why forecasting methods have not generally been adopted in the past. Virtually no use of a variable attends to the mechanisms by which rain or dew affects pathogens in crops. Although some variables take account of infection, none are used to exploit dispersal. This section examines how improvements in forecasting might be made through understanding some of the important ways in which water behaves in crops and thereby influences pathogen behaviour.

As precipitation occurs, the amount and distribution of water in the canopy is constantly changing. It is important to know how water is distributed since, although liquid water may be present in the crop, it may not be located at the site of the pathogen. Movement of water is important for transport of spores from sources of inoculum, and to or from sites of potential infection. This section explains some of the complexities of the wetting of canopies and the movement of water in relation to spore dispersal.

The dynamics of wetting in relation to spore dispersal

At the onset of rain most interception occurs on the uppermost leaves, although some drops will penetrate through gaps in the foliage. Drops impinging on leaves will be redistributed; some splash droplets are held on the leaf, others are transferred to adjacent leaves, some fall to the ground and a few escape from the canopy in air currents. Drops which penetrate through gaps in the foliage may reach a source of inoculum at the base of the canopy and secondary splash may then carry spores onto leaves in the crop. If rain stops, most water movement stops shortly afterwards and the subsequent persistence of leaf wetness at any location in the canopy will depend on the amount of water held in that location. If rain continues, drops coalesce and some reach a size sufficient to drip from the leaf. These drips may fall onto lower leaves and be redistributed in a similar way to rain drops. This initial phase of wetting vegetation is important because (a) redistributed water may carry spores to new locations, and (b) the distribution of surface water in the canopy is changing and this will influence the persistence of wetness after precipitation. The amount of water held in the canopy increases non-linearly towards an upper limit, known as the canopy storage capacity. According to an equation given by Couturier & Ripley (1973), the amount of rainfall necessary to reach 90% of the canopy storage capacity is about 2.3 times its value. However, this assumes that there are no gaps in the canopy and that initially all rain is captured by the canopy. Spores previously deposited on the surface of leaves may be carried in splash droplets onto adjacent leaves. Spores of some pathogens are normally only dispersed in water but may become airborne if they are carried in small droplets which leave the canopy and evaporate (Faulkner & Colhoun, 1976; see also Ch. 5). Drops which coalesce and drip from leaves may carry spores away from potential infection sites.

The effectiveness of this washing-off process will depend on the tenacity of the spores, which is determined by the extent to which they are adapted to the splash-dispersal process. Conidia of pathogens like *Septoria nodorum* are born in mucilage and appear to be much more tenacious on wet leaves than spores from pathogens such as the downy mildews which, although transported by rain in crops, are regarded mainly as air dispersed. For these pathogens, infection from spores which are splashed onto lower and relatively protected leaf surfaces could be highly significant. A few pathogens, such as *Pseudoperonospora humuli* on hop, only enter the host through stomata of hypostomatous leaves, and therefore are well adapted to exploit the natural deposition of water onto the lower surfaces.

In continued rain, canopy saturation is reached when the rate at which water is lost from the canopy is equal to the rate at which it is gained by rainfall interception. The amount of wash-off is at a maximum in this dynamic situation, and the amount of water held in the canopy is greater than the equilibrium water-holding capacity of the canopy by an amount which depends on the rainfall rate. The equilibrium water-holding capacity of the canopy is the amount of water held after saturation when both rain and drips from the leaves have ceased. This strongly affects the duration of surface water on the vegetation.

Where gaps exist between the top of a canopy and an inoculum source, rain drops which penetrate directly may produce secondary droplets which carry spores to infection sites. This depends on the primary drops having sufficient kinetic energy to carry the secondary droplets to a particular site. The extent to which splash dispersal occurs is dependent on canopy density and the size of falling rain drops. Canopy structure and density often differ between varieties of a crop species and can also be manipulated by agronomic practice, and so an understanding of their effects on dispersal is important. The size of rain drops is related to rainfall intensity (Laws & Parsons, 1943) but, as noted earlier, rainfall intensity data are often not available from standard meteorological stations. This fact, together with the inherent complexity of dispersal, are reasons why dispersal has so far not been included in forecasting methodology.

Results from some recent work at Long Ashton (D. J. Royle & M. W. Shaw, unpubl.) illustrate the importance of factors affecting splash dispersal in the interpretation of epidemic patterns of *Septoria nodorum* and *S. tritici* in winter wheat. From the start of stem extension, in spring, sudden outbreaks of these diseases can occur when inoculum residing in the senescent leaves at the base of tillers is transported vertically upwards in splash droplets produced in heavy rainstorms. Lesions later appear simultaneously and for the first time on leaves at all heights. Inoculum is transported in a single rainstorm up to at least 50 cm from the ground within the crop, a height that has also been achieved in splash experiments with *S. nodorum* in the laboratory (Brennan, Fitt, Taylor & Colhoun, 1985). When crop density is manipulated by row spacing or seed rate, the efficiency of dispersal varies. Significantly more disease arises in sparse than in dense crops, probably due to the proportion of large rain drops at terminal velocity which pass directly through the canopy and strike inoculum sources. These examples demonstrate that dispersal may dominate the disease pattern and factors which affect both dispersal and infection should be considered concurrently. Longer persistence of surface water in dense crops will tend to favour infection, but more frequent gaps in sparse crops will allow greater vertical inoculum transport; this is an interesting conflict between factors that should be a challenge in developing more realistic forecast methods.

Wetness duration

For the majority of pathogens, the period that surface water remains on vegetation determines whether infection is possible. Many pathogens also require a minimum period of free water for sporulation. Hence, disease forecasting requires an estimate of the persistence of water on leaves, and it is convenient to use a simple measure of 'leaf wetness duration', but this avoids the questions of how much water is on the crop and where it is.

Water capture by crop canopies. The length of time that vegetation remains wet depends on the time of onset of precipitation, its duration and the time that water held on the plants after precipitation takes to dry. Persistence of wetness after precipitation depends on the evaporation rate, but it is often dominated by the amount of water held in the canopy. To estimate the persistence of water on leaves in a canopy after rain, a measure of interception is therefore required.

The efficiency with which rainfall is intercepted by a plant canopy depends on the degree of ground cover achieved by the vegetation, the rainfall intensity, wind speed and the water-holding capacity of the canopy. Light rainfall is associated with small drops which are captured efficiently whereas large drops in heavy rainfall are redistributed and their impact may shake water from the leaves. Water is also shaken from the leaves by strong wind. If the ground cover is complete, the interception efficiency at the onset of light rain in calm conditions will be very close to unity. A value of one was assumed by Couturier & Ripley (1973) in deriving an equation to describe the wetting of a crop canopy. However, the initial interception efficiency will frequently be less than one, so the amount of intercepted rain is difficult to predict. The curve in Fig. 8.1 assumes an initial interception efficiency of 0.7 to describe the wetting of a wheat canopy (Butler, 1983).

The water-holding capacity of the canopy is also affected by rainfall intensity and wind speed, but depends primarily on the leaf area index and the contact angle between water and the leaf cuticle. This is the angle included between the tangent plane to the surface of the water and the tangent plane to the surface of the leaf at any point along their line of contact. The contact angle between the leaf surface and water varies widely between species. With small contact angles, water spreads over the surface and total wetting is readily achieved but gives a small amount of water per unit leaf area. With large contact angles, surface water accumulates as discrete drops and, after rain, a large proportion of the surface of 'saturated' leaves may be dry. However, the equivalent depth of water per unit leaf area is much greater for discrete drops than for a uniform film. Where drops form, prostrate leaves can hold much more water than erect ones. The significance to the infection process of different patterns of water distribution on leaves is largely unknown.

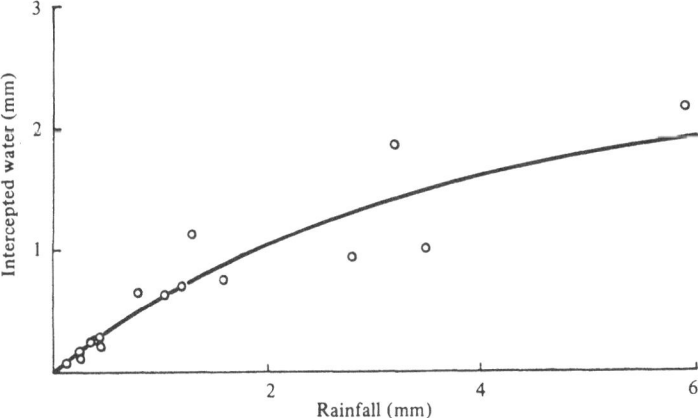

Fig. 8.1. Intercepted water in a wheat canopy plotted against rainfall. The points were obtained from the difference between measured rainfall and throughfall and the curve from an equation given by Butler (1983).

Wetness from rain generally results in a large range of drop sizes, giving a non-uniform cover of water on the scale of a single leaf. Measurements of drop diameters on leaves of cereals shortly after rain show that the drop population has a log-normal distribution, and the most frequently occurring diameters are between 1 and 2 mm (Fig. 8.2).

For a short time after the onset of rain, most water will be held on the upper leaves, but after prolonged rain the distribution of water in the canopy will be more or less uniform with height. Mist and fog produce a uniform cover of water both on leaves and throughout the canopy. In contrast, dew is usually restricted to a uniform film on exposed leaves at the top of the canopy.

Estimating wetness persistence. Estimates of the persistence of wetness can be obtained by calculating the rate of evaporation from surface water in the canopy provided that the amount of water held after precipitation is known. This approach permits the effect of drop shape and size on wetness persistence to be considered and can give an estimate of the

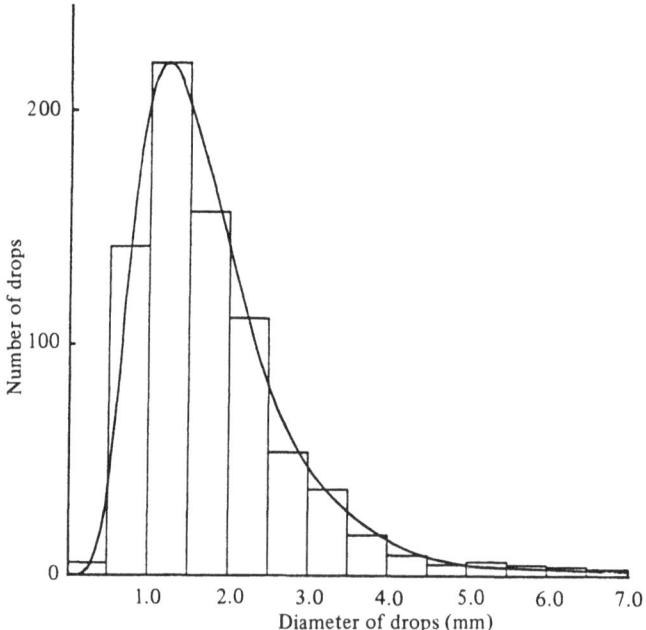

Fig. 8.2. The number of raindrops on wheat leaves in 0.5-mm diameter size classes with a fitted log-normal distribution. The diameters were measured from photographs taken within 1.5 h of the end of a rain shower. Reproduced from Brain & Butler (1985).

proportion of the leaf area covered by water at any time (Brain & Butler, 1985). The time taken for leaves to dry depends to a large extent on whether water is held as a continuous film or as discrete drops. If discrete drops are present their size distribution is important to the drying time, as illustrated in Fig. 8.3 where theoretical drying curves are shown. In this example, the time to reach complete dryness was about 40 h for the truncated log-normal size distribution, although over 95% of the volume had evaporated in 10 h. The significance to infection of small amounts of water, which persist for long periods, seems never to have been considered. The effect of the canopy microclimate on the evaporation rate from leaves with different exposures can also be taken into account and so, for example, the persistence of water on leaves deep in a canopy can be differentiated from that on upper leaves (Butler, 1983).

As noted previously, the recognised importance of leaf wetness duration on pathogen behaviour has lead to an interest in its measurement. There is a variety of instruments that detect water from dew or precipitation on an exposed surface. These are mostly based on measurements of changes in weight, length or electrical impedance. A balance, described by Hirst (1957), where the exposed surface was an expanded polystyrene block shaped to give similar results to a shoot from an apple tree, has been widely used in the UK. Other mechanical recorders mea-

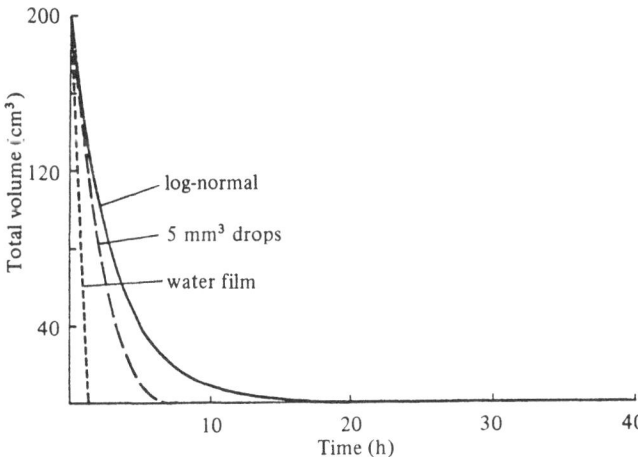

Fig. 8.3. The change in total volume of water on a surface with a constant evaporation rate per unit surface area of water. Drops with an initial log-normal distribution (———); drops with an initial volume of 5 mm³ (— —); a uniform film of water (- - - - -). Reproduced from Brain & Butler (1985).

sure the decrease in length of hemp or cotton thread when it becomes wet (Post, 1959; MacHardy, 1979) or make use of a surface marked with a water-soluble pencil (Taylor, 1956). More recently, a number of designs of printed circuit grids or 'mock leaves' have been described for which the electrical impedance between two interlocking electrodes changes when the surface is wet (e.g. Davis & Hughes, 1970; Gillespie & Kidd, 1978; Sparks & Wass, 1983). The behaviour of different designs of these sensors depends mainly on their water-holding capacity and whether water is held on their surface as discrete drops or as a continuous film, and they give different relative results in different weather conditions (Huband & Butler, 1984).

In dense canopies much of the vegetation normally remains dry when dew occurs. Dew usually occurs as a uniform film of water which ensures that wetness is provided to spores on the tissue, no matter where their location. Since free water usually is confined to exposed leaves, this is an ideal mechanism for providing water to spores which had been previously air-dispersed and deposited on upper leaves by impaction or sedimentation. There are exceptions to the normal pattern of dew however, which may have important epidemiological implications. The temperature of parts of the plant, such as fruits which have a large thermal capacity, lags behind the air temperature when it is changing. This can cause the organ to remain dry during the night (when it is warmer than the air) and water to condense on the surface after sunrise (when it is cooler than the air). This phenomenon has been observed with cocoa pods (Monteith & Butler, 1979) and is likely to occur in a variety of other tree crops.

The precipitation of water on all plant surfaces as a uniform film in mist or fog is ideal for infection by spores previously deposited on the plants, since wetness is guaranteed. The film of water may reach sufficient thickness to cause drips which will be redistributed in a similar way to those which occur during rain, with the same implications for dispersal.

Conclusions

The majority of existing proposals for forecasting fungal diseases make use of water-related variables. They either relate to the infection phase alone or are entirely empirical with no defined link to any particular phase of the pathogen's life-cycle. For certain diseases, progress in producing reliable forecast methods is possible if aspects of the environment which affect dispersal, sporulation and infection are all considered. Rainfall intensity may indicate events when dispersal of splash-spread disease is likely to occur. Information on leaf wetness in different parts of a crop canopy has much significance for interpreting phases of life-cycles at different sites of pathogen activity. Artificial sensors offer a feasible way of obtaining a measure of leaf wetness, but their performance needs interpreting in relation to a particular crop and should also be related to the pathogen's perception of wetness on the plant. Siting them within a canopy cannot be justified for routine monitoring; many sensors would be needed to represent the large spatial variation within the canopy, and errors in their performance from plant and other debris would be serious. Relating observations from sensors exposed outside a crop to conditions within it is necessary, but not straightforward, especially after rain. The distribution of water on leaves varies with the source of water and age of leaf, so no single type of sensor is suitable for all forms of wetness (Huband & Butler, 1984). An alternative is to calculate wetness from weather data. This may in the long term offer the best solution but is feasible only if continuous records are available. Calculation of the persistence of rain drops requires knowledge of rain interception by canopies. However, dew is more easily predicted, and some methods have already been proposed, e.g. Pedro & Gillespie (1982).

Recent technical developments are now permitting the routine collection of weather data which are relevant to disease forecasting. In the UK, a network of automatic weather stations is being established in areas of agricultural importance (Thompson, 1984). These will provide the continuous information required to make calculations about the environmental conditions in individual crop canopies. However, although factors other than rain may be interpolated in this manner, regional rainfall data are inadequate for forecasting at the farm level. One interesting recent development is the routine collection of rain data by radar which provides frequent information for 5 × 5 km squares within the UK. On-farm weather stations provide more precise local rainfall information and at the same time allow other factors to be integrated into forecast models which can provide advice for quick responses in farmers' day-to-day decision making. The first commercial automatic weather station for farm use is now available and others are likely to follow.

Although these and other developments in disease-management schemes (Royle, 1985) are encouraging, new disease forecast criteria, which recognise the new technologies available, still need to be produced. We argue that they must embrace the mechanistic meaning of

154   *D. J. Royle and D. R. Butler*

water-related, biological and other physical factors to the pathogen's response in crops.

References

Beaumont, A. (1940). Potato blight and the weather. *Transactions of the British Mycological Society*, **24**, 266.
Bourke, P. M. A. (1970). Use of weather information in the prediction of plant disease epiphytotics. *Annual Review of Phytopathology*, **8**, 345–70.
Brain, P. & Butler, D. R. (1985). A model of drop size distribution for a system with evaporation. *Plant, Cell and Environment*, **8**, 247–52.
Brennan, R. M., Fitt, B. D. L., Taylor, G. S. & Colhoun, J. (1985). Dispersal of *Septoria nodorum* pycnidiospores by simulated rain drops in still air. *Phytopathologische Zeitschrift*, **112**, 281–90.
Butler, D. R. (1983). Microclimate, rainfall interception and wetness persistence within a cereal canopy. Ph.D. Thesis, University of Bristol.
Butt, D. J. (1978). Epidemiology of powdery mildews. In *The Powdery Mildews*, ed. D. M. Spencer, pp. 51–81. London: Academic Press.
Colhoun, J. (1973). Effect of environmental factors on plant disease. *Annual Review of Phytopathology*, **11**, 343–64.
Couturier, D. E. & Ripley, E. A. (1973). Rainfall interception in mixed grass prairie. *Canadian Journal of Plant Science*, **53**, 659–63.
Davis, D. R. & Hughes, J. E. (1970). A new approach to recording the wetting parameter by the use of electrical resistance sensors. *Plant Disease Reporter*, **54**, 474–9.
Eisensmith, S. P. & Jones, A. L. (1981). Infection model for timing fungicide applications to control cherry leaf spot. *Plant Disease*, **65**, 955–8.
Faulkner, M. J. & Colhoun, J. (1976). Aerial dispersal of pycnidiospores of *Leptosphaeria nodorum*. *Phytopathologische Zeitschrift*, **86**, 357–60.
Fehrmann, H. & Schrödter, H. (1972). Ökologische Untersuchungen zur Epidemiologie von *Cercosporella herpotrichoides*. IV. Erarbeitung eines praxisnahen Verfahrens zur Bekämpfung der Halmbruchkrankheit des Weizens mit systemischen Fungiciden. *Phytopathologische Zeitschrift*, **74**, 161–4.
Gillespie, T. J. & Kidd, G. E. (1978). Sensing duration of moisture retention using electrical impedance grids. *Canadian Journal of Plant Science*, **58**, 179–87.
Gillespie, T. J. & Sutton, J. C. (1979). A predictive scheme for timing fungicide applications to control *Alternaria* leaf blight in carrots. *Canadian Journal of Plant Pathology*, **1**, 95–9.
Hirst, J. M. (1957). A simplified surface-wetness recorder. *Plant Pathology*, **6**, 57–61.
Huband, N. D. S. & Butler, D. R. (1984). A comparison of wetness sensors for use with computer or microprocessor systems designed for disease forecasting. In *Proceedings of the 1984 British Crop Protection Conference – Pests and Diseases*, vol. 2, pp. 633–8. British Crop Protection Council.
Hyre, R. A. (1959). The development of a method for forecasting downy mildew of lima bean. *Plant Disease Reporter Supplement*, **257**, 179–80.
Jones, A. L., Fisher, P. D., Seem, R. C., Kroon, J. C. & VanDeMotter, P. J. (1984). Development and commercialization of an in-field microcomputer delivery system for weather-driven predictive models. *Plant Disease*, **68**, 458–63.
King, J. E. & Polley, R. W. (1976). Observations on the epidemiology and effect on grain yield of brown rust in spring barley. *Plant Pathology*, **25**, 63–73.

Krause, R. A., Massie, L. B. & Hyre, R. A. (1975). Blitecast, a computerised forecast of potato blight. *Plant Disease Reporter*, **59**, 95–8.

Kremheller, H. T. & Diercks, R. (1983). Epidemiologie und Prognose des Falschen Mehltaues (*Pseudoperonospora humuli*) an Hopfen. *Zeitschrift fur Pflanzenkrankheiten und Pflanzenschutz*, **90**, 599–616.

Laws, J. O. & Parsons, D. A. (1943). The relation of raindrop size to intensity. *Transactions of the American Geophysical Union*, **24**, 452–60.

MacHardy, W. E. (1979). A simple, quick technique for determining apple scab infection periods. *Plant Disease Reporter*, **63**, 199–204.

MacKenzie, D. R. & Schimmelpfennig, H. (1978). Development of a microcomputer unit for forecasting potato late blight using Pensylvania State University's Blitecast system. *American Potato Journal*, **55**, 384.

Mangstl, A., Englert, G., Anderl, A., Reiner, L. & Rössler, S. (1982). SEPTPROG: Ein Modell zur frühzeitigen Prognose von *Septoria nodorum* bei Weizen. *Pflanzenschutz-Praxis*, **2**, 7–10.

Mills, W. D. & LaPlante, A. A. (1954). Diseases and insects in the orchard. Apple scab. *Cornell University Extension Bulletin* no. 711, 20–8.

Monteith, J. L. & Butler, D. R. (1979). Dew and thermal lags: a model for cocoa pods. *Quarterly Journal of the Royal Meteorological Society*, **105**, 207–15.

Pedro, M. J. & Gillespie, T. J. (1982). Estimating dew duration. II. Utilising standard weather station data. *Agricultural Meteorology*, **25**, 297–310.

Polley, R. W. (1971). Barley leaf blotch epidemics in relation to weather conditions with observations on the overwintering of the disease on barley debris. *Plant Pathology*, **20**, 184–90.

Ponchet, J. (1958). La prévision des épidémies du piètin-verse, *Cercosporella herpotrichoides* Fron. *Phytiatrie et Phytopharmacie*, **7**, 133–44.

Populer, C. (1981). Epidemiology of downy mildews. In *The Downy Mildews*, ed. D. M. Spencer, pp. 57–105. London: Academic Press.

Post, J. J. (1959). De hydrograaf als bladnatschrijver. *Versl. Koninklijk Nederlands Meteorologisch Instituut*, V–51, RIII, 234.

Preece, T. F. & Smith, L. P. (1961). Apple scab infection weather in England and Wales, 1956–60. *Plant Pathology*, **10**, 43–51.

Richter, J. & Haussermann, R. (1975). Ein elektronisches Schwarngerät. *Anz. Schädlingskde., Pflanzenschutz, Umweltschutz*, **48**, 107–9.

Rotem, J., Cohen, Y. & Bashi, E. (1978). Host and environmental influences on sporulation *in vivo*. *Annual Review of Phytopathology*, **16**, 83–101.

Royle, D. J. (1973). Quantitative relationships between infection by the hop downy mildew pathogen, *Pseudoperonospora humuli*, and weather and inoculum factors. *Annals of Applied Biology*, **73**, 19–30.

Royle, D. J. (1985). Rational use of fungicides on cereals in England and Wales. In *Fungicides for Crop Protection: 100 Years of Progress*, vol. 1, pp. 171–80. British Crop Protection Council.

Schrödter, H. & Ullrich, J. (1965). Untersuchungen zu Biometeorologie und Epidemiologie von *Phytophthora infestans* (Mont.) de By. auf mathematisch-statistischer Grundlage. *Phytopathologische Zeitschrift*, **54**, 87–103.

Shaner, G. R. E. & Finney, R. E. (1976). Weather and epidemics of *Septoria* leaf blotch of wheat. *Phytopathology*, **66**, 781–5.

Smith, L. P. (1956). Potato blight forecasting by 90% humidity criteria. *Plant Pathology*, **5**, 83–7.

Sparks, W. R. & Wass, S. N. (1983). Development of equipment to observe the weather and calculate crop disease risk. *EPPO Bulletin*, **13**, 27–31.

Sutton, J. C., James, T. D. W. & Rowell, P. M. (1984). BOTCAST – a forecaster for timing fungicides to control botrytis leaf blight of onions. *University of Guelph Circular* no. EBO784, pp. 1–15.

Taylor, C. R. (1956). A device for recording the duration of dew deposits. *Plant Disease Reporter*, **40**, 1025–8.

Thompson, N. (1984). Automatic acquisition of meteorological data for crop protection. In *Proceedings 1984 British Crop Protection Conference – Pests and diseases*, vol. 2, pp. 647–54. British Crop Protection Council.

Tyldesley, J. B. & Thompson, N. (1980). Forecasting *Septoria nodorum* on winter wheat in England and Wales. *Plant Pathology*, **29**, 9–20.

Yarwood, C. E. (1978). Water and the infection process. In *Water Deficits and Plant Growth*, vol. 5, ed. T. T. Kozlowski, pp. 141–73. London: Academic Press.

Zadoks, J. C. (1981). EPIPRE: a disease and pest management system for winter wheat developed in the Netherlands. *EPPO Bulletin*, **11**, 365–9.

9

Water stress predisposition to disease – an overview

DONALD F. SCHOENEWEISS

Illinois Natural History Survey, 607 East Peabody Drive, Champaign, Illinois 61820, USA

Free moisture is required for growth and development of both vascular plants and their associated fungal pathogens, thus variations in water availabilty before, during, and after infection may have pronounced effects on disease development. Unfortunately, it is often difficult or impossible to separate the effects of water stress on the host from those on the parasite or on host–parasite interaction. The term 'predisposition', however, implies an effect on the 'disposition' or 'proneness' of the host plant to attack by a pathogen (Schoeneweiss, 1975a; Yarwood, 1976). Yarwood (1976) defines predisposition as the tendency of treatments and conditions acting prior to inoculation to affect disease susceptibility, and includes both increases and decreases in susceptibility in his definition. Whether the predisposing effect of water stress occurs before or after inoculation may be irrelevant to the present discussion, since diseases such as stalk rots of grain crops and canker diseases of woody hosts may develop from infections that occurred prior to the onset of stress. Therefore, reports of water-stress effects on fungal diseases will be included regardless of when inoculation took place. For broader treatments of environmental stresses as predisposing factors in plant disease, the reader is referred to several previous reviews (Cook & Papendick, 1972; Schoeneweiss, 1975a, 1978; Yarwood, 1976; Bell, 1982).

The preponderance of published reports on water-stress predisposition to fungal pathogens involve root rots, stalk rots, and stem cankers, with occasional papers reporting on vascular wilts and other diseases. Quantitative estimates of the level or severity of water stress range from numbers of days or weeks of below normal rainfall, to percentages of soil moisture-holding capacity (MHC) and available soil moisture (ASM), or to thermodynamic terms such as water potential of soil (ψ_{soil})

or plant (ψ). Quantitative estimates of the results of predisposition to disease include yield reductions, rate and extent of lesion development or tissue colonisation, and arbitrary disease indices. Thus it is difficult to summarise or compare published data on water-stress predisposition. In most cases, data are presented from plants that were naturally or artificially subjected to severe water stress and compared with well-watered plants. Only rarely have reports included data on disease development over a range of water-stress levels. Most of the available data showing valid correlations between water stress and predisposition to fungal pathogens have appeared in previous reviews (Schoeneweiss, 1975a, 1978) and are only summarised here (Fig. 9.1).

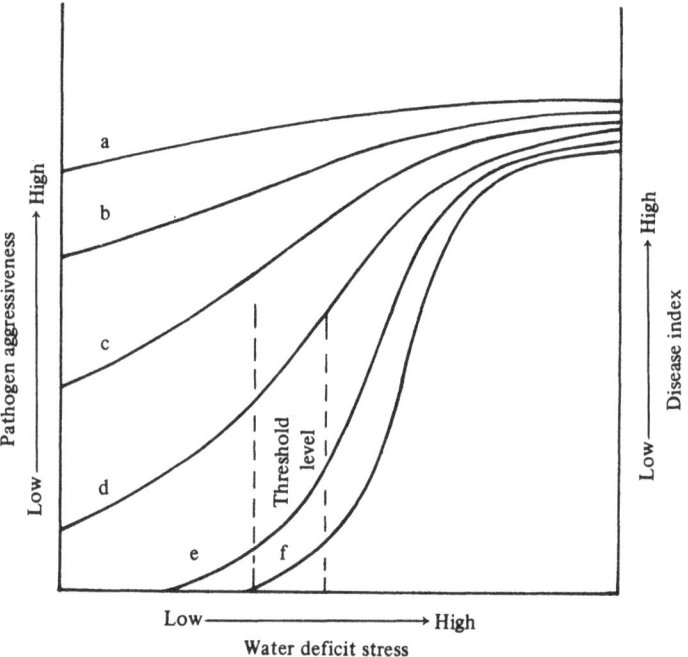

Fig. 9.1. Water stress predisposition: relation between pathogen aggressiveness, level of water deficit stress prior to inoculation, and disease development (disease index). Curves adapted from: a – Cytospora canker of sugar maple (McPartland, 1983); b – Sclerotinia dollar spot of Kentucky bluegrass (Couch & Bloom, 1960); c – Macrophomina charcoal rot of grain sorghum (Jordan *et al.*, 1984); d – Cytospora canker of French prune (Bertrand *et al.*, 1976); e – Botryosphaeria canker of red-osier dogwood (Schoeneweiss, 1975a); f – Botryosphaeria canker of white birch (Crist & Schoeneweiss, 1975).

The role of pathogen aggressiveness

The expression of the genetic potential of a microorganism for pathogenicity on a particular host is referred to as virulence or aggressiveness (Schoeneweiss, 1975*a*). Since biotrophic or obligate parasites derive all of their essential nutrients from living cells and characteristically cause maximum damage on vigorously growing plants, they are commonly considered aggressive pathogens (Yarwood, 1976). Water stress usually results in increased susceptibility or predisposition to necrotrophic or facultative parasites, which may vary considerably in aggressiveness. Research on woody plants subjected to controlled environmental stresses provides convincing evidence that threshold levels of water stress are required for predisposition to nonaggressive stem canker fungi (Bier, 1964; Crist & Schoeneweiss, 1975; Schoeneweiss, 1975*a*, 1978, 1983; Wene, 1979; McPartland, 1983). Unfortunately, very few data are available on more highly aggressive fungal pathogens of woody or nonwoody hosts. In general, as pathogen aggressiveness increases, the influence of threshold levels of stress becomes less evident (Fig. 9.1). However, much additional evidence is needed to substantiate or refute this relationship. Many alterations in metabolism occur in higher plants under water stress, particularly near those threshold levels, -1.2 to $-1.5\,\mathrm{MPa}\,\psi$, required for predisposition to nonaggressive canker fungi (Schoeneweiss, 1978). Which of these alterations is involved in predisposition is not clear at present.

This discussion on water-stress predisposition will include flooding stress where the evidence indicates that the effect of flooding on disease development may involve a change in host susceptibility and is not merely an effect on growth and development of the pathogen.

Diseases associated with water-stress predisposition
Root rots

The influence of soil moisture and aeration on development of root rots is often attributed (as discussed by Carlile, Ch. 6, and Duniway & Gordon, Ch. 7) to effects on the ecology of soil fungi, while the influence of soil water and aeration on host predisposition is less clear. Bateman (1961) subjected rooted poinsettias to moisture levels ranging from 30 to 90% MHC by placing potted cuttings at various heights above a water table in a continuous sand–vermiculite medium. After soil moisture levels approached equilibrium, pots were infested with *Pythium ultimum*, *Rhizoctonia solani*, or *Thielaviopsis basicola*. Rhizoctonia root and stem rots were most severe at moisture levels below 40% MHC and disease severity decreased with increasing soil moisture. Pythium root rot became severe only above 70% MHC, while *Thielaviopsis* caused appreciable damage between 36% and 70% MHC. Bateman interpreted the results as an effect of soil aeration on the pathogens, but speculated that host susceptibility may also have been altered. Other workers have reported increased root rot damage under saturated soil conditions and commonly attribute effects to oxygen stress or reduced soil aeration. Phipps & Beute (1977) found that *Cylindrocladium crotalariae* was most aggressive on peanut at soil moisture levels near field capacity. Pre-inoculation drought treatment reduced root rot severity, while drought following a post-inoculation wet period enhanced above-ground symptom development because of a lack of healthy roots. Progressive colonisation of celery roots by *Fusarium oxysporum* f. sp. *apii* was accelerated by periods of oxygen stress resulting from heavy irrigation (Correll & Schneider, 1982). Stoltzy, Letey, Klotz & Labanauskas (1965) reported less decay by *Phytophthora* in citrus roots if saturated soil was artificially aerated; a low oxygen supply in soils prevented growth and regeneration of roots, and prolonged soil saturation enhanced zoospore production, which together resulted in higher root decay. Brief periods of oxygen stress greatly increased root damage on bean by *Fusarium solani* f. sp. *phaseoli*, a common bean pathogen, but also resulted in damage by *F. solani* f. sp. *pisi*, which is not normally a pathogen of bean (Miller, Burke & Kraft, 1980).

In the above examples, the influence of excess soil moisture or oxygen depletion stress on the pathogen was emphasised, but the experiments were not designed to separate effects on the pathogen from those on host susceptibility. This was attempted by Kuan & Erwin (1980) who grew alfalfa seedlings in saturated or unsaturated soil for one week prior to inoculation with *Phytophthora megasperma* f. sp. *medicagnis* and found that saturation predisposed alfalfa roots by increasing root damage. The exudation of nutrients from damaged roots in turn increased the chemotaxic attraction of zoospores to the roots. Kuan & Erwin concluded that lack of oxygen or nutrients could account only in part for the predisposition effect and suggested that soil saturation also affected resistance. Rhododendron seedlings subjected to pre-inoculation drought stress or flooding stress developed severe symptoms of root and crown rot when soil was infested with zoospores of *Phytophthora cinnamomi* (Blaker & McDonald, 1981). Since soils drained freely and were adequately aerated at the time of inoculation, and increasing inoculum density did not increase disease severity, the authors concluded that predisposition was probably due to impairment of host resistance mechanisms, possibly through disruption of metabolic pathways.

Drought or water deficit stress has often been associated with increased damage caused by root rot fungi but, as with flooding or oxygen deficit stress, more attention has been paid to effects on the pathogen than on host predisposition. In most cases, claims of predisposition were based on observations or on results of experiments that did not separate effects on host susceptibility from those on the pathogen or on root regeneration (Schoeneweiss, 1978). Thus, Leaphart & Stage (1971) analysed growth rings of *Pinus elliotii* affected by pole blight and concluded that failure of root regeneration to keep pace with root rot over a prolonged drought period from 1916 to 1940, particularly on sites with low water-holding capacity, ultimately resulted in pole blight damage. However, they did not attempt to confirm these conclusions experimentally. As a result of conflicting reports that *Fomes annosus* root rot of forest trees was most severe under soil moisture conditions ranging from waterlogging to drought, Towers & Stambaugh (1968), conducted a series of controlled water-stress experiments in the field and greenhouse. Drought conditions, induced by covering the soil, decreased the resistance of loblolly pines to root colonisation by *F. annosus*. Predisposition was confirmed in greenhouse tests in which seedlings subjected to soil moisture-depletion cycles showed a higher incidence of infection and colonisation by the fungus than in soils maintained at field capacity.

The influence of water deficit on root and foot rot of wheat, caused by *Fusarium roseum* f. sp. *cerealis* 'Culmorum', has been described in detail by Cook & Papendick (1972), Cook (1973), and Papendick & Cook (1974). Although the fungus is able to grow at low ψ_{soil} where most antagonists are inhibited, Papendick & Cook (1974) found that the pathogen population was less important than water stress in determining disease severity. The influence of water stress was on disease development after infection, and they concluded that water stress predisposes host plants to aggressive attack by the pathogen, which would otherwise be a slow-decay organism. Ghaffar & Erwin (1969) reported similar results with *Macrophomina phaseolina* root rot of cotton growing in sand. Root rot damage occurred regardless of temperature or the addition of nutrients, but developed only in plants subjected to water stress before or after inoculation.

Phytophthora root rots are usually associated with wet soils, but Duniway (1977) presented convincing evidence that water stress prior to inoculation predisposed safflower to *P. cryptogea*. Although root rot developed following inoculation in a well watered, susceptible cultivar, significant predisposition was demonstrated following depression of ψ_{leaf} to -1.3 MPa or lower. Visible symptoms of root rot and significant reduction in fresh weight of roots occurred in a resistant cultivar when ψ_{leaf} were depressed to -0.9 to -1.2 MPa before inoculation; no root rot occurred in well watered controls. Induction of water stress after inoculation did not increase severity of root rot in safflower. Predisposition appeared to be related to the extent to which established infections could develop.

Stalk rots

Stalk rots of grain crops, such as maize and sorghum, cause extensive damage in plants under drought stress and may become limiting factors in grain production, particularly in areas of the semi-arid tropics where hot, dry weather during critical periods of the growing season is common. The association of soil water deficits with increased severity of stalk and root rots, caused by normally weak or nonaggressive fungi such as *Diplodia*, *Fusarium*, *Gibberella*, and *Macrophomina*, has been well established, although the underlying mechanisms remain unclear (Jordan, Clark & Seetharama, 1984). In most cases, the stalk rot phase develops from root rot infections if host plants become predisposed during grain filling. Edmunds (1964) reproduced typical field symptoms of charcoal rot (*Macrophomina phaseolina*) on grain sorghum in controlled greenhouse experiments where ASM in soil beds was maintained at 25% prior to inoculation. At 25% ASM, severe stalk rot developed rapidly after bloom, while there was no infection in plants at 80% ASM. Trimboli & Burgess (1983) reported similar results with *Fusarium moniliforme* on grain sorghum. Basal stalk and root rot were produced when plants were grown in the greenhouse under optimum soil moisture until flowering, then subjected to gradual development of severe moisture depletion stress until mid-dough stage before rewetting. Stalk rot did not develop on plants grown to maturity at optimal soil moisture. In irrigated field-grown sweet corn, significant increases in stalk rot and reduction in seed quality due to infection by *F. moniliforme* occurred when plants were subjected to water stress by withholding irrigation after silking (Chotena, Makus & Simpson, 1980). Odvody & Dunkle (1979) presented evidence that water stress promoted initial infection of sorghum roots by the charcoal rot fungus *M. phaseolina*. Nonstressed plants growing in infested soil remained relatively free of infection, while susceptible male-fertile and resistant male-sterile sorghum lines both had high rates of root infection when water stressed during the soft-dough stage.

The frequent association of stalk rots with water stress during grain filling led Dodd (1980) to advance the 'photosynthetic stress–translocation balance' concept of predisposition to stalk rot. According to the concept, root and lower stalk tissues in plants under stress are decayed as the tissues lose their metabolically-dependent defence system because of an increase in cellular senescence. Carbohydrate depletion resulting from insufficient photosynthesis and unbalanced translocation to the grain sink weakens the plant and is followed by root rot and subsequent stalk rot. However, Schneider & Pendery (1983) reported that stalk rot of maize, caused by *F. moniliforme*, was strongly enhanced at the end of the season by mild water deficit during early growth stages. Even though ψ_{leaf} were only about 0.2 MPa lower than in well watered controls, root senescence was accelerated and root infection increased in infested treatments prior to flowering. It seems unlikely that simple change in carbohydrate status of roots or stalks is sufficient to explain predisposition phenomena in stalk rot diseases (Jordan *et al.*, 1984).

Duncan (1984) discussed the association between plant senescence and root and stalk rots in sorghum. Nonsenescent genotypes, and genotypes with delayed senescence characteristics, are generally less susceptible to root- and stalk-rot fungi. Associations between water stress, senescence, and predisposition are discussed below.

Stem cankers

The term 'canker' is used to describe a local lesion on a stem, which is usually but not always accompanied by an eventual sloughing off of tissue, leaving an open wound. Nearly all cankers have been described on woody hosts and the vast majority are caused by facultative or obligately necrotrophic fungi. The fungi grow saprotrophically but some may become parasitic, depending upon their virulence or aggressiveness toward a particular host. Aggressive canker fungi cause cankers on nonstressed, susceptible hosts and the rate and extent of canker development commonly increase under water stress (Bagga & Smalley, 1974; Schipper, McNabb & Haywood, 1977; Lewis & VanArsdel, 1978; McPartland, 1983). Weak or nonaggressive canker fungi form cankers only on hosts predisposed by stresses such as drought (see Schoeneweiss, 1978, for references; also Schoeneweiss, 1981, 1983; Wene, 1979; McPartland, 1983; Appel & Stipes, 1984). Outbreaks of canker diseases, such as bleeding necrosis of sweetgum (Neely, 1968) caused by *Botryo-*

sphaeria ribis (*B. dothidea*), ash dieback caused by *Cytophoma pruinosa* and *Fusicoccum* sp. (Ross, 1964; Silverborg & Ross, 1968), and honey locust decline caused by *Thyronectria austro-americana* (Stim & Hime-lick, 1981) appeared following several years of extended drought stress.

In a summary of many years of research on the influence of water stress on development of canker disease of forest trees caused by faculta-tive fungal parasites, Bier (1964) concluded that most canker fungi were present in or on bark tissue of healthy trees, but formed cankers only if bark relative turgidity fell below a critical threshold level. For example, *Hypoxylon pruinatum* caused cankers on trembling aspen when bark turgor fell below 76%, while *Septoria musiva* required a threshold turgor below 71% for infection. Bier felt that bark turgor was associated with host vigour and that Hypoxylon canker would appear on more vigorous hosts than *Septoria*, which was indeed the case. However, Bagga & Smalley (1974) reported Hypoxylon canker formation on well watered aspens, with canker development increasing in proportion to increasing water stress. They stated that any factor that contributed to moisture stress in the host increased susceptibility to the pathogen. Bloomberg (1962) and Landis & Hart (1967) reported threshold levels of bark turgor for canker formation by *Cytospora chrysosperma* on poplars and *Physalospora obtusa* on crabapples, respectively, but Filer (1967) found no simple, direct relationship between relative bark turgidity of cotton-wood and susceptibility to *C. chrysosperma*, *Phomopsis macrospora*, and *Hypomyces solani*. Bertrand, English, Uriu & Schick (1976) reported good correlation between ψ_{leaf}, measured with a pressure chamber, and canker formation on French prune by *Cytospora leuco-stoma* following late-season water stress, but much poorer correlation with bark moisture values.

Bier (1964) suggested that the influence of bark moisture on sapro-trophic microflora antagonistic to canker fungi may play a role in canker development. When moisture content of aspen bark was lowered, sapro-trophs would, thus, be ineffective in preventing infection and disease spread by *H. pruinatum*. He reported that development of bark sapro-trophs inactivated the pyrocatechol in aspen bark which was inhibitory to the canker fungus. However, most canker fungi are wound parasites (Schoeneweiss, 1981), so bark saprotrophs may have little effect if the fungus enters the plant through wounds. No cankers formed on white birch stems surface-inoculated with *Botryosphaeria dothidea* and sub-jected to predisposing water, freezing, and defoliation stresses or on nonstressed wound-inoculated stems (Crist & Schoeneweiss, 1975). All three stresses resulted in canker formation on wound-inoculated stems, provided that stress severity exceeded threshold levels for predisposition. Tobiessen & Buchsbaum (1976) suggested that physiological changes in white ash bark due to reduced CO_2 fixation under water stress were involved in predisposing ash to fungal infection. In general, however, the role of bark factors in predisposition to canker fungi remains unclear.

Since most reports on water-stress predisposition to canker fungi are based on field observations of disease outbreaks following droughts, or at best on comparisons of disease development on unstressed control plants with that on plants subjected to severe water stress in the field or greenhouse, valid data on the predisposing effect of different levels of water stress is meagre (Schoeneweiss, 1978). Crist & Schoeneweiss (1975) subjected white birch seedlings to various levels of water stress by withholding irrigation and allowing plants to deplete soil moisture over different lengths of time prior to wound inoculation with *B. dothidea*. The plants were then placed in a cabinet under temperature and humidity conditions conducive to stabilisation of ψ, as described by Schoeneweiss (1975*b*). No cankers developed until ψ exceeded a threshold level, beginning at approximately -1.2 MPa, after which the rate of canker expansion increased rapidly with increasing stress severity. Similar results were reported with the same pathogen on red-osier dogwood and sweetgum (Schoeneweiss, 1975*a*). Although the rate of canker development and stem colonisation by *B. dothidea* varied depending upon the host species, the predisposing threshold level remained constant. *B. dothidea* is typical of many nonaggressive canker fungi in that the fungus is often present as a saprotroph in or on dead twigs and branch stubs but does not attack live tissues until the host is stressed beyond a predisposing threshold level (Schoeneweiss, 1981). From the very limited amount of data available, it appears that the relationship between water-stress severity and canker development is more linear with aggressive fungi, such as *H. pruinatum* on aspen (Bagga & Smalley, 1964) and *Nectria cinnabarina* and *Cytospora* sp. on sugar maple (McPartland, 1983), and may or may not involve threshold levels (Fig. 9.1).

The duration of water stress required for predisposition to canker fungi remains unresolved. Reports of canker appearance in the field following droughts of varying lengths provide little useful information for understanding the role of stress duration, and few studies have been reported where duration was an experimental variable. No cankers were formed on white birch inoculated with *B. dothidea* and maintained for

20 d at ψ approaching, but not exceeding, the threshold level of −1.2 MPa, whereas canker formation was initiated within 4 d at more negative potentials (Schoeneweiss, 1978). Wene (1979) inoculated stems of red-osier dogwood with *B. dothidea* at 3-d intervals before, during, and after potted plants were subjected to a water-stress regime. Plants were held without irrigation until ψ were between −1.2 and −1.8 MPa, then incubated for 6 d at constant ψ before rewatering. The extent of colonisation of stem tissues by the pathogen was significantly greater than in watered controls only if ψ remained below the threshold level of −1.2 MPa for more than 3 d. Plants inoculated at the time stress was relieved by watering were not colonised more than well water controls. According to Levitt (1980), stresses such as water deficits exert strains on plants. Elastic strains may be alleviated by removal of the stress, whereas plastic strains are irreversible. Crist & Schoeneweiss (1975) found that the rate of canker expansion decreased and finally ceased in predisposed white birch after stress was relieved by watering, indicating that predisposition was an elastic strain. However, Schreiber & Dochinger (1967) reported that cankers caused by *Fusarium solani* on water-stressed paper mulberry continued to expand after seedlings were watered. Thus water stress may exert either an elastic or a plastic strain, the type possibly depending on stress level and duration.

Wilts

Vascular wilt fungi are capable of growing at ψ of −9.0 to −11.0 MPa, where growth of antagonists is inhibited, and infection may even occur in drier soils. However, Cook (1973) concluded that low ψ make the host plant more resistant to disease. Also, spread of the fungus within the plant is more rapid during periods of maximum water flow. Thus, vascular wilts are characteristically more severe in wet than in dry soils, although wilt symptoms develop most rapidly under hot, dry weather.

The effect of water stress on host predisposition to wilt fungi appears complex. Otazu, Gudmestad & Ziuk (1978) reported that the black dot organism, *Colletotrichum atramentarium*, was an important component of the potato wilt complex in North Dakota. The fungus was normally a 'superficial colonizer' of healthy potato vines or root systems, or both, but could colonise the vascular system of those plants predisposed by either water stress or infection by other wilt fungi. Bell (1982) stated that moisture stress before inoculation may increase resistance to wilts caused by *Verticillium* spp., but it usually aggravates wilt when it occurs after inoculation. He did not, however, indicate whether the effect was on the pathogen or on host susceptibility.

Correll & Schneider (1982) observed that yellows of celery caused by *Fusarium oxysporum* f. sp. *apii* was more severe where irrigation water accumulated and oxygen stress developed. They subjected field-grown plants to schedules of heavy and light irrigation and measured oxygen diffusion rates. Progression of the disease from roots to crowns was accelerated by periods of oxygen stress. Greenhouse experiments also indicated that root colonisation by the pathogen was increased following brief periods of oxygen stress. In contrast, Melchior & Morehart (1979) studied the progress of vascular wilt caused by *Verticillium* sp. in yellow poplar under continuously high or low soil moisture levels. They reported that the fungus was substantially more aggressive in host tissue stressed by periods of low moisture; at no time did diseased, flooded plants exhibit wilt symptoms, nor could the pathogen be re-isolated from root tissue. The influence of water stress on wilt diseases and the effects on host predisposition remain controversial.

Other diseases

The influence of water stress as a predisposing factor has not been investigated in many diseases other than root and stalk rots, cankers, and wilts. In the case of foliar diseases where moisture films or high humidity are required for germination and/or penetration by fungal propagules, studies have concentrated on the effects of moisture levels on the pathogen. Couch & Bloom (1960), however, found that soil moisture stress heightened disease proneness of Kentucky bluegrass to *Sclerotinia homeocarpa*, dollar spot, and disease development increased with decreasing soil moisture content, regardless of nutrient level or soil pH. Similar results were reported with *Helminthosporium sativum* leaf blight of Kentucky bluegrass (Atalla & Capellini, 1977). Leaf blight increased significantly in water-stressed plants and sporulation was higher on wilted than on turgid leaves. In contrast, although leaves of tobacco plants kept under a low soil moisture regime before inoculation were more highly predisposed to infection by *Peronospora tabacina* than those given a more abundant water supply, this occurred only if leaves were turgid at the time of inoculation (Rotem, Cohen & Spiegel, 1968). Negligible infection occurred on wilted leaves regardless of the pre-inoculation moisture regime. Drought-enhanced resistance to powdery mildew in barley is described by Ayres & Paul (Ch. 15).

Pre-infection water stress may predispose sugarbeet roots to storage-rot fungi. Harvested roots aged under moisture stress were more resistant to storage rot caused by *Botrytis cinerea* and *Phoma betae* than those provided with adequate moisture (Bugbee, 1979). When stored under identical conditions, however, roots of plants grown under adequate moisture in autumn were less liable to rot than those grown under moisture stress. Water stress, induced by withholding irrigation during silking to late dough stage of grain development in maize, increased infection and aflatoxin production by *Aspergillus flavus* in developing kernels (Jones, Duncan & Hamilton, 1981).

Water stress, senescence, and predisposition

Water stress is known to have significant effects on hormonal balance, cell-wall and protein synthesis, and many other processes in plants (Levitt, 1980; Hanson & Hitz, 1982). Although it is not worthwhile to attempt to make valid cause-and-effect correlations between metabolic changes under water stress and predisposition to fungal pathogens, given our present state of knowledge, the promotion of premature or accelerated senescence by water stress appears to be directly involved in the appearance of many diseases. Leopold (1980) defines senescence as the 'deteriorative processes that are the natural causes of death', while aging refers to 'processes of accruing maturity with the passage of time'. Farkas (1978), however, regards senescence as a developmental phase and states that pathogens do not 'recognize' a sharp boundary between aging and senescence. Although many organisms may colonise older plant tissues, we are more concerned with the effects of water stress on premature or accelerated senescence that leads to increased damage by fungal pathogens which otherwise remain ineffective or weak colonisers of healthy plant tissue. Much has been written on the physiological processes in natural senescence (Leopold, 1980; Nooden, 1980) and disease-induced senescence (Farkas, 1978), but less is known about senescence-induced diseases caused by pathogens.

The importance of cellular senescence in root and stalk rots of sorghum is well documented (Duncan, 1984). Water stress, which accelerates senescence of pith parenchyma cells in sorghum, also accelerates the development of stalk rots. Standard sorghum genotypes undergo sequential senescence in which the older leaves at the base of the stalk die first. Under water stress, senescence is followed by plant death at physiological maturity, whereas nonstressed plants continue to grow from tillers at the base until killed by frost. The phenomenon of green leaf retention after the grain has reached physiological maturity is termed nonsenes-cence (Duncan, 1984). In general, nonsenescent genotypes have been shown to be more resistant to root- and stalk-rot fungi.

A similar association between cellular senescence and stalk rot was found in maize. Death of parenchyma cells was preceded by nuclear and nucleolar degeneration, loss of *t*RNA methylase activity, abnormal protein synthesis, and increased synthesis and activity of nucleases and proteases; cell death was accelerated by water stress (Pappelis & BeMiller, 1984). Even mild water stress appears to increase the rate of parenchyma-cell death in maize; Schneider & Pendery (1983) concluded that symptoms of stalk rot may be caused by early-season water stress, which predisposes roots and eventually stalks by death of parenchyma cells. Since the mechanisms of resistance to root- and stalk-rot fungi in maize and sorghum are unknown, it is also not known whether predisposition results from altered metabolism associated with senescence or whether the fungi only grow saprotrophically in dead parenchyma cells. Cellular senescence reflects a loss in energy that is available to play a major role in host defence responses, such as repair or synthesis of cell wall components in cells under attack.

Cellular senescence associated with stress predisposition has not been described for diseases such as stem cankers of woody plants. Wene (1979) found that hyphae of *B. dothidea* remain viable for weeks in open xylem vessels of wound-inoculated, vigorous woody stems. No physical barriers to colonisation were observed and the hyphae were thin, contorted, unbranched, and stained poorly with vital stains as compared to robust, branched, heavily stained hyphae in stems predisposed by water stress. Xylem parenchyma cells appeared alive in predisposed stems, and uninoc-ulated plants recovered when stress was relieved without the appearance of dead or discoloured parenchyma cells. Subsequent studies with the scanning electron microscope (McPartland & Schoeneweiss, 1984) showed a significantly higher percentage of swelling and bursting of hyphal tips of *B. dothidea* in nonstressed, resistant stems compared with predisposed stems. Removal of amorphous cytoplasm revealed distinct holes in the cell walls of burst hyphal tips, indicating the possible action of lytic enzymes. Hyphal tips in stems predisposed by water stress remained intact and were often coated with a two-layered hyphal sheath, which may serve as a protective mechanism against enzymatic degrada-tion. Wargo (1975) reported that chitinase and β-1-3-glucanase from forest trees caused *in vitro* lysis of cell walls of *Armillaria mellea*. He hypothesised that host-produced lytic enzymes were involved in resistance of forest trees to fungal pathogens. Pegg & Young (1981) also reported lysis of cell walls of *Verticillium albo-atrum* in tomato by glycosidases that increased in activity in infected hosts and were associated with a sharp reduction in the amount of viable pathogen in resistant plants. In summary of earlier work on lytic enzymes, Pegg (1977) suggested that synthesis of glucanhydrolases in response to infection is a viable alternative to phytoalexin production as a disease-resistance mechanism. Since water-stress induced senescence is associated with changes in plant hormonal balance (Nooden, 1980), and hormones such as ABA and ethylene have pronounced effects on synthesis of both glucanhydrolase enzymes and phytoalexins, Ayres (1984) suggested that ethylene might have a dual, regulatory role in defence.

Conclusions

Water stress causes many changes in the metabolism of higher plants, some of which are well known, while others are poorly understood. It influences the ability of plants to defend themselves against attack by pathogens, yet our understanding of the processes involved is rudimentary. Host defence mechanisms, such as lignification and synthesis and accumulation of phytoalexins and lytic enzymes, have received much attention, yet their role in disease resistance remains controversial. Unless we know how a plant resists attack by a fungus, attempts to elucidate the changes in metabolism that are responsible for predisposition must be considered hypothetical. As Ayres (1984) has stated, both abiotic and biotic stresses cause defence reactions that draw upon common pools of energy and upset the energy balance in plants. These stresses interact, and it is reasonable to assume that abiotic stresses, such as water stress, either cause the plant to expend energy in adapting to the stress or reduce the ability of the plant to produce the energy needed in active defence against biotic stress.

Plants subjected to mild water stress adapt or 'drought-harden' through osmoregulation or other means (Levitt, 1980). If plants are able to adapt to water stress, are they also able to maintain a higher level of effective defence response under stress than plants that are not able to adapt? If this is so, it may be that threshold levels of water stress required for predisposition to nonaggressive fungal pathogens may be lowered in adapted plants, as hypothesised by Schoeneweiss (1978), and that imposition of mild water stress by cultural practices may be an effective means of reducing the predisposing effects of drought.

References

Appel, D. N. & Stipes, R. J. (1984). Canker expansion on water-stressed pin oaks colonized by *Endothia gyrosa*. *Plant Disease*, **68**, 851–3.

Atalla, N. S. & Cappellini, R. A. (1977). Effect of water stress on *Helminthosporium sativum* infection and development in Kentucky blue grass. *Proceedings of the American Phytopathology Society*, **4**, 182 (Abstr.).

Ayres, P. G. (1984). The interaction between environmental stress injury and biotic disease physiology. *Annual Review of Phytopathology*, **22**, 53–75.

Bagga, D. K. & Smalley, E. B. (1974). The development of Hypoxylon canker of *Populus tremuloides*: role of interacting environmental factors. *Phytopathology*, **64**, 658–62.

Bateman, D. F. (1961). The effect of soil moisture upon development of poinsettia root rots. *Phytopathology*, **51**, 445–51.

Bell, A. A. (1982). Plant pest interaction with environmental stress and breeding for pest resistance: plant disease. In *Breeding Plants for Less Favorable Environments*, ed. M. N. Christiansen & C. F. Lewis, pp. 353–63. New York: Wiley Press.

Bertrand, P. F., English, H., Uriu, K. & Schick, F. J. (1976). Late season water deficits and development of Cytospora canker of French prune. *Phytopathology*, **66**, 1318–20.

Bier, J. E. (1964). The relation of some bark factors to canker susceptibility. *Phytopathology*, **54**, 272–5.

Blaker, N. S. & McDonald, J. B. (1981). Predisposing effects of soil moisture extremes on the susceptibility of *Rhododendron* to Phytophthora root and crown rot. *Phytopathology*, **71**, 831–4.

Bloomberg, W. J. (1962). Cytospora canker of poplars: factors affecting the development of the disease. *Canadian Journal of Botany*, **40**, 1271–80.

Bugbee, W. M. (1979). The effect of plant age, storage, moisture, and genotype on storage rot evaluation of sugarbeet. *Phytopathology*, **69**, 414–16.

Chotena, M., Makus, J. D. & Simpson, W. R. (1980). Effect of water stress on production and quality of sweet corn seed. *Journal of the American Society for Horticultural Science*, **105**, 289–93.

Cook, R. J. (1973). Influence of low plant and soil water potentials on diseases caused by soilborne fungi. *Phytopathology*, **63**, 451–8.

Cook, R. J. & Papendick, R. I. (1972). Influence of water potentials of soils and plants on root diseases. *Annual Review of Phytopathology*, **10**, 349 74.

Correll, J. C. & Schneider, R. W. (1982). The effect of frequency of irrigation on severity of Fusarium yellows of celery. *Phytopathology*, **72**, 948 (Abstr.).

Couch, H. B. & Bloom, J. R. (1960). Influence of environment on diseases of turfgrasses. II. Effect of nutrition, pH, and soil moisture on Sclerotinia dollar spot. *Phytopathology*, **50**, 761–3.

Crist, C. R. & Schoeneweiss, D. F. (1975). The influence of controlled stresses on susceptibility of European white birch stems to attack by *Botryosphaeria dothidea*. *Phytopathology*, **65**, 369–73.

Dodd, J. L. (1980). The role of plant stresses in development of corn stalk rots. *Plant Disease*, **64**, 533–7.

Duncan, R. R. (1984). The association of plant senescence with root and stalk rot diseases in sorghum. In *Sorghum Root and Stalk Rots: A Critical Review, Proc. Consultative Group Discussion on Research Needs and Strategies for Control of Sorghum Root and Stalk Rot Diseases*, pp. 99–109. Bellagio, Italy. Patancheru,

India: International Crops Research Institute for the Semi-Arid Tropics (ICRISAT).

Duniway, J. M. (1977). Predisposing effect of water stress on the severity of Phytophthora root rot in safflower. *Phytopathology*, **67**, 884–9.

Edmunds, L. K. (1964). Combined relation of plant maturity, temperature, and soil moisture to charcoal stalk rot development in grain sorghum. *Phytopathology*, **54**, 514–17.

Farkas, G. L. (1978). Senescence and plant disease. In *Plant Disease, An Advanced Treatise*, vol. 3: *How plants Suffer From Disease*, ed. J. G. Horsfall & E. B. Cowling, pp. 391–412. New York: Academic Press.

Filer, T. H., Jr (1967). Pathogenicity of *Cytospora, Phomopsis*, and *Hypomyces* on *Populus deltoides. Phytopathology*, **57**, 978–80.

Ghaffar, A. & Erwin, D. C. (1969). Effect of soil water stress on root rot of cotton caused by *Macrophomina phaseolina. Phytopathology*, **59**, 795–7.

Hanson, A. D. & Hitz, W. D. (1982). Metabolic responses of mesophytes to plant water deficits. *Annual Review of Plant Physiology*, **33**, 163–203.

Jones, R. K., Duncan, H. E. & Hamilton, D. B. (1981). Planting date, harvest date, and irrigation effects on infection and aflatoxin production by *Aspergillus flavus* in field corn. *Phytopathology*, **71**, 810–16.

Jordan, W. R., Clark, R. B. & Seetharama, N. (1984). The role of edaphic factors in disease development, pp. 81–96 (ref. see Duncan, 1984).

Kuan, T-L. & Erwin, D. C. (1979). Predisposition effect of water saturation of soil on Phytophthora root rot of alfalfa. *Phytopathology*, **70**, 981–6.

Landis, W. R. & Hart, J. H. (1967). Cankers of ornamental crabapples associated with *Physalospora obtusa* and other microorganisms. *Plant Disease Reporter*, **51**, 230–4.

Leaphart, C. D. & Stage, A. R. (1971). Climate: a factor in the origin of the pole blight disease of *Pinus monticola* Dougl. *Ecology*, **52**, 229–39.

Leopold, A. C. (1980). Aging and senescence in plant development. In *Senescence in Plants*, ed. K. V. Thimann, pp. 1–12. Boca Raton: CRC Press.

Levitt, J. (1980). *Responses of Plants to Environmental Stresses*, vol. **2**: *Water, Radiation, Salt, and Other Stresses*, 2nd edn. New York: Academic Press.

Lewis, R. Jr & VanArsdel, E. P. (1978). Vulnerability of water-stressed sycamores to strains of *Botryodiplodia theobromae. Plant Disease Reporter*, **62**, 62–3.

McPartland, J. M. (1983). Stress predisposition and histopathology of canker diseases in woody hosts. M.Sc. Thesis, University of Illinois, Urbana. 60 pp.

McPartland, J. M. & Schoeneweiss, D. F. (1984). Hyphal morphology of *Botryosphaeria dothidea* in vessels of unstressed and drought-stressed stems of *Betula alba. Phytopathology*, **74**, 358–62.

Melchior, G. L. & Morehart, A. L. (1979). Effect of water stress on the development of Verticillium wilt of yellow poplar. *Phytopathology*, **69**, 536 (Abstr.).

Miller, D. E., Burke, D. W. & Kraft, J. M. (1980). Predisposition of bean roots to attack by the pea pathogen, *Fusarium solani* f. sp. *pisi* due to temporary oxygen stress. *Phytopathology*, **70**, 1221–4.

Neely, D. (1968). Bleeding necrosis of sweetgum in Illinois and Indiana. *Plant Disease Reporter*, **52**, 223–5.

Nooden, L. D. (1980). Senescence in the whole plant. In *Senescence in Plants*, ed. K. V. Thimann, pp. 219–58. Boca Raton: CRC Press.

Odvody, G. N. & Dunkle, L. D. (1979). Charcoal stalk rot of *Sorghum bicolor*, effect of environment on host–parasite relations. *Phytopathology*, **69**, 250–4.

Otazu, V., Gudmestad, N. C. & Ziuk, R. T. (1978). The role of *Colletotrichum atramentarium* in the potato wilt complex in North Dakota. *Plant Disease Reporter*, **62**, 847–51.

Papendick, R. I. & Cook, R. J. (1974). Plant water stress and development of Fusarium foot rot in wheat subjected to different cultural practices. *Phytopathology*, **64**, 358–63.

Pappelis, A. J. & BeMiller, J. N. (1984). The maize root rot, stalk rot, lodging syndrome, pp. 155–71 (ref. see Duncan, 1984).

Pegg, G. F. (1977). Glucanhydrolases of higher plants: a possible defense mechanism against parasitic fungi. In *Cell Wall Biochemistry Related to Specificity in Host–Plant Pathogen Interactions*, ed. B. Solheim & J. Raa, pp. 304–45. Oslo: Universitetsforlaget.

Pegg, G. F. & Young, F. H. (1981). Changes in glycosidase activity and their relationship to fungal colonization during infection by *Verticillium albo-atrum*. *Physiological Plant Pathology*, **19**, 371–82.

Phipps, P. M. & Beute, M. K. (1977). Influence of soil temperature and moisture on the severity of Cylindrocladium black rot in peanut. *Phytopathology*, **67**, 1104–7.

Ross, E. W. (1964). Cankers associated with ash dieback. *Phytopathology*, **54**, 272–5.

Rotem, J., Cohen, Y. & Spiegel, S. (1968). Effect of soil moisture on the predisposition of tobacco to *Peronospora tabacina*. *Plant Disease Reporter*, **52**, 310–13.

Schipper, A. L. Jr, McNabb, H. S. Jr & Haywood, W. F. (1977). *Dothichiza populae* and *Phomopsis macrospora* cankers of hybrid poplar in Iowa. *Proceedings of the American Phytopathological Society*, **4**, 109–11.

Schneider, R. W. & Pendery, W. E. (1983). Stalk rot of corn: mechanism of predisposition by an early season water stress. *Phytopathology*, **73**, 863–71.

Schoeneweiss, D. F. (1975a). Predisposition, stress, and plant disease. *Annual Review of Phytopathology*, **13**, 193–211.

Schoeneweiss, D. F. (1975b). A method for controlling plant water potentials for studies on the influence of water stress on disease susceptibility. *Canadian Journal of Botany*, **53**, 647–52.

Schoeneweiss, D. F. (1978). Water stress as a predisposing factor in plant disease. In *Water Deficits and Plant Growth*, vol. **5**, ed. T. T. Kozlowski, pp. 61–99. New York: Academic Press.

Schoeneweiss, D. F. (1981). The role of environmental stress in diseases of woody plants. *Plant Disease*, **65**, 308–14.

Schoeneweiss, D. F. (1983). Drought predisposition to Cytospora canker in blue spruce. *Plant Disease*, **67**, 383–5.

Schreiber, L. P & Dochinger, L. S. (1967). Fusarium canker on paper mulberry (*Broussonetia papyrifera*). *Plant Disease Reporter*, **51**, 531–2.

Silverborg, S. B. & Ross, E. W. (1968). Ash dieback disease development in New York State. *Plant Disease Reporter*, **52**, 105–7.

Stim, J. A. & Himelick, E. B. (1981). Honey locust decline in urban areas. *Phytopathology*, **71**, 906 (Abstr.).

Stoltzy, L. H., Letey, J., Klotz, L. J. & Labanauskas, C. K. (1965). Water and aeration as factors in root decay of *Citrus sinensis*. *Phytopathology*, **55**, 270–5.

Tobiessen, P. & Buchsbaum, S. (1976). Ash dieback and drought. *Canadian Journal of Botany*, **54**, 543–5.

Towers, B. & Stambaugh, W. J. (1968). The influence of induced soil moisture stress upon *Fomes annosus* root rot of loblolly pine. *Phytopathology*, **58**, 269–72.

Trimboli, D. S. & Burgess, L. W. (1983). Reproduction of *Fusarium moniliforme* basal stalk rot and root rot of grain sorghum in the greenhouse. *Plant Disease*, **67**, 891–4.

Wargo, P. M. (1975). Lysis of the cell wall of *Armillaria mellea* by enzymes from forest trees. *Physiological Plant Pathology*, **5**, 99–105.

Wene, E. G. (1979). Stress predisposition of woody plants to stem canker caused by *Botryosphaeria dothidea*. Ph.D. Thesis, University of Illinois, Urbana. 60 pp.

Yarwood, C. E. (1976). Modification of the host response-predisposition. In *Physiological Plant Pathology, Encyclopedia of Plant Physiology*, New Series, vol. **4**, ed. R. Heitefuss & P. H. Williams, pp. 703–18. New York: Springer-Verlag.

10
Drought, irrigation and fungal diseases of tropical crops

J. M. WALLER

Commonwealth Mycological Institute, Ferry Lane, Kew, Surrey, TW9 3AF, UK

Introduction

Water availability, more than any other factor, determines plant growth in the tropics both in absolute terms and by influencing seasonal fluctuations. In contrast to temperate areas, where day-length variation is the constant regulator of plant growth patterns, it is the seasonal differences of rainfall which determine cropping patterns in the tropics (Waller, 1976). Large areas of the tropics are characterised by seasonally arid climates but even in humid equatorial areas, short dry seasons control the annual growth cycle of plants.

Traditional agricultural systems have developed to cope with seasonal moisture fluctuations especially in the semi-arid regions of the tropics, but aridity and drought differ in that the former is normal and the latter abnormal. Aridity is a permanent feature of a region characterised by low rainfall and limited water availability whereas drought is a temporary state caused by a deficiency of rainfall significantly below the average. Drought may be defined in purely meteorological terms of rainfall and/or temperature or, more realistically from an agricultural point of view, in terms of evapotranspiration, soil moisture availability and a minimum duration of dry conditions (see WMO, 1975). It is the apparently unpredictable frequency and severity of droughts which causes the greatest concern. However, there is evidence to indicate long-term cyclic fluctuations in weather patterns, for example, Lumb (1966) showed that fluctuations in rainfall and lake levels in E. Africa coincided with cyclical fluctuations in sunspot activity.

Generally, droughts are synonymous with unusual rainfall deficiencies occurring over periods long enough to have adverse effects on ecosystems, but drought is an emotive word and provides a convenient scape-

175

goat for many of man's self-imposed ills. In an agricultural context we are concerned with plants suffering from water stress but this can be caused by factors other than unusual rainfall deficiency. These include edaphic factors causing soil moisture deficits or restricting root growth, biotic factors which may hinder water uptake and transport by the plant causing 'physiologic drought' and, possibly the most important factor associated with water stress in tropical crops, improper matching of the crop or cropping system to the environment.

Effects of drought on hosts
Direct effects
The high levels of solar radiation and temperature characteristic of most tropical areas mean that water stress is a common phenomenon and it could be claimed that erratic rainfall is in itself a characteristic of many tropical climates. Plants able to cope with this, especially when exposed to other stresses such as those imposed by diseases, will be more fit to thrive in the tropics. Natural plant communities have evolved to meet these environmental challenges; their natural phenotypic plasticity allows them to adapt to dry conditions. Such locally adapted plants have been incorporated into traditional agriculture and this natural adaptability is greatest in locally evolved traditional land races of tropical crops. Their yields, however, are often too low for man's expanding agricultural requirements. Modern high-yielding varieties, often bred and selected to take full advantage of ideal or at least 'normal' ecological conditions, have often lost much of this phenotypic drought adaptability and thus are more susceptible to the effects of drought and more readily predisposed to associated diseases.

As rainfall irregularity must be considered a major characteristic of many tropical climates, particularly in Africa, this loss of adaptability is a high price to pay for increased yield. Buddenhagen (1983a) has recently reviewed breeding strategies for stress and disease resistance in developing countries and advocates 'balanced ecosystem adaptation'. Yield increases must be sought under the existing adversity complex and selection for resistance to adverse physical factors cannot be separated from interacting disease effects. Selection for drought tolerance often results in lower attainable yields in non-stress situations, but the resulting cultivars are able to give stable yields over a wide range of ecological conditions.

Although the most serious direct effects of drought far outweigh the effects of most plant pathogens, sublethal water stress does influence many tropical plant pathosystems. Hard data to indicate the magnitude of these effects, particularly in relation to the prevalence and severity of associated diseases on plant productivity are difficult to find for those areas of the tropics where the effects of drought are most critical. Nevertheless it is possible to group the types of interactions between drought and plant pathogens into those which primarily affect the resistance of the host, usually by predisposing it to infection, those which have a more direct effect on the pathogen, and those in which diseases influence the drought tolerance of the host.

Predisposition to disease

To soilborne pathogens. Premature senescence of drought-stressed tissues has been implicated as the major mechanism accounting for predisposition. Schneider & Pendery (1983) showed that water stress early in the life of maize plants subsequently predisposed them to stalk rots even though the stress was removed; there was greater resistance to water flow between root and leaves in prestressed plants, even when water was not limiting and premature senescence ensued. These mechanisms are discussed in more detail by both Schoeneweiss and Hall (Chs 9 and 14).

Water-stress predisposition is especially prevalent in root and stalk diseases caused by a range of soilborne fungi and these effects are very evident on sorghum – a widely grown cereal crop in the semi-arid tropics (ICRISAT, 1984).

Macrophomina phaseolina causing charcoal root and stalk rot is probably the most well known of the tropical drought-favoured pathogens, attacking a wide range of herbaceous crops but particularly important in tropical cereals, cotton and grain legumes. *M. phaseolina* is ubiquitous in tropical soils, inhabiting plant rhizospheres and surviving as small resistant sclerotia. It is isolated frequently from the roots of a wide range of plants, whether or not they are obviously diseased or recognised as susceptible hosts of the pathogen. This phenomenon led Garrett (1956) to refer to *M. phaseolina* as an 'imposter' pathogen. Water stress apparently upsets the balance of this cohabitation by impairing the host's defence mechanisms. The fungus then takes over, spreading up the roots to the stem base and frequently causing premature death of its host.

Aspergillus niger which can tolerate very low moisture deficits, can cause severe crown rot symptoms on young groundnuts stressed by hot, dry conditions. This problem, among others, was widespread during the ill-fated groundnut scheme of the Overseas Food Corporation in

Tanzania in the late 1940s (Gibson, 1953) – a prime example of where a crop was improperly matched to the variability of the local environment.

Diseases caused by *Fusarium* spp. can also be exacerbated by drought stress and the foot rot pathogens, such as *Fusarium solani, F. culmorum*, and less-aggressive species such as *F. equiseti* are commonly encountered on cereals and other annual crops in the tropics (Giha, 1976). The higher incidence of these diseases may be due as much to enhanced pathogen activity as to host predisposition – effects which are difficult to partition without careful research. The vascular wilt pathogen, *F. oxysporum*, is often reported to cause less damage under dry conditions and it has been hypothesised that reduced transpiration apparent under conditions of water stress delays the spread of the pathogen through the vascular tissue (Cook & Papendick, 1972). However, this situation does not always occur in the tropics as will be discussed later. Many of the stalk and root rots associated with water stress predisposition are of apparently complex etiology and are often associated with normally weak pathogens.

To pathogens of shoots. Most pathogens of shoots may be expected to be at a disadvantage in drought conditions due to the prevailing low humidity, lack of leaf wetness and the hardening of drought-adapted plant surfaces. Yet certain diseases, particularly those induced by normally non-aggressive inhabitants of the bark of perennial plants, do become important. Coffee bark disease caused by *Fusarium stilboides* is an example. The fungus is widespread throughout the world on coffee and some other perennial hosts such as citrus where it exists saprotrophically or as a weak pathogen of damaged or senescent tissue. However, on stressed plants, the fungus can invade the bark phellogens of coffee causing flaking and eventual canker formation which kills the tree. In most coffee-producing areas the disease is of sporadic occurrence flaring up only when drought or injury predisposes the plant to infection. In areas marginal for coffee production where rainfall amount and distribution are erratic, the disease can be more persistent. In Malawi, coffee bark disease restricted coffee development for many years until resistant cultivars were selected (Siddiqi, 1980). Similar drought-favoured canker pathogens are *Phomopsis coffea* on coffee and *Phomopsis theae* on tea (Shanmuganathan & Rodrigo, 1967).

The effect of moisture stress on predisposition of trees to infection by *Botryosphaeria dothidea* is well established (Schoeneweiss, 1975) and the tropical counterpart of this is *B. ribis* which causes considerable canker damage to tree crops in the drier areas of Africa (J. M. Waller, unpubl.). *Hendersonia toruloidea* is another example of a fairly non-aggressive pathogen that is able to cause severe damage to fruit trees under water-stress conditions, although there is evidence that sun-scorch damage may be the primary predisposing mechanism (Punithalingam & Waterston, 1970). *Botryodiplodia theobromae* is a very common 'imposter pathogen' of plant shoots in the tropics occurring mostly as a secondary invader except of fruit where it is a primary storage pathogen. However, it can cause a severe dieback gummosis of citrus on trees which have been predisposed by drought, saline soils or the citrus root nematode *Tylenchulus semipenetrans*, all of which induce water stress in citrus shoots (Waller & Bridge, 1978). Another citrus disease exacerbated by drought is black spot caused by *Guignardia citricarpa* (Kotze, 1971) which is more prevalent on fruit from trees exposed to water stress just before harvesting.

Not unexpectedly, drought can affect some diseases of rice, the most widely irrigated tropical crop. Brown spot (*Cochliobolus myabeanus*) can cause a severe seedling blight of dry sown broadcast rice in Bangladesh and also in Cuba where irrigation otherwise prevents this loss (Herrera & Seidel, 1978). Blast (*Pyricularia oryzae*) is most severe on upland (rainfed) rice, particularly under dry conditions, and part of the difficulty of understanding the highly variable rice–blast pathosystem is that the genetic interactions of the system are greatly modified in their expression by environmental effects, particularly water stress. Old rice varieties showing the most stable resistance to blast, such as Tetep, are upland varieties which evolved under conditions very conducive to blast and there is evidence that resistance to water stress and blast are correlated (Buddenhagen, 1983*b*).

Effects of drought on pathogens

Most fungal and bacterial pathogens require ample moisture to infect plants, and dry conditions clearly put the vast majority of pathogens at a disadvantage during this stage of their life cycles, particularly those which infect plant shoots. However, in the soil, the amount of water deficit which can limit plant growth (-1.0 MPa) does not generally limit the growth of root-infecting fungi which can grow at deficits down to -3.0 MPa (Griffin, 1969), and growth of some soil fungi is stimulated by low water potentials. Many tropical pathogens, especially those adapted to seasonally dry areas have effective drought-resistance survival mechanisms, oospores, sclerotia, etc. and are not greatly affected by water stress. Nevertheless many of the more virulent soilborne pathogens, especially the oomycetes, are greatly restricted by drought. Water relations and pathogen activity in the soil are discussed in more detail by Duniway & Gordon (Ch. 7).

Indirect effects occur through the influence of soil moisture and temperature on the development of antagonistic microflora. Although actinomycetes appear to flourish in dry soils, many bacterial antagonists do not and the reduction of their activity during dry conditions may allow soilborne pathogens to become more aggressive. Such a reduction of natural microbial control has been proposed as the mechanism which allows certain *Fusarium* spp. causing foot rot diseases to become more active in dry conditions (Cook, 1973). This inhibition of microbial activity may also allow longer survival of pathogen propagules in dry soils.

Effect of diseases on drought resistance

Plants resist the effects of drought in several ways. In seasonally arid climates, rapid growth and early maturity, to make full use of the short rainy season, allow plants to escape drought. Drought endurance can be achieved by rapid root extension, particularly in depth, allowing plants to tap potentially large water supplies. Other in-built mechanisms affect the transpiration rate of plants in arid areas and include various morphological modifications to leaves, stomata, etc. characteristic of a xerophytic habit. The efficiency of these mechanisms is frequently impaired by the physiological effect of diseases so that plants can be rendered less tolerant to the effect of drought. For example, pathogens which reduce the growth of roots through the soil, or otherwise impair their exploration for soil moisture, will clearly have larger effects in dry conditions. Relatively little work has been done on this aspect with foliar diseases of tropical crops although work with temperate cereals has shown that foliar pathogens occurring early in the season will reduce root growth.

On coffee, berries are a particularly powerful physiological sink and heavy cropping is induced by exposure to full sunlight. This in turn increases susceptibility to coffee rust, *Hemileia vastatrix*, and the combined effect of defoliation caused by leaf rust and the high carbohydrate demand of the developing berries leads to a condition known as overbearing dieback in which both young shoots and roots are killed. Coffee trees suffering from this condition are much more susceptible to the effects of drought due to their reduced root activity.

'Cacar daun' (leaf blister) disease of cloves caused by a *Phyllosticta* sp. is a widespread but normally mild disease in Sumatra, but a drought in Lumpung Province in 1982/3 resulted in the death of 15 000 ha, or 20%, of the clove acreage. Healthy trees, including those on which the disease had been controlled with fungicides, were unaffected by the drought. The cacar daun pathogen disrupts the cuticle and epidermis of the leaf and probably induces excessive transpiration. A similar effect is known to occur with *Fusicoccum amygdali* on almonds. This produces a phytoalexin-like compound which induces stomatal opening; other pathogens may induce stomatal closure and this aspect is discussed in more detail by Ayres & Paul (Ch. 15).

Root pathogens have very significant effects on drought tolerance. Blast of young oil palm seedlings is caused by root infection by *Pythium splendens* and *Rhizoctonia lamellifera* coupled with excessive evapotranspiration in hot dry conditions, and can be controlled either by preventing root infection or by shading and irrigation (Rajagopalan, 1974). Crops infested with root knot nematodes, which are a major problem on light tropical soils, are also more susceptible to drought.

The interactions between drought and vascular pathogens are difficult to unravel. On the one hand these diseases have been reported to be less severe under dry conditions due to the reduced transpiration stream limiting the spread of the pathogen through the vascular tissues of the plant, but on the other the wilting observed is often more severe on cotton and tomato in the tropics when both drought and vascular wilt pathogens are present than when only one factor exists. Often, however, root knot nematodes are also present and the interacting effects of nematodes and *Fusarium oxysporum* f.sp. *lycopersici* have been shown to be more pronounced under dry conditions (Hienlein & Kumar, 1982). Growth of sugarcane is much reduced on droughted cane infected with ratoon stunting disease and the effects are additive (Rossler, 1974).

Irrigation

Measures used to combat the effects of drought can also cause plant health problems. Irrigation is of particular significance; others, such as mulching and sowing density, although worthy of further discussion, cannot be dealt with here. In the tropics, irrigation can be used to extend both the *area* and the *duration* of cultivation. There are also many different techniques of irrigation. Much work has been done on the effects of overhead irrigation (sprinkler) and low volume surface irrigation (trickle) particularly with regard to high volume crops in Mediterranean climates (Rotem & Palti, 1969). These techniques, however, require fairly high initial investment and are best suited to intensive agricultural enterprises. Various techniques of surface irrigation such as furrow, basin and flood irrigation (including paddy and deep water rice) are more common and applicable to developing countries in the tropics. In areas where water availability is the main determining factor for crop production, irrigation can be of great significance. FAO data show that during the last decade there has been an approximate 20% increase in the area of land under irrigation. This figure is applicable to both developed and developing regions but is proportionally smaller in the African continent where benefits would be largest, partly because suitable sources of irrigation water are fewer (Abernethy, 1985).

Irrigation in the tropics is not the panacea it may seem. Some large-scale schemes have been ruined by deteriorating soil, while others have run into trouble through pest and disease problems, brought about largely through the difficulties of maintaining effective control strategies under the severe and continuous pest and disease pressure that is characteristic of intensive agriculture in the tropics. Nevertheless, irrigation does generally remove the constraints imposed by water stress and consequently tends to have effects opposite to those associated with drought. Thus plant vigour is improved so that defence reactions are more active and the plant can cope more successfully with disease stress. However, the external moisture supply often favours many stages of the pathogen's life cycle; there is usually an abundance of succulent host tissue to infect and in some irrigated circumstances in the tropics, little seasonal constraint to epidemic development. Which of these apparently opposing factors dominates depends as much on the type of irrigation used, the efficiency of its management and its ecological effects on cropping patterns as on the particular pathosystem involved.

Effects on host susceptibility

As well as increasing the succulence and abundance of host tissue, making it more suitable as a substrate for many biotrophic pathogens, irrigation may impose stresses on plant tissues, particularly if it is poorly managed. Such indirect and unintentional predisposition effects may be similar to those normally associated with drought stress and are commonly encountered in the small-scale informal irrigation seen in many developing countries.

Waterlogging and resultant root oxygen starvation are particularly evident on heavy soils and predisposes citrus to Phytophthora root rot for example. Basin irrigation of lucerne in Oman leads to a rapidly debilitating crown rot of complex etiology which eventually halves the useful life of the crop (Waller & Bridge, 1978).

Salinity problems bedevil many irrigation projects in the arid tropics, particularly on soils where there is often an already high mineral salt content. High evaporation rates can lead to increasing salt concentrations in the top layers of the soil unless appropriate management techniques are used. This problem also occurs in coastal areas where excessive groundwater use can result in subterranean encroachment of seawater or tidal incursions up rivers as flow rates fall during the dry season. Salinity predisposes plants to many of the diseases characteristic of drought stress such as charcoal rot (*Macrophomina phaseolina*) (El Mahjoub *et al.*, 1979).

Erratic irrigation is a major problem of small-scale areas where meagre water supplies may be used beyond their optimal capacity. Village gardens irrigated by hand from wells during the dry season are particularly prone to these problems. As groundwater levels fall, well recharge times increase and irrigation becomes increasingly sporadic. A survey of dry-season vegetable production in the Gambia (J. M. Waller & J. Bridge, unpubl.) showed that problems such as blossom end rot of cucurbits and solanaceous fruits increase, and basal rot of onions (*F. oxysporum* f.sp. *cepae*) can become devastating. Onion is a major trade commodity during this time of year and losses commonly reach 20–30%.

Effects on pathogens

Water made available by irrigation for plant growth is also available for pathogens and microclimatic conditions for inoculum production, dispersal and infection are improved, particularly by overhead irrigation. The magnitude of these effects depends greatly on the conditions under which sprinkler irrigation is practised, as discussed by Rotem & Palti (1969). In very arid situations the increases in humidity and leaf wetness duration within crop canopies are minimal and the risks of infection by foliar pathogens are not greatly increased. Where prevailing conditions are less hostile, the risks are appreciably larger, particularly where sprinkler irrigation is used to extend the duration of the cropping season and the list of foliage and fruit diseases that are exacerbated under these conditions is extensive. However, where rates of application are high, overhead irrigation may succeed in washing inoculum away. On coffee in Kenya, irrigation is usually applied at 70–100 mm in 6 h, and the quantity of spores of *Colletotrichum coffeanum*, the coffee

berry disease pathogen, remaining in water drops on berries after irrigation is extremely small. Even with natural rainfall, numbers of spores
left on berries decrease rapidly after more than 15 mm rain (Waller,
1972). As well as dispersing fungal spores, overhead irrigation can be
used to disperse fungicides and Phytophthora rots of citrus fruit have
been controlled in this way (Oren & Solel, 1978).

Surface irrigation has a much smaller effect on the humidity within
crop canopies but it does have a significant effect on the movement
of soilborne inoculum of many pathogens. The zoospores of many *Phytophthora* species causing root and collar rots of many tropical crops
are widely dispersed in surface irrigation water which itself may be the
source of many waterborne pathogens (Shokes & McCarter, 1979).
Other examples of inoculum spread include sclerotia of *Corticium* spp.
causing sheath blight and stem rot of rice (Ou, 1985), and of *Sclerotinia
sclerotiorum* (Schwartz & Steadman, 1978).

Irrigation has a major effect on the survival of many pathogens during
dry seasons. The microbial activity of soils is increased when they are
wet and this enhances the destruction of crop residues and pathogens
surviving on them. For example *Alternaria solani* survives well on host
remains in dry conditions, but is rapidly destroyed in wet soils (Rotem,
1968). Flood irrigation is known to destroy the inoculum of many soilborne pathogens such as *Fusarium oxysporum* f.sp. *cubense* (Panama
disease of bananas) and teliospores of *Puccinia carthami* (Safflower rust)
(Kleisiewicz, 1977). Yields of cotton were substantially increased when
the crop was rotated with paddy rice and this was attributed to the
destruction of soilborne inoculum of *Verticillium dahliae* when the soil
was flooded for rice cultivation (Pullman & DeVay, 1981). The building
of the Aswan dam prevented the seasonal flooding of agricultural land
along the Nile valley and since then white rot of onions (*Sclerotinia
cepivorum*) has become a major problem. Apparently, flooding killed
the overseasoning sclerotia; these now survive and controlled irrigation
favours development of the disease (Rushdi *et al.*, 1974).

Effects on crop ecology

Irrigation can allow crops to be grown continuously with no
off-season breaks and can have major epidemiological impacts upon
pathogens and pests as the natural constraining influences of a dry season
are removed and disease epidemics can proceed unchecked over many
seasons. Frequently in large organised irrigation schemes, cropping flexibility is restricted and, even if land is fallowed, vegetation persisting

along the sides of irrigation ditches, paths, etc. allows ideal overseasoning sites for many pathogens and pests. Obligate pathogens, including viruses, are particularly favoured by the continuing presence of living hosts. Short-season rice cultivars now permit the growing of three crops during the year in some areas; furthermore adjacent crops often overlap in maturity and pathogens spread rapidly from maturing or senescing crops to younger ones.

The cultivation of dry-season irrigated cereals is becoming increasingly widespread in Africa and the natural off-season constraints to the continuing development of pathogen populations have been partly removed. This has allowed obligate pathogens such as the wheat rusts and maize streak virus to become more prevalent.

Irrigation can, however, be used to avoid diseases. Firstly, by extending the area rather than the duration of cropping, crops can be grown in arid regions well away from the inoculum sources of most pathogens. Secondly, short-season crops grown in semi-arid areas need not necessarily allow pathogens to bridge the off-season gap provided that the crops to be grown and their planting times are carefully selected. Thirdly, irrigation can be used to manipulate cropping times. Muller (1973) showed that early irrigation of coffee produced an early flowering and berries had passed their most susceptible growth stage by the time conditions were most suitable for coffee berry disease.

To summarise, irrigation requires careful use if diseases of tropical crops are not to be exacerbated. Surface irrigation is less likely to favour pathogens of plant shoots, but it requires careful management if predisposing soil conditions are to be avoided. The greatest dangers accrue when irrigation is used to extend seasons and where it bridges the off-season gap allowing disease epidemics to proceed unchecked between seasons, thus increasing the selection pressure on pathogens to adapt. When used to extend the area of land available for agriculture, irrigation has fewer epidemiological disadvantages. The desert can bloom with healthy crops under irrigation, but the subsequent spread of pathogens to these new areas must be prevented.

References

Abernethy, C. L. (1985). Irrigation in Africa: present situation and prospects. In *Advancing Agricultural Production in Africa*, ed. D. L. Hawksworth, pp. 342–6. Slough: Commonwealth Agricultural Bureaux.

Buddenhagen, I. W. (1983a). Breeding strategies for stress and disease resistance in developing countries. *Annual Review of Phytopathology*, 21, 385–409.

Buddenhagen, I. W. (1983b). Disease resistance in rice. In *Durable Resistance in Crops*, ed. F. Lamberti, J. M. Waller & N. A. Van der Graaff, pp. 401–28. NATO ASI Series A. vol. 55. New York: Plenum.

Cook, R. J. (1973). Influence of low plant and soil water potentials on diseases caused by soil borne fungi. *Phytopathology*, **63**, 451–8.

Cook, R. J. & Papendick, R. F. (1972). Influence of water potential of soils and plants on root diseases. *Annual Review of Phytopathology*, **10**, 349–74.

El Mahjoub, M., Bouzaidi, A., Jouhri, A., Hamrouni, A. & El Beji. (1979). Influence de la salinité des eaux d'irrigation sur la sensibilité du tournesol au *Macrophomina phaseolina* (Maulbl.) Ashby. *Annales de Phytopathologie*, **11**, 61–7.

Garrett, S. D. (1956). *Biology of Root-Infecting Fungi*. Cambridge: Cambridge University Press.

Gibson, I. A. S. (1953). Crown rot, a seedling disease of groundnut caused by *Aspergillus niger*. *Transactions of the British Mycological Society*, **36**, 198–209.

Giha, O. H. (1976). Fungi associated with foot and root rot diseases of irrigated wheat in the northern states of Nigeria. *PANS*, **22**, 479–87.

Griffin, D. M. (1969). Soil water in the ecology of fungi. *Annual Review of Phytopathology*, **7**, 289–310.

Herrera, L. & Seidel, D. (1978). On the injurious effect of *Cochliobolus miyabeanus* (Ito & Kuribayashi) Drechsler ex Dastur in rice growing in Cuba. *Archiv für Phytopathologie und Pflanzenschutz*, **14**, 285–90.

Hienlein, M. & Kumar, J. (1982). Interaction of *Meloidogyne incognita* and *Fusarium oxysporum* f.sp. *lycopersici* on 'Alton' variety tomatoes. *Fiji Agricultural Journal*, **44**, 17–20.

ICRISAT (1984). Sorghum Root and Stalk Rots, a Critical Review: *Proceedings of the consultative group discussion on the research needs and strategies for control*. Bellagio, 1983. Patancheru, India: ICRISAT.

Kleisiewicz, J. M. (1977). Effects of flooding and temperature on incidence and severity of safflower seedling rust and viability of *Puccinia carthami* teliospores. *Phytopathology*, **67**, 787–90.

Kotze, J. M. (1971). The effect of drought on the development of blackspot of citrus. *Citrus and Subtropical Fruit Journal*, **452**, 19.

Lumb, F. E. (1966). Variation of rainfall over Lake Victoria catchment since 1899 and over E. Africa since 1934. *Kenya Coffee*, **31**, 347–50.

Muller, R. A. (1973). L'anthracnose des baies du caféier d'Arabie (*Coffea arabica* L.) due à une forme virulente du *Colletotrichum coffeanum* Noack. *Café, Cacao Thé*, **17**, 281–312.

Oren, Y. & Solel, Z. (1978). Control of brown rot of citrus fruit by application of fungicides via sprinkler irrigation systems. *Phytoparasitica*, **6**, 65–70.

Ou, S. H. (1985). *Rice Diseases*. Slough: Commonwealth Agricultural Bureaux.

Pullman, G. S. & DeVay, J. E. (1981). Effect of soil flooding and paddy rice culture on the survival of *Verticillium dahliae* and incidence of Verticillium wilt of cotton. *Phytopathology*, **71**, 1285–9.

Punithalingam, E. & Waterston, J. M. (1970). *Hendersonula toruloidea. CMI Descriptions of Pathogenic Fungi and Bacteria*, No. 274.

Rajagopalan, K. (1974). Influence of irrigation and shading on the occurrence of blast disease of oil palm seedlings. *Journal of the Nigerian Institute of Oil Palm Research*, **5**, 23–31.

Rossler, L. A. (1974). The effects of ratoon stunting disease on three sugarcane varieties under different irrigation regimes. *Proceedings of the 15th Congress of*

International Society of Sugarcane Technologists, ed. J. Dick & D. J. Collingwood, pp. 250–7. Durban: ISSCT.

Rotem, J. (1968). Thermoxerophytic properties of *Alternaria porri* f.sp. *solani*. *Phytopathology*, **58**, 1284–7.

Rotem, J. & Palti, J. (1969). Irrigation and plant disease. *Annual Review of Phytopathology*, **7**, 267–88.

Rushdi, M., Shatla, M. N., Abd-El-Razik, A., Darwish, F. A., Ali, A. & El-Yamani, E. (1974). Effect of cultural practices and fungicides on control of white rot of onion. *Zeitschrift für Pflanzenkrankheiten und Pflanzenschutz*, **81**, 337–40.

Schneider, R. W. & Pendery, W. E. (1983). Stalk rot of corn – mechanism of predisposition by an early season water stress. *Phytopathology*, **73**, 863–71.

Schoeneweiss, D. F. (1975). Predisposition, stress and plant disease. *Annual Review of Phytopathology*, **13**, 193–211.

Schwartz, H. F. & Steadman, J. R. (1978). Factors affecting sclerotium populations of, and apothecium production by *Sclerotinia sclerotiorum*. *Phytopathology*, **68**, 383–8.

Shanmuganathan, N. & Rodrigo, W. R. F. (1967). Studies on collar and branch canker of young tea (*Phomopsis theae* Petch). II. Influence of soil moisture on the disease. *Tea Quarterly*, **38**, 320–30.

Shokes, F. M. & McCarter, S. A. (1979). Occurrence, dissemination and survival of plant pathogens in surface irrigation ponds in southern Georgia. *Phytopathology*, **69**, 510–16.

Siddiqi, M. A. (1980). The selection of arabica coffee for Fusarium Bark Disease resistance at Bvumbwe. *Kenya Coffee*, **45**, 55–9.

Waller, J. M. (1972). Water-borne spore dispersal in coffee berry disease and its relation to control. *Annals of Applied Biology*, **71**, 1–18.

Waller, J. M. (1976). The influence of climate on the incidence and severity of some diseases of tropical crops. *Review of Plant Pathology*, **55**, 185–94.

Waller, J. M. & Bridge, J. (1978). Plant diseases and nematodes in the Sultanate of Oman. *PANS*, **24**, 313–26.

WMO (1975). *Drought and Agriculture*. Technical Note No. 138, pp. 127. Geneva: World Meteorological Organization.

11

Interaction of host stress and pathogen ecology on *Phytophthora* infection and symptom expression in nutrient film-grown tomatoes

M. HOLDERNESS and G. F. PEGG

Department of Horticulture, University of Reading, Earley Gate, Reading, RG6 2AU, UK

The nutrient film technique of plant culture (NFT or alternatively NFC) is an important method of commercial glasshouse crop production in Britain. It comprises a shallow stream of nutrient solution, flowing down an inclined channel, usually of polyethylene sheet, over the bare roots of plants, to a catchment tank from which solution is recirculated by pump to the channel heads. Nutrient concentration and pH are monitored regularly and adjusted as required to maintain constant conditions.

The technique has led to substantial increases in yields over those obtained in soil or peat, by providing a more uniform root environment and avoiding the problems of erratic or suboptimal water or nutrient supply (Graves, 1983). The need for substrate sterilisation is obviated, avoiding the risk of toxic residues where chemical methods are used. Although this method of plant culture is relatively hygienic and the incidence of disease has been generally less than was originally anticipated, severe root and foot-rot diseases of tomato caused by the zoosporic pathogens *Phytophthora nicotianae* var. *parasitica* and *P. cryptogea* have still occurred. Possible sources of inoculum include the introduction of diseased plant material, soil splash into the channels, zoospore contamination of the water supply and wind-blown oospores in plant debris.

While zoospores may be disseminated freely throughout a recirculatory system, the pattern and frequency of *Phytophthora* outbreaks are irregular, with ostensibly healthy and diseased plants juxtaposed in the same channel. This type of disease distribution in a plant population of uniform genotype may be indicative of host predisposition (Yarwood, 1959), or modified disease proneness (Gäumann, 1950).

Several factors in the NFC environment could modify host susceptibility to infection and, or, disease expression. Root growth may be retarded by reduced oxygen levels due to inadequate solution flow rates or a limiting gradient of oxygen availability down the channel (Jackson, 1980). Root desiccation and the possible enhancement of infection may occur during plant establishment, through the cultural practice of reducing the water supply to restrict root growth and enhance early flower induction (Graves & Hurd, 1983), or from the effect of a hydrophobic channel surface leading to meandering of the solution stream and drying of isolated roots. Roots may also undergo osmotic water stress caused by high nutrient concentrations of up to 9.5 mS, used experimentally to induce flowering (Richardson, 1982).

Environmental conditions may either enhance infection or lead to increased disease expression, either of which may be further affected by the growth stage of the plant and corresponding innate physiological stresses. Leonard & Head (1958) described a decline in root growth in soil-grown tomato plants coincident with fruit development. This was attributed to a redistribution of assimilate from roots to fruit (Khan & Sagar, 1966) leading to the physiological death of roots. A similar but more severe root die-back ('root death') leading to the collapse of nutrient film-grown plants has also been associated with fruiting (Hurd, Gay & Mountifield, 1979). A wide range of organisms has been isolated from such moribund roots, including *Phytophthora erythroseptica* (Price, 1976), *P. cryptogea, P. nicotianae* var. *parasitica, Colletotrichum coccodes* and *Pythium* spp. (Evans, 1979). Although pathogens have been isolated from diseased plants showing aerial symptoms, the re-establishment of infection experimentally, using the same isolate and cultivar, has not always been possible. In the case of suspected physiological root death, it is not known whether recognised pathogens or avirulent 'rhizosphere' fungi, frequently found in association with the condition, are causally involved or are exploiting a decaying root substrate as saprotrophs.

Disease severity may be regarded as a function of the quantity and virulence of inoculum at the root surface, affected by host and environment, and the susceptibility of the plant to infection and disease expression as determined by host physiology, itself subject to environmental influence. Thus,

Disease severity = inoculum potential × disease potential
(Baker, Maurer & Maurer, 1967).

Propagules and lesion production can be monitored regularly in NFC without damage to the root or the root–pathogen interface. NFC may therefore provide a valuable technique to aid the construction of interpretive models of pathogen and host root interactions in the more complex soil environment.

The literature on plant disease in hydroponics is limited. Following the preliminary observations of Price (1979), and recognising both the importance of assimilate for root growth and the dependence of the pathogen on a supply of carbohydrate, experiments were mainly confined to a manipulation of assimilate supply and its effect on host–pathogen relations. Treatments included defoliation, shading, enhanced illumination and fruit removal. Experiments were carried out with young tomato plants, cv. Sonato, in model NFC systems and with larger fruiting plants (up to 25 flower trusses) in small commercial-scale units. *P. nicotianae* var. *parasitica* L. was used as a representative pathogen.

Inoculum motility

Turbulence in the circulatory pump induced the encystment of all zoospores without any reduction in viability after ten passages at 2600 rpm. This suggested that zoospore motility would be of limited duration in NFC and might only be of significance in localised infection and disease spread within a root mat. The concentration of nutrient solution also affected zoospore motility (Fig. 11.1). Zoospores were motile for longer periods in deionised water than in the nutrient solution normally used for plant culture (2.5 to 3 mS). At a concentration of 9 mS, encystment of all spores occurred within 2 min. Similar effects of violent agitation and nutrient concentration on encystment are described by Carlile (1983) (see also Ch. 6). The use of high solution conductivities to promote flowering would severely reduce the motility of zoospores entering such an environment.

Effect of zoospore motility on dispersal through the root mat

Motile and encysted zoospores behaved differently after entering root mats from a point source at the head of each channel. Encysted spores were deposited by sedimentation more rapidly than motile ones, resulting in steeper dispersal gradients (Table 11.1). No differences in gradients were found between the root densities tested.

Surface tension restricted the flowing solution into a narrow stream between newly established plants, but with the development of a root mat the solution formed a continuous film. The introduction of fluores-

Table 11.1. *Log:log gradient coefficients ('b') for numbers of zoospores dispersed linearly along tomato root mats*

Spore motility	Plants per channel			Mean
	3	4	6	
Encysted	−1.188	−0.857	−1.159	−1.068
Motile	−0.447	−0.348	−0.468	−0.451
Mean	−0.817	−0.602	−0.859	
Encysted/motile	2.7	2.5	2.5	
Mean root number cm^{-2}	16	22.5	30.4	

Means of gradients from two replicate channels per treatment, based on 24 samples per channel.
s.e. for comparison of motility means = 0.193.
s.e. for comparison of plant density means = 0.236.

cent dyes showed that flow was most rapid along the channel edges and underside of the root mat and through minor rivulets caused by differences in root size and distribution. Other areas within the root matrix received nutrient solution by greatly reduced flow or by diffusion into stagnant solution. Such regions could be exploited by zoospores

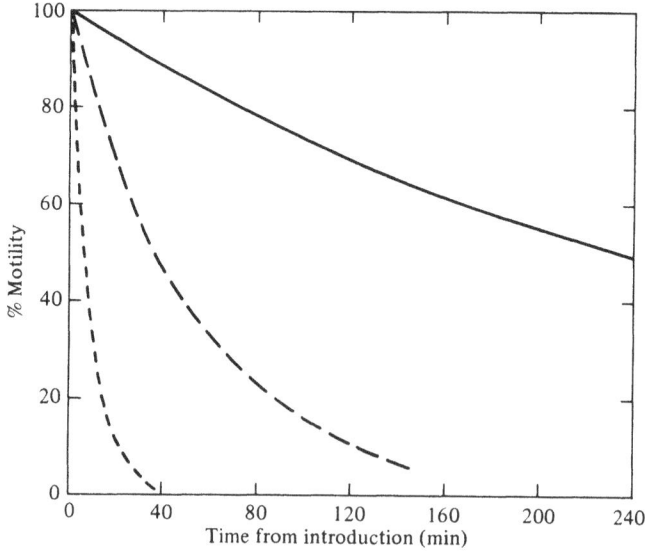

Fig. 11.1. Zoospore motility in NFC nutrient solution at different concentrations: (———) distilled water; (— —), 3mS; (-----), 4mS.

where encysted spores could not gain access. An advantage of zoospore motility in the spread of infection in NFC therefore lies in the possibility of spore deposition on roots which would be unavailable to non-motile cysts at a distance from the inoculum source. Zoospores are maintained longer in suspension and are less likely to be filtered by root mats such as exist in the long channels used commercially.

This dispersal pattern may be compared with zoospore dissemination in soil. The filtering effect of soil particles prevents the passive dispersal of non-motile soil inocula over distances of more than a few centimetres (Wallace, 1978). Motility, influenced by rheotaxis (Katsura & Kiyata, 1966), reduces the propensity for deposition on non-plant surfaces and entrapment in dead-end spaces, permitting the exploration of a greater volume of soil (Duniway, 1983). Passive spore dispersal in percolating water is predominantly downwards, particularly in the larger pores, but zoospore motility is considered sufficiently rapid, with negative geotaxis (Cameron & Carlile, 1977), to prevent spores being flushed too deeply into the soil (Young, Newhook & Allen, 1979).

Zoospore dispersal in nutrient-film culture is paralleled by the distribution of *Spongospora subterranea* zoospores through watercress beds. Tomlinson (1958) showed that zoospores were carried away from infected plants near the water inlet by fast-flowing water and sedimented in areas of reduced flow rate, resulting in disease dispersal distal from the inlet. The dissemination of inoculum in nutrient-film culture may be more analogous to that of airborne particles at a constant wind speed than to soilborne propagules; spores deposited from a turbulent dispersal medium are filtered through a matrix consisting almost entirely of plant material, the standing crop or, in NFC, the root mat. In soil however, the propagule has greatly increased likelihood of deposition on non-plant material and infection may depend on root growth into the vicinity of stationary inoculum (see Duniway & Gordon, ch. 7). Host growth may, therefore, play a more central role in the infection process in soil.

Dispersion of inoculum at microsites

Microscopic examination of excised roots from plants in solutions to which motile or encysted zoospores had been introduced showed that spores were frequently enmeshed among root hairs, but no significant accumulation occurred in the elongation zone, or at sites of lateral root emergence, both of which are known to be regions of high exudation. When uninoculated roots were excised and placed in zoospore suspensions, however, accumulation occasionally occurred at the cut

Table 11.2. *Factors affecting the water-soluble carbohydrate content of nutrient-film grown tomato roots*

(a) Shading and supplementary light

	Light	Shade	L.S.D.
PPFD	$134\,\mu mol\,m^{-2}s^{-1}$	$17.5\,\mu mol\,m^{-2}s^{-1}$	
	(mg glucose equiv. g^{-1} f.wt)		
Total water-soluble carbohydrate	2.34	0.49	1.73*
Total reducing sugars	1.57	0.42	1.09*

(b) Fruit and leaf removal

	Present	Removed	L.S.D.
	($\log_{10}$ mg glucose equiv. g^{-1} d.wt)		
Fruit	0.948	1.172	0.147**
Leaf	1.135	0.985	0.147**

*, **, significant at $P = 0.05, 0.01$, respectively.

ends. Dukes & Apple (1961) found no response of *P. parasitica* var. *nicotianae* (*P. nicotianae* var. *nicotianae*) zoospores to tomato roots. Chemotaxis was shown only towards wounded roots of tobacco, whereas Troutman & Wills (1964) described aggregation of zoospores of the same fungus in the elongation zone of uninjured tobacco roots. These differences were attributed by Hickman & Ho (1966) to variations in host physiology affecting root exudation.

A rhizosphere in NFC differs from that in soil. Significant gradients of exudates from roots would probably exist only in stagnant microsites, host substances otherwise dispersing in the nutrient stream. Similarly, organisms of the root microflora may have no fixed spatial arrangement in relation to the root mat. The microorganism population and host exudate content of the recycled solution will therefore change with time but it will not be affected by physical damage caused by attrition from soil particles (Barber & Gunn, 1974), or the larger microfauna.

Zoospore cysts from both motile and encysted inoculum were irregularly clustered along the root surface. Variance:mean ratios, calculated for spore numbers deposited per $350\,\mu m$ microscope field along individual roots, were greater than unity, implying an aggregated, non-Poisson distribution with no differences between spore types or root densities.

Although no gross spore aggregation was apparent, the random cluster-ing of small numbers of spores may enhance infection synergistically.

Inoculum effectiveness may also be heightened by an increase in its nutritional status (Garrett, 1956). Sloughed-off root debris and changes in the rate and composition of root exudation may affect germination and growth of the pathogen. The maintenance of root exudation requires a difference in concentration between the free pool of soluble sugars in the root cell and the external medium (Hale, Moore & Griffin, 1978). Roots of tomato plants given enhanced and extended illumination had much higher levels of total water-soluble carbohydrates than those from shaded plants (Table 11.2a). Similar but less consistent differences were found in roots from plants with fruit removed and the converse was seen in roots of defoliated plants (Table 11.2b). These differences would be expected to be reflected in exuded root sugars.

Germ-tube growth from zoospore cysts responded to increases in both glucose supply and nutrient solution concentrations (Fig. 11.2). These

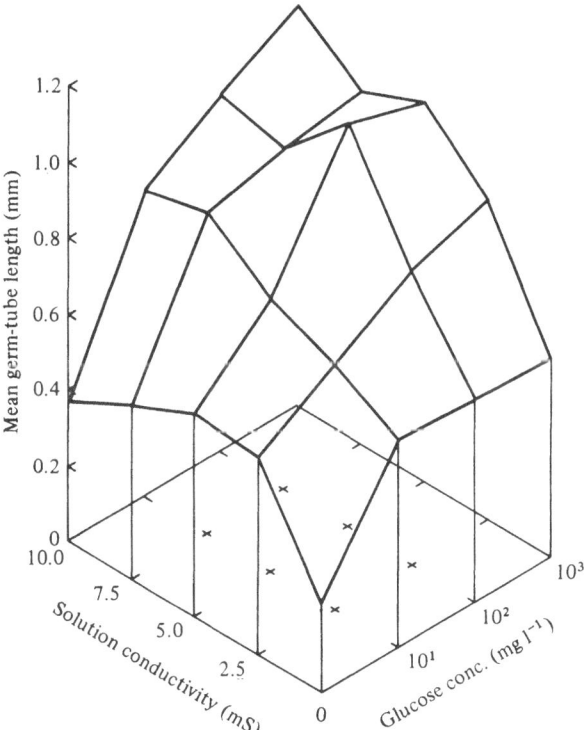

Fig. 11.2. Influence of glucose and ionic concentrations on zoospore germ-tube growth (means of 45 observations).

factors showed a positive interaction in relation to growth, but the increase was not proportionate to that of glucose concentration. Barash, Klisiewicz & Kosuge (1965) similarly reported that the addition of various sugars to the medium significantly increased the rate of germ-tube elongation in *P. drechsleri*, but had little effect on zoospore motility and germination.

Indirect germination to produce microsporangia occurred in distilled water and at low glucose levels, consistent with the findings of Thomson & Allen (1976). Appressorium-like structures were produced more frequently in solutions with glucose than in non-amended solutions, several commonly forming on each germ-tube.

Primary infection

Infection arising from zoospore inoculum was greatest towards the channel edges, where available inoculum was increased by a relatively free flow of solution alongside an area of high root density, and at the channel outlet where roots funnelled, creating an effective spore trap.

No difference was observed in the overall percentage of infection resulting from motile or encysted inoculum, probably because motility did not increase the likelihood of spore to root contact. Kliejunas & Ko (1974) working with *P. palmivora* on papaya, and Hwang & Ko (1978) with *P. cinnamomi* on ohia (*Metrosideros* (*Callistemon*) *collina* subsp. *polymorpha*), similarly showed that where the filtering effect of soil was eliminated by mixing inoculum with the soil, encysted spores were at least as effective as motile spores in colonising buried stem segments.

When zoospores were introduced into the root mats of illuminated or shaded plants, more primary lesions were produced on the former (Table 11.3a). (A similar increase was observed where fruit was removed from mature plants, the converse was obtained by defoliation treatments.) Increased infection reflected in part a more successful deposition on the denser lateral roots of those plants with a higher capacity for root growth (Table 11.3b). The results also suggested that roots supplied with high levels of assimilate were more susceptible than those deficient in carbohydrate, but when such roots were excised and inoculated *in vitro* there was a small but statistically insignificant increase in the rate of lesion spread. Since tomato roots do not normally function as storage organs (Nightingale *et al* 1928; Ludwig & Withers, 1977), this result may indicate the importance, in relation to infection, of the rate of assimilate supply and its turnover, rather than the transient level *per se*.

Table 11.3. *Effect of shading on development of infection*

(a) 3 d from inoculation

	Light 134 μmol m^{-2}s^{-1}	Shade 17.5 μmol m^{-2}s^{-1}	L.S.D.
PPFD			
Lesion number, 420 cm^{-2}	17.67	0.67	16.23**
Total length of infected root (cm)	24.23	0.5	16.89*
Number of roots with symptoms, 420 cm^{-2}	12.67	0.67	7.29*

(b) 4 d from inoculation (cm^{-2})

	Light		Shade		L.S.D. for either pair
	Control	Infected	Control	Infected	
% Infection main roots[a]		3.955		1.712	
arc sine transformed		11.31		6.12	9.46 n.s.
% infection lateral roots		1.437		0	
arc sine transformed		6.75		0	6.32**
Number of infected roots, cm^{-2}		0.287		0.021	0.102*
Number of healthy main roots, cm^{-2}	5.89	3.11	2.56	1.10	1.409*
Number of healthy lateral roots, cm^{-2}	22.64	13.70	5.64	1.72	5.56*

Means from 3 channels.
*, **, significant at $P = 0.05, 0.01$, respectively. n.s., not significant.
[a], roots >0.5 mm thick.

Production of secondary inoculum

The production of zoosporangia and release of zoospores are essential to the rapid multiplication of infection loci throughout the nutrient-film system. Serial dilutions of nutrient solutions plated into Benomyl, Nystatin, PCNB, Prifamycin, Ampicillin agar (BNPRA) (Masago, Yoshikawa, Fukada & Nakanishi, 1977), showed that the initial release of zoospores from roots of illuminated plants was greater than that from shaded (Table 11.4). This difference, initially correlated with the number of primary lesions, diminished 14 d after inoculation, through increased variability caused by several cycles of inoculum pro-

Table 11.4. *Effect of shading on production of secondary inoculum*

Days from inoculation	Zoospores cm^{-3} of nutrient solution		L.S.D.
	Light PPFD 134 μmol m^{-2}s^{-1}	Shade 17.5 μmol m^{-2}s^{-1}	
0	85[a]	85	
9	27.4	1.8	12.08***
14	63.89	19.5	150.8 n.s.
17	22.22	12.22	22.33 n.s.

Means from three channels.
[a] = initial inoculum concentration.
***, significant at $P = 0.001$. n.s., not significant.

duction and infection. Greater numbers of zoospores were also produced from the infected roots of plants from which fruit had been removed (statistically significant ($P = 0.05$) 20 and 35 d after inoculation), again reflecting the increased number of roots and zoosporangial sites (Fig. 11.3).

Studies on uniform young mycelia of *P. nicotianae* var. *parasitica* derived from zoospores showed that the solution used in NFC stimulated sporangial formation, compared with distilled water (Table 11.5), an

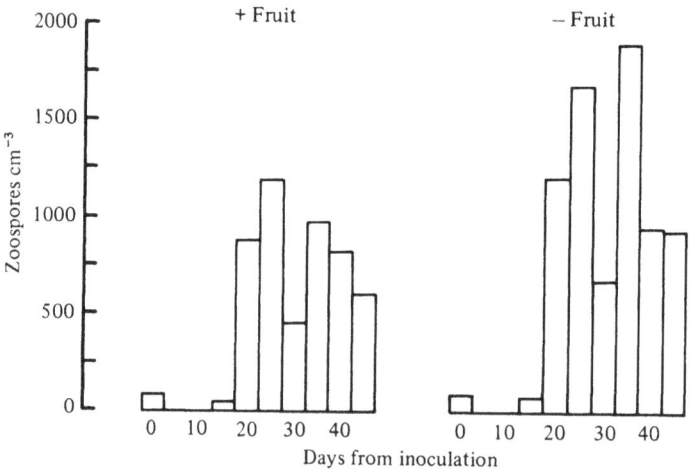

Fig. 11.3. Effect of fruit removal on production of secondary inoculum (means from five channels).

Table 11.5. *Sporangial production in nutrient solutions (mean number per microscope field)*

Solution source		Plant treatment	
		Light (±s.e.)	Shade (s.e.)
Sterile	Control	88.85 (±14.16)	51.78 (± 3.15)
	Infected	69.32 (±16.51)	70.73 (±14.6)
Non-sterile	Control	124.37 (±20.56)	114.95 (±15.58)
	Infected	153.0 (±20.65)	127.95 (±32.89)
2.5 mS Stock sterile		35.66 (± 4.01)	
Glass-distilled water		8.4 (± 2.75)	

s.e. for a comparison of pairs light shade, control infected = 8.91.
s.e. for comparison sterile vs. non-sterile = 9.53.

effect frequently found with individual nutrient ions (Gooding & Lucas, 1959; Allen & Nandra, 1975; Halsall & Forrester, 1977). Solutions that were filter-sterilised after plants had been growing for five weeks increased zoosporangial formation compared with fresh stock nutrient solution, indicating the influence of root exudate or breakdown products, or the solution microflora. Solutions from infected plants, or those receiving supplementary illumination, showed no consistent effect. Sporangial production was enhanced further when mycelial wefts were placed in the same, but unsterilised, nutrient solutions. Stimulation by the microflora has been observed in soil extracts (Mehrlich, 1935) and is considered to result from depletion of potential fungal nutrients by microorganisms, or the production of specific bacterial metabolites which stimulate sporangial formation (Marx & Haasis, 1965; Ayers, 1971; Ayers & Zentmeyer, 1971; Ribeiro, 1983). Microflora antagonistic to sporulation of *P. cinnamomi* have been reported in some Australian soils (Broadbent & Baker, 1974), but none of the non-sterile NFC solutions suppressed sporangial formation.

Progress of infection and disease outcome

As the epidemic progressed the production of secondary inoculum reduced the initial differences in infection found between plants with a high capacity for assimilate supply (illuminated or with fruit removed) and those with a low capacity (shaded or with fruit) (Table 11.6). No difference in percentage infection was found 30 d after inocula-

Table 11.6. *Effect of shading on development of infection 18 d from inoculation* (cm^{-2})

PPFD	Light 134 μmol m^{-2}s^{-1}		Shade 17.5 μmol m^{-2}s^{-1}		L.S.D.
	Control	Infected	Control	Infected	
% Infection main roots[a]		36.88		35.60	8.17 n.s.
% Infection lateral roots		70.66		47.70	36.25 n.s.
Number of infected roots		22.92		4.11	14.01**
Number of healthy main roots	6.804	2.78	3.33	1.75	1.239*
Number of healthy lateral roots	39.53	8.70	7.0	3.14	7.195*

[a], roots >0.5 mm thick.
*, **, significant at $P = 0.05, 0.01$, respectively. n.s., not significant.

tion, but the total number of infected roots was higher in the illuminated or de-fruited plants.

The overall balance of growth in a plant is widely believed to be determined by the existence of a functional equilibrium between the size and activity of the shoot, and that of the root (White, 1937; Brouwer, 1962). The specific activities of both root and shoot are further dependent upon the environment (Reynolds & Thornley, 1982).

Since neither water nor nutrients are limiting in nutrient film culture, and a greater part of the root system may be physiologically more active than in soil, the plant is able to sustain a considerable loss of root volume through infection, before aerial symptoms are expressed. Various authors have assessed the effect of excising portions of the root system on the growth of tomato in hydroponic culture. Went (1943) found that less than 10% of the root system of young plants was responsible for more than 50% of stem growth rate. Similarly, Humphries (1960) showed that up to 40% of the lateral roots could be removed without affecting root growth. Shoot growth, while unaffected by removal of up to 26% of lateral roots, was sharply reduced by further excision. Removal of all laterals reduced shoot growth by 54%, but growth, supported by the tap root, nevertheless still occurred. This is indicative of the capacity of suberised older roots to absorb water and nutrients, albeit generally at a reduced rate relative to that in unsuberised tissue at the root tips (Graham, Clarkson & Sanderson, 1974; Atkinson &

Wilson, 1979). The tomato plant therefore has a capacity for adequate uptake of nutrients, even after substantial root loss.

Root metabolism may also be affected by the availability of assimilate. The uptake of specific ions is energy dependent and, while little energy is required relative to the amounts used in respiration and growth, ion uptake is often sensitive to a reduction in assimilate supply (Ayres, 1984). Crapo & Ketellapper (1981) found a close relationship between the concentration of photosynthate in tomato roots, light intensity and respiration, growth and potassium uptake, suggesting that the supply of assimilate is the intermediary between light intensity and root metabolism. Farrar (1981), however, concluded that daily variations in root respiration of barley are not a simple function of root carbohydrate content and so may not be directly mediated by the rate of assimilate supply from the shoot.

Where assimilate is lacking, the plant gives a higher priority to the maintenance and function of existing root, at the expense of root growth (Crapo & Ketellapper, 1981). As the progress curve for root infection approached the upper asymptotic phase, the influence of assimilate supply on the plant's capacity for new root production was found to determine the expression of aerial symptoms. Root regrowth was primarily determined by the amount of inter-sink competition for assimilate within the plant; thus, whereas fruit removal markedly inhibited root regrowth, defoliation, which reduced the overall amount of assimilate produced, did not significantly inhibit regrowth (Fig. 11.4). The developing tomato fruit was a powerful competitive sink, acquiring assimilate to the detriment of root growth such that by 65 d after inoculation, manipulation of the source–sink relationships resulted in marked differences in symptom expression and survival of infected plants. Plants from which fruit had been removed showed little or no foliar necrosis and no wilt, whereas 17% of plants with fruit, together with 53% of those additionally stressed by leaf removal, were dead.

Murneek (1926) and Ho (1979) also showed that the competitive sink demand of fruit development in tomato was of overriding importance in the regulation of assimilate distribution and movement. Assimilate supply therefore has diametrically opposite effects on infection and disease. The tolerance to disease, shown by tomato plants which were more susceptible to infection, may be compared with the effect of soil nitrogen supply on the 'disease escape mechanism' of cereal plants infected by *Gaeumannomyces graminis* as described by Garrett (1956).

The location of infection was also of importance in NFC. Several

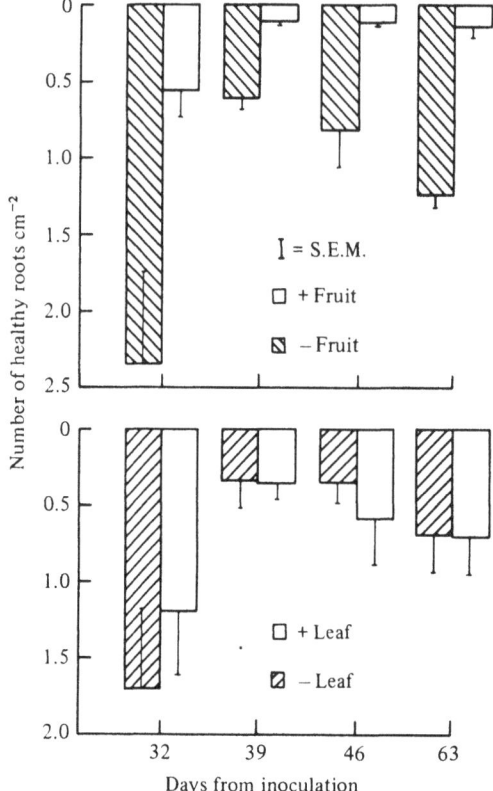

Fig. 11.4. Effects of fruit and leaf removal on root regrowth. Upper: □ with fruit; □ without fruit. Lower: □ with leaf; □ without leaf. (Bars represent s.e.m.)

shaded plants showed a dramatic early collapse, without recovery, where infection occurred at the base of the stem. The etiolated plant could not outgrow such a strategic infection by compensatory root production so that these plants, through being susceptible to infection, were *ipso facto* susceptible to disease.

In both NFC and soil, the development and outcome of disease caused by root pathogens depend on the relative rates of root loss from infection, and root initiation and growth. Such interactions may well be commonplace for a wide range of root pathogens, but are obscured in soilborne diseases by the difficulties of assessing inoculum and measuring the volume of functional root in the soil.

References

Allen, D. J. & Nandra, S. S. (1975). Effects of pH and calcium concentration on the sporulation of *Phytophthora* isolates from *Agave*. *Plant Disease Reporter*, **59**, 555–8.

Atkinson, D. & Wilson, S. A. (1979). The root–soil interface and its significance for fruit tree roots of different ages. In *The Soil–Root Interface*, ed. J. L. Harley & R. Scott Russell, pp. 259–71. New York: Academic Press.

Ayers, W. A. (1971). Induction of sporangia in *Phytophthora cinnamomi* by a substance from bacteria and soil. *Canadian Journal of Microbiology*, **17**, 1517–23.

Ayers, W. A. & Zentmyer, G. A. (1971). Effect of soil solution and two soil Pseudomonads on sporangium production by *Phytophthora cinnamomi*. *Phytopathology*, **61**, 1188–93.

Ayres, P. G. (1984). Effects of infection on root growth and function; consequences for plant nutrient and water relations. In *Plant Diseases: Infection, Damage and Loss*, ed. R. K. S. Wood & G. J. Jellis, pp. 105–17. Oxford: Blackwell Scientific Publications.

Baker, R., Maurer, C. L. & Maurer, R. A. (1967). Ecology of plant pathogens in soil. VII Mathematical models and inoculum density. *Phytopathology*, **57**, 662–6.

Barash, I., Klisiewicz, J. M. & Kosuge, T. (1965). Utilization of carbon compounds by zoospores of *Phytophthora drechsleri*, and their effect on motility and germination. *Phytopathology*, **55**, 1257–61.

Barber, D. A. & Gunn, K. B. (1974). The effect of mechanical forces on the exudation of organic substances by the roots of cereal plants grown under sterile conditions. *New Phytologist*, **73**, 39–45.

Broadbent, P. & Baker, K. F. (1974). Behaviour of *Phytophthora cinnamomi* in soils suppressive and conducive to root rot. *Australian Journal of Agricultural Research*, **25**, 121–37.

Brouwer, R. (1962). Distribution of dry matter in the plant. *Netherlands Journal of Agricultural Science*, **10**, 361–76.

Cameron, J. N. & Carlile, M. J. (1977). Negative geotaxis of zoospores of the fungus *Phytophthora*. *Journal of General Microbiology*, **98**, 599–602.

Carlile, M. J. (1983). Motility, taxis and tropism in *Phytophthora*. In *Phytophthora, its Biology, Taxonomy, Ecology and Pathology*, ed. D. C. Erwin, S. Bartnicki-Garcia & P. H. Tsao. St Paul: American Phytopathological Society.

Crapo, N. L. & Ketellapper, H. J. (1981). Metabolic priorities with respect to growth and mineral uptake in roots of *Hordeum, Triticum* and *Lycopersicon*. *American Journal of Botany*, **68**, 10–16.

Dukes, P. D. & Apple, J. L. (1961). Chemotaxis of zoospores of *Phytophthora parasitica* var. *nicotianae* by plant roots and certain chemical solutions. *Phytopathology*, **51**, 195–7.

Duniway, J. M. (1983). Role of physical factors in the development of *Phytophthora* diseases. In *Phytophthora, its Biology, Taxonomy, Ecology and Pathology*, ed. D. C. Erwin, S. Bartnicki-Garcia & P. H. Tsao, pp. 175–87. St Paul: American Phytopathological Society.

Evans, S. G. (1979). Susceptibility of plants to fungal pathogens when grown by the nutrient film technique (NFT). *Plant Pathology*, **28**, 45–8.

Farrar, J. F. (1981). Respiration rate of barley roots: its relation to growth, substrate supply and the illumination of the shoot. *Annals of Botany*, **48**, 53–63.

Garrett, S. D. (1956). *Biology of Root-infecting Fungi*. Cambridge: Cambridge University Press.

Gäumann, E. A. (1950). *Principles of Plant Infection*. London: Crosby Lockwood.

Gooding, G. V. & Lucas, G. B. (1959). Factors influencing sporangial formation and zoospore activity in *Phytophthora parasitica* var. *nicotianae*. *Phytopathology*, **49**, 277–81.

Graham, J., Clarkson, D. T. & Sanderson, J. (1974). Water uptake by the roots of marrow and barley plants. *ARC Letcombe Laboratory, Annual Report* 1973, pp. 9–12.

Graves, C. J. (1983). The nutrient film technique. *Horticultural Reviews*, (June), pp. 1–44.

Graves, C. J. & Hurd, R. G. (1983). Intermittent solution circulation in the nutrient film technique. *Acta Horticulturae*, **133**, 47–52.

Hale, M. G., Moore, L. D. & Griffin, G. J. (1978). Root exudates and exudation. In *Interactions Between Non-pathogenic Soil Microorganisms and Plants*, ed. Y. R. Dommergues & S. V. Krupa, pp. 163–203. Amsterdam: Elsevier Scientific Publishing Co.

Halsall, D. M. & Forrester, R. I. (1977). Effects of certain cations on the formation and infectivity of *Phytophthora* zoospores. I. Effect of calcium, magnesium, potassium and iron ions. *Canadian Journal of Microbiology*, **23**, 994–1001.

Hickman, C. J. & Ho, H. H. (1966). Behaviour of zoospores in plant-pathogenic Phycomycetes. *Annual Review of Phytopathology*, **4**, 195–220.

Ho, L. C. (1979). Regulation of assimilate translocation between leaves and fruits in the tomato. *Annals of Botany*, **43**, 437–48.

Humphries, E. C. (1960). Effects of mutilation of the root on subsequent growth. *Scientific Horticulture*, **14**, 42–8.

Hurd, R. G., Gay, A. P. & Mountfield, A. C. (1979). The effect of partial flower removal on the relation between root, shoot and fruit growth in the indeterminate tomato. *Annals of Applied Biology*, **93**, 77–89.

Hwang, S. C. & Ko, W. H. (1978). Biology of chlamydospores, sporangia and zoospores of *Phytophthora cinnamomi* in soil. *Phytopathology*, **68**, 726–31.

Jackson, M. B. (1980). Aeration in the nutrient film technique of glasshouse crop production and the importance of oxygen, ethylene and carbon dioxide. *Acta Horticulturae*, **98**, 61–78.

Katsura, K. & Kiyata, Y. (1966). Movement of zoospores of *Phytophthora capsici*. III Rheotaxis. *Scientific Report Kyoto Prefecture University of Agriculture*, **18**, 51–6.

Khan, A. A. & Sagar, G. R. (1966). Distribution of ^{14}C labelled products of photosynthesis during the commercial life of the tomato crop. *Annals of Botany*, **30**, 727–43.

Kliejunas, J. T. & Ko, W. H. (1974). Effect of motility of *Phytophthora palmivora* zoospores on disease severity in Papaya seedlings and substrate colonization in soil. *Phytopathology*, **64**, 426–8.

Leonard, E. R. & Head, G. C. (1958). Technique and preliminary observations on growth of the roots of glasshouse tomatoes in relation to that of the tops. *Journal of Horticultural Science*, **33**, 171–85.

Ludwig, L. J. & Withers, A. C. (1977). Root respiration. In *Annual Report of Glasshouse Crops Research Institute*, 1976, pp. 51–2. Littlehampton, England.

Marx, D. H. & Haasis, F. A. (1965). Induction of aseptic sporangial formation in *Phytophthora cinnamomi* by metabolic diffusates of soil microorganisms. *Nature*, **206**, 673–4.

Masago, H., Yoshikawa, M., Fukada, M. & Nakanishi, N. (1977). Selective inhibition of *Pythium* spp. on a medium for direct isolation of *Phytophthora* spp. from soils and plants. *Phytopathology*, **67**, 425–8.

Mehrlich, F. P. (1935). Nonsterile soil leachate stimulating to zoosporangia production by *Phytophthora* sp. *Phytopathology*, **25**, 432–5.

Murneek, A. E. (1926). Effects of correlation between vegetative and reproductive functions in the tomato (*Lycopersicon esculentum* Mill.). *Plant Physiology*, **1**, 3–56.

Nightingale, G. T., Schemerhorn, L. G. & Robbins, W. R. (1928). The growth of the tomato as correlated with organic nitrogen and carbohydrates in roots, stems and leaves. *Bulletin of the New Jersey Agricultural Experimental Station* No. 461.

Price, D. (1976). Nutrient-film culture. In *Annual Report of the Glasshouse Crops Research Institute* 1975, p. 114. Littlehampton, England.

Price, D. (1979). Effect of etridiazole and *Pythium* on tomatoes grown in nutrient film. *Proceedings 1979 British Crop Protection Conference – Pests and Diseases*, pp. 421–4.

Reynolds, J. F. & Thornley, J. H. M. (1982). A shoot:root partitioning model. *Annals of Botany*, **49**, 585–97.

Ribeiro, O. K. (1983). Physiology of asexual sporulation and spore germination in *Phytophthora*. In *Phytophthora – Its Biology, Taxonomy, Ecology and Pathology*, ed. D. C. Erwin, S. Bartnicki-Gracia & P. H. Tsao. St Paul: American Phytopathological Society.

Richardson, S. J. (1982). Crop nutrition in nutrient film culture, 46 pp. *Fertiliser Society*, London, Dec. 1981.

Thomson, S. V. & Allen, R. M. (1976). Mechanisms of survival of zoospores of *Phytophthora parasitica* in irrigation water. *Phytopathology*, **66**, 1198–202.

Tomlinson, J. A. (1958). Crook root of watercress. 1. Field assessment of the disease and the role of calcium carbonate. *Annals of Applied Biology*, **46**, 593–607.

Troutman, J. L. & Wills, W. H. (1964). Electrotaxis of *Phytophthora parasitica* zoospores and its possible role in infection of tobacco by the fungus. *Phytopathology*, **54**, 225–8.

Wallace, H. R. (1978). Dispersal in time and space:soil pathogens. In *Plant Disease; an Advanced Treatise*, vol. 2, *How Disease Develops in Populations*, ed. J. G. Horsfall & E. B. Cowling. New York: Academic Press.

Went, F. W. (1943). Effect of the root system on tomato stem growth. *Plant Physiology*, **18**, 51–65.

White, H. L. (1937). The interaction of factors in the growth of *Lemna*. XII The interaction of nitrogen and light intensity in relation to root length. *Annals of Botany*, **1**, 649–54.

Yarwood, C. E. (1959). Predisposition. In *Plant Pathology*, vol. 1, ed. J G. Horsfall & A. E. Dimond, pp. 521–62.

Young, B, R., Newhook, F. J. & Allen, R. N. (1979). Motility and chemotactic response of *Phytophthora cinnamomi* zoospores in 'ideal soils'. *Transactions of the British Mycological Society*, **72**, 395–401.

12
Phytoalexins, water stress and stomata

C. M. WILLMER and A. M. PLUMBE
*Department of Biological Science, University of Stirling, Stirling FK9 4LA,
Scotland, UK*

Much research has been directed to finding methods of artificially controlling stomatal movements because of the many potential applications
of considerable agronomic importance. For example, in areas of the
world where it is possible to forecast the development of a drought
period fairly reliably, application of compounds capable of closing stomata ('antitranspirants') prior to the onset of the drought may slow
the development of plant water stress and therefore give crops a better
chance of survival. A practice already used is to apply antitranspirants
to transplanted trees and other plants to improve their chances of survival
by protecting them from excessive water loss until the root systems are
established. The entry of potentially damaging aerial pollutants through
stomata could also be restricted by antitranspirant compounds. 'Pro-
transpirants', compounds capable of enhancing stomatal apertures, could
be used to extend the period of stomatal opening (provided light and
water were not limited), which might result in increased crop yields.
Most success has been experienced with antitranspirants which are sprays
of either naturally occurring or synthetic compounds that directly affect
stomata. Since the discovery of the role of abscisic acid (ABA) in protecting water-stressed plants from desiccation by closing their stomata, many
other naturally occurring compounds have been found to accumulate
to relatively low concentrations in water-stressed leaf tissues and protect
the plant by directly or indirectly bringing about stomatal closure. They
include all *trans* farnesol (Ogunkanmi, Wellburn & Mansfield, 1974;
Fenton, Mansfield & Wellburn, 1976; Fenton, Davies & Mansfield,
1977), certain short-chain fatty acids of carbon chain length between
6 and 11 (Willmer, Don & Parker, 1978) and the unsaturated longer-
chain fatty acids, linolenic and linoleic acids (Fenton *et al.*, 1976; Wilson
& Davies, 1979).

Other substances like proline (Kemble & MacPherson, 1954), gluta-
mine, asparagine (Chen, Kessler & Monselisi, 1965; Barrett & Naylor,
1966) and betaines accumulate to relatively high concentrations in leaf
tissue under water stress and are considered to be primarily involved
in cellular osmotic adjustment as cell water potential decreases.

Although there are no reports that naturally occurring phenolic compounds accumulate in leaf tissue under water stress, there are a number
of reports that some of them affect stomata. Certain phenolic compounds, such as chlorogenic acid and scopoletin, also accumulate in
some plants in response to pathogen attack.

Phytoalexins are compounds produced by plants in response to pathogen attack. The term is generally restricted to antimicrobial compounds
synthesised from remote precursors. Antimicrobial compounds which
already exist in tissue before infection (e.g. catechol), or antimicrobial
compounds which are released from immediate inactive precursors upon
infection of tissue (e.g. tuliposides), are not generally considered to
be phytoalexins.

Phytoalexins and some of the phenolic compounds may be loosely
defined as microbial stress compounds since they accumulate in response
to the stress of microbial attack. Often plants under attack lose control
of gas exchange between the leaf and its environment, mainly because
stomata are directly affected. In some cases stomata remain wide open
with uncontrollable loss of water followed by severe wilting and possibly
death of the plant (see Ayres & Paul, Ch. 15). There is also a report
(Ayres, 1980) that a phytoalexin, pisatin, affects stomatal responses.

These observations and the fact that some phytoalexins (e.g. rishitin)
and ABA have common precursors to their biosynthesis prompted investigations to determine whether certain phytoalexins (microbial stress
compounds) and certain phenolics could mimic water-stress compounds
(compounds which accumulate in leaves under water stress and may
close stomata) such as ABA. Investigations were also made to test
whether water-stress compounds possessed antifungal properties.

Occurrence of phytoalexins in water-stressed leaf tissue

In order to test whether phytoalexins accumulate in water-
stressed leaf tissue, pea and broad (faba) bean plants were stressed in
one of two ways. In one set of experiments ('stress only' treatment)
plants were stressed by withholding water from a sub-sample of plants
for 7 d and then leaf samples were taken. This mild water-stress treatment
caused the plants to wilt, but there was no other visible damage to the

leaves. A second set of experiments ('stress–recovery' treatment) involved withholding water for a much longer period until the plants were severely wilted and the leaves showed signs of damage along their margins, after which time the plants were rewatered and allowed to recover to full turgor. A 10-d drought followed by a 5-d or 2-d recovery period was used for pea and broad bean, respectively. The 'stress–recovery' treatment resulted in leaves of broad bean developing necrotic areas round the leaf margin as turgor was regained, while the central portion of the leaf recovered fully from the stress. With pea leaves, no browning of damaged areas of the leaf margins occurred on recovery of turgor, but the delineation of the recovered and damaged areas was quite distinct. Control plants were watered twice daily.

After the appropriate stress treatment, leaf samples were taken and assayed for antifungal activity using extraction methods similar to those of Hargreaves, Mansfield & Rossall (1977) and detection procedures (thin layer chromatography (TLC) plate bioassay) originally devised by Klarman & Sanford (1968). The antifungal compounds were detected by spraying a suspension of *Cladosporium herbarum* spores on to the TLC plates. White regions of the silica gel persisted where the fungal growth was inhibited. The procedures are fully described in Plumbe & Willmer (1985).

Antifungal substances were not detected in pea or broad bean leaves subjected to mild or moderate water stress ('stress' only treatment), but severe water stress ('stress–recovery' treatment) resulted in detection of a number of new substances in the leaves of both species (Fig. 12.1a,b). Most of the antifungal activity occurred in extracts from damaged and necrotic patches of leaf tissue, but some antifungal activity was also detected in undamaged areas. No antifungal activity was detected in unstressed leaves.

The area of inhibition of fungal growth observed on the TLC plates of pea leaf extracts corresponded with that produced by a sample of pure pisatin. Numerous inhibitory areas were observed on the TLC plates of broad bean leaf extracts, two of which corresponded with those produced by samples of wyerone and wyerone acid. However, further investigations are needed to confirm the identity of pisatin, wyerone and wyerone acid. Under less severe water stress, antifungal compounds were not detected by the TLC plate bioassay; these are conditions under which conventional water-stress compounds such as ABA would be evident in the tissue at physiologically high levels.

These results support the widely held view that cell damage or death is necessary for induction of phytoalexins in plant tissues (Bailey, 1981). Although the greatest amounts of the phytoalexins were found in the damaged or dead tissue, considerable amounts of the compounds were also detected in the central, recovered areas of the leaves, supporting

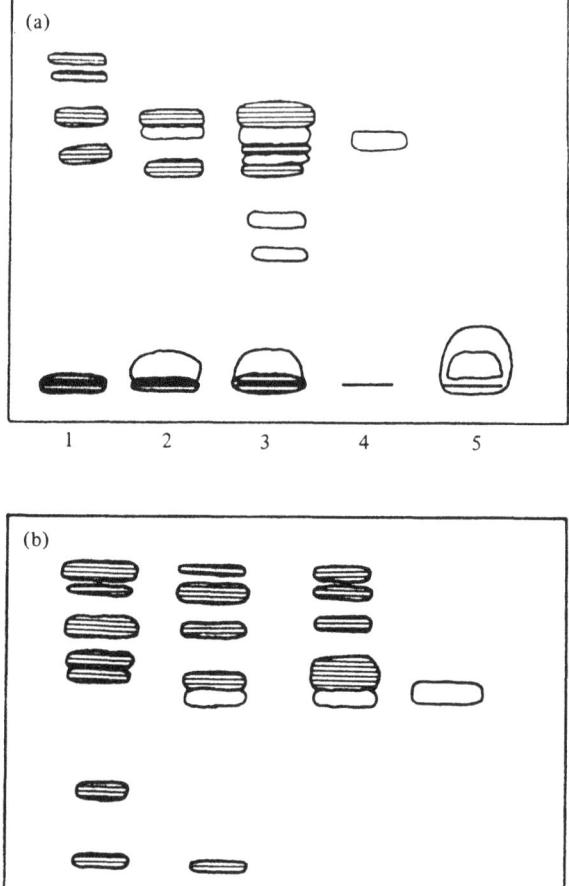

Fig. 12.1a, b. TLC plate bioassay of 80% methanol extracts from leaves of (a) broad bean and (b) pea, subjected to severe water stress. The horizontal line at the base of the plate represents the origin. The hatched areas represent leaf pigments and the encircled white areas represent regions of fungal inhibition. Key: 1, unstressed leaves; 2, central recovered areas of water stressed leaves; 3, damaged margins of water-stressed leaves; 4, wyerone (50 μg); 5, wyerone acid (50 μg); 6, pisatin (50 μg).

the current view that phytoalexins accumulate in dead tissues but that their site of synthesis is in the adjacent living tissue (Bailey, 1981).

Effects of phytoalexins and some phenolics on stomatal behaviour

A standard epidermal strip technique was used to observe the effect of the compounds (all at $0.1 \, mol \, m^{-3}$) on stomatal behaviour. Essentially, epidermal strips from the lower leaf surface of *Commelina communis* were incubated under constant conditions of temperature, photon flux density and CO_2 concentration, in a buffered medium containing the test compounds, and after a fixed time interval of usually 1.5 h stomatal apertures were measured. Stomata in the epidermal strips were initially either wide open (Fig. 12.2b) or completely closed (Fig. 12.2a). The method is described in detail in Plumbe (1983). Phaseollin and wyerone acid were about as potent as ABA (also at $0.1 \, mol \, m^{-3}$) at preventing opening. Both substances caused visible damage to the guard cells. Pisatin had an intermediate inhibitory effect on opening and rishitin had no effect.

The effects of the phytoalexins and of ABA were not so great on open stomata (Fig. 12.2b) as they were on closed stomata. This phenomenon has been observed before and is believed to be related to the relatively high concentration of KCl ($100 \, mol \, m^{-3}$) which is needed to maintain opening in the control medium. ABA was most effective at closing open stomata, although phaseollin also, ultimately, decreased apertures to the level of the ABA treatment. Effects of rishitin were not greatly different from the controls and, although apertures increased initially in the presence of wyerone acid, after 5 h incubation they were similar to control values. Pisatin tended to enhance stomatal opening although this effect was not statistically significant. No damage to the guard cells was observed with any of the treatments when stomata were initially open.

Figure 12.3 shows the effects of a selection of naturally occurring phenolic compounds (all at $1 \, mol \, m^{-3}$) and of ABA (at $0.1 \, mol \, m^{-3}$) on stomatal opening and on wide-open stomata of *C. communis*. Sinapic acid was as effective as ABA at preventing opening; caffeic, salicylic and *p*-coumaric acids were also fairly effective at preventing opening, scopoletin, protocatechuic acid and ferulic acid were less so.

The general pattern of the effectiveness of phenolic compounds at closing open stomata was similar to that observed for inhibition of stomatal opening, but the effects were less marked. Ferulic acid, however,

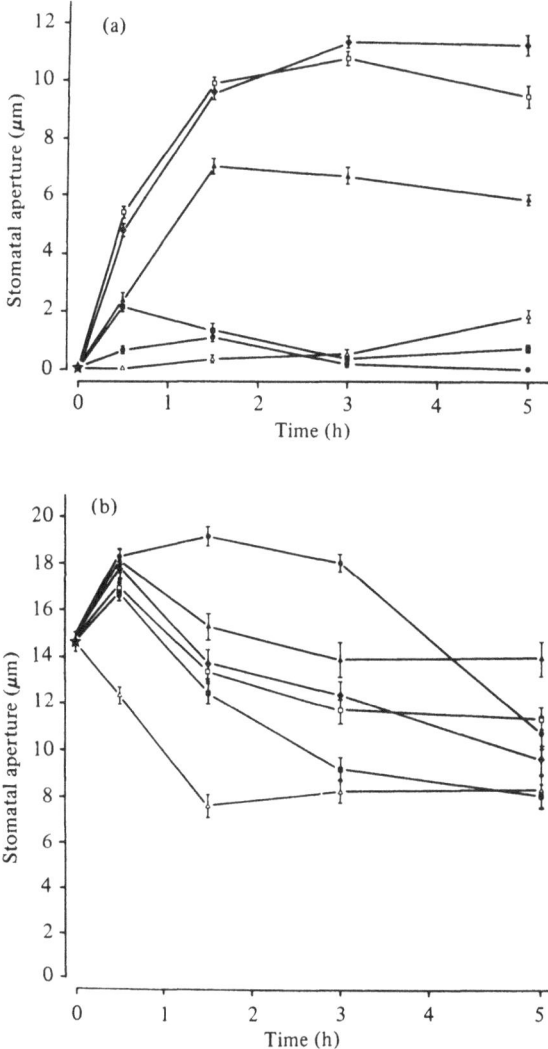

Fig. 12.2a, b. Stomatal apertures in epidermal strips of *Commelina communis* incubated for up to 5 h in various phytoalexin solutions or ABA (all at 0.1 mol m^{-3}). Stomata were either closed (a) or open (b) at the start of the experiment. Control, □; ABA, △; phaseollin, ■; pisatin, ▲; rishitin, ◆; wyerone acid, ●. All compounds were in 20 mol m^{-3} MES buffer, pH 5.5 containing 100 mol m^{-3} KCl. CO$_2$-free air was bubbled through the medium and the photon flux density (400–700 nm) in the region of the tissue was 195 μmol m^{-2} s^{-1}. Temperature was maintained at 27 °C. Each point is associated with its standard error of the mean.

caused significant further opening of already open stomata. Unlike phaseollin and wyerone acid, none of the phenolics caused any obvious cell damage.

No previous studies have been made of the effects of phytoalexins on stomatal responses in epidermal strips, but one study has been made of the effects of a phytoalexin on stomata in intact leaves. In that study Ayres (1980) observed that a $0.1 \, mol \, m^{-3}$ application of pisatin to the surface of pea leaves reduced transpiration rates. A number of studies have been made of the effects of phenolic compounds on stomata and results have been conflicting. The findings of this study that chlorogenic

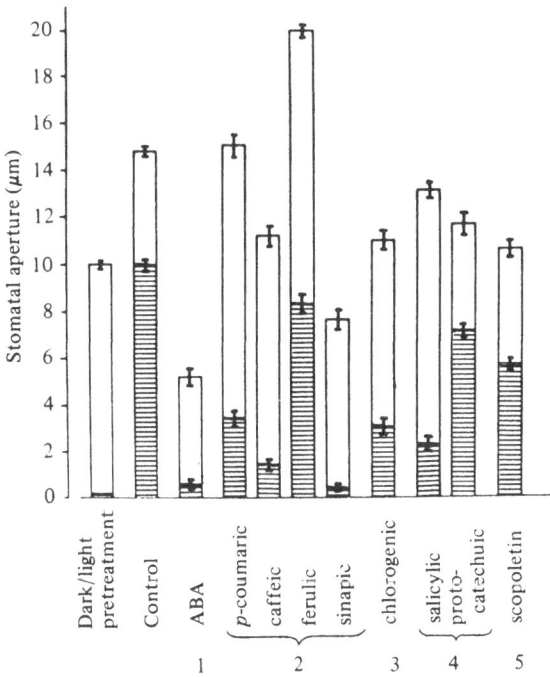

Fig. 12.3. Stomatal apertures in epidermal strips of *Commelina communis* incubated for 2 h in various simple phenolic solutions (all at $1 \, mol \, m^{-3}$) or ABA (at $0.1 \, mol \, m^{-3}$). Stomata were either closed (hatched columns) or open (open columns) at the start of the experiment. The compounds have been grouped in the following manner: 1, sesquiterpenoid compound; 2, cinnamic acids; 3, benzoic acid; 4, coumarins. All compounds were in $20 \, mol \, m^{-3}$ MES buffer, pH 6.0 containing $75 \, mol \, m^{-3}$ KCl. Air was bubbled through the medium and the photon flux density (400–700 nm) in the region of the tissue was $195 \, \mu mol \, m^{-2} s^{-1}$. Temperature was maintained at $27 \, °C$. Associated with each column is the standard error of the mean.

acid inhibits stomatal opening agree with the findings of Ogunkanmi, Tucker & Manfield (1973), and Stevens (1977). However, the slight closure of open stomata in the presence of chlorogenic acid contrasts with the observation of Stevens (1977) who found chlorogenic acid had no effect on open stomata.

The slight inhibition of stomatal opening and closure of open stomata observed with 1.0 mol m^{-3} scopoletin contrasts with the findings of Ogunkanmi *et al.* (1973) who observed no inhibition of stomatal opening (effects on open stomata were not studied). The effects of 1.0 mol m^{-3} ferulic acid on stomatal responses observed in this study were considerably different from those obtained by Stevens (1977). In the present study, ferulic acid did not inhibit stomatal opening and enhanced opening of already open stomata. In contrast, Stevens (1977) found ferulic acid inhibited stomatal opening and had no effect on open stomata. However, a report that ferulic acid increases the transpiration rate from leaves of *Phaseolus vulgaris* (Michniewicz & Rozej, 1974) tends to support the findings of the present study rather than those of Stevens (1977).

Other simple phenolic compounds used here which were found to be effective at inhibiting stomatal opening (with most also causing closure of open stomata) were *p*-coumaric, salicylic, caffeic and sinapic acids. These compounds have not previously been tested on stomata but acetyl-salicylic acid (aspirin), a compound related to salicylic acid, was found to be almost as effective as ABA at closing open stomata in epidermal strips of *C. communis* (Larqué-Saavedra, 1978, 1979).

Thus, the phytoalexins, wyerone acid and phaseollin (and to some extent pisatin) behave like the water-stress compound ABA in that they inhibit stomatal opening and may also cause stomatal closure at relatively low concentrations. However, the mode of action of the phytoalexins on stomata is very different from ABA. Unlike ABA, phaseollin and wyerone acid at a concentration of 0.1 mol m^{-3} are phytotoxic and their effects on stomatal behaviour are a result of cell damage. It is possible, however, that lower concentrations of the phytoalexins may still be effective without producing phytotoxic responses and future work should establish whether this is the case.

The phenolic compounds were not apparently phytotoxic at the higher concentration of 1 mol m^{-3} and a few of them such as salicylic acid, *p*-coumaric acid, caffeic acid and chlorogenic acid, and particularly sinapic acid, were effective at inhibiting stomatal opening and causing stomatal closure. Thus, these compounds deserve further attention to ascertain whether they will be useful as antitranspirants.

The fungitoxicity of water-stress compounds

Two methods were used to ascertain the antifungal properties of water-stress compounds. They were, (A) a TLC plate spot bioassay, and (B) a spore-germination and germ-tube growth bioassay in liquid culture. In (A) a spore suspension of *Cladosporium herbarum* was sprayed onto a plate on which the compounds to be tested had been spotted. White areas of silica gel remained where the spores did not germinate, indicating the presence of antifungal substances. In (B) a Czapek Dox medium containing the compound under test and the fungal spores was inspected at regular intervals and germination and growth were compared with suitable controls. Both methods are fully described in Plumbe (1983).

Compounds which inhibited growth of the fungus *C. herbarum* in (A) tended to do so in one of two ways (see Fig. 12.4). The compounds either caused a 'discrete' type of inhibition, characterised by a central area totally devoid of fungal growth (corresponding almost exactly to where the compound had been applied) bounded by a dark band where fungal growth appeared to be enhanced, or caused a 'diffuse' type of inhibition where the area in which fungal growth was inhibited was much greater (often extending 10 to 15 mm beyond the area on which the compound was applied) but there was no dark band bounding the inhibitory zone.

Of the water-stress compounds tested, the saturated short-chain fatty acids, nonanoic acid (C_9), decanoic acid (C_{10}) and undecanoic acid (C_{11}), were increasingly effective at preventing fungal growth and showed a 'discrete' type of inhibition. The unsaturated longer-chain fatty acids, oleic acid, linoleic acid and α-linolenic acid were also increasingly effective at preventing fungal growth but caused a 'diffuse' type of growth inhibition. ABA had no effect on fungal growth at any of the levels tested, while farnesol caused a 'diffuse' type of inhibition, but only at the highest amount tested (1.0 μmol). Proline stimulated fungal growth at 1.0 μmol but smaller amounts had no effect on growth.

Of the other compounds tested, the C_3 saturated short-chain fatty acid, propionic acid, stimulated fungal growth at 1.0 μmol while pisatin and wyerone acid showed 'diffuse' growth inhibition at all levels tested. The ethanol and distilled water control treatments had no effect on fungal growth.

In the liquid culture bioassay all the water-stress compounds tested, except linoleic acid, initially delayed germination with wyerone acid being particularly effective. However, after 17 h, 100% spore germina-

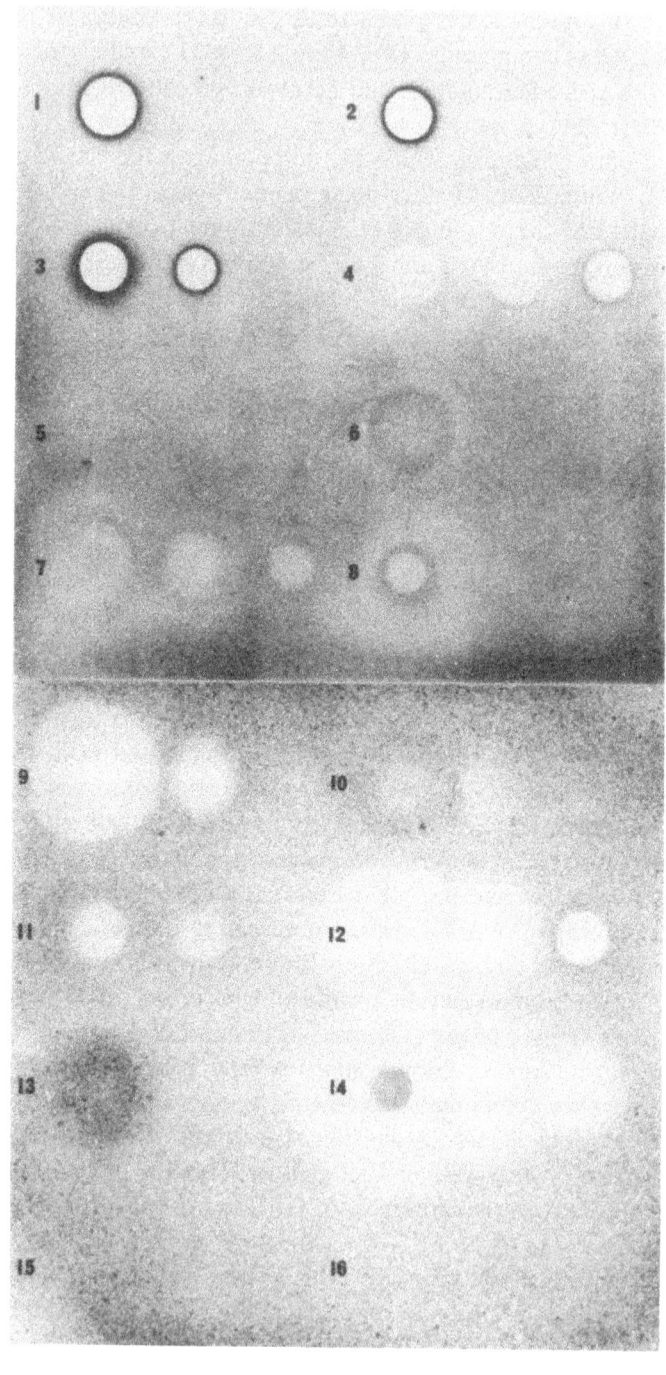

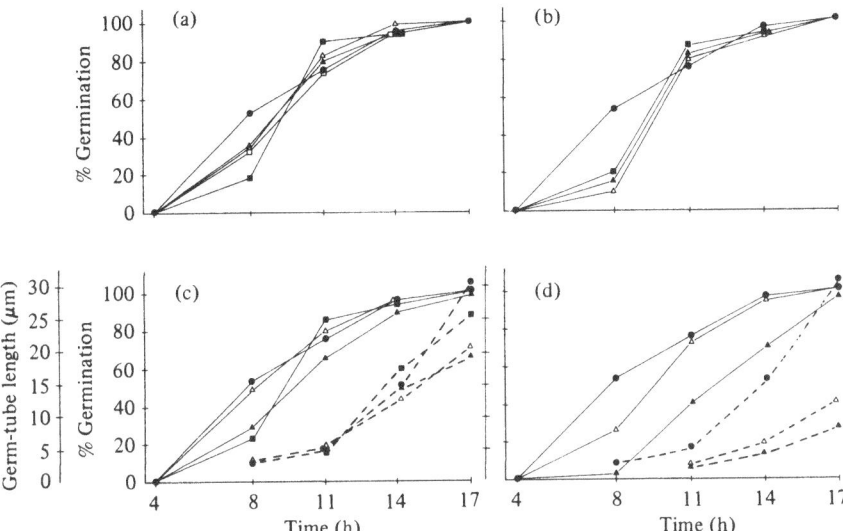

Fig. 12.5a–d. Effect of various water-stress compounds and phytoalexins (all at 0.1 mol m^{-3}) on germination of spores (solid lines) and growth of germ-tubes (dashed lines) of *Cladosporium herbarum* at $21.5 \pm 1.5\,°C$. (a) Short-chain saturated fatty acids: ●, control; ▲, nonanoic acid; △, decanoic acid; ■, undecanoic acid; □, propionic acid. (b) Sesquiterpenoids and proline: ●, control; ▲, ABA; △, farnesol; ■, proline. (c) Unsaturated long-chain fatty acids. ●, control; ▲, α-linolenic acid; △, linoleic acid; ■, oleic acid. (d) Phytoalexins: ●, control; ▲, wyerone acid; △, pisatin. In the presence of all short-chain fatty acids (a), and the sesquiterpenoids and proline (b), germ-tube growth was not significantly different from controls and therefore is not presented.

tion occurred in the presence of all compounds, including the phytoalexins (Fig. 12.5). Once the spores had germinated, the rate of germ-tube growth was not affected by the short-chain fatty acids, ABA, farnesol or proline, but was severely inhibited by the phytoalexins, wyerone acid and pisatin, and was also retarded by α-linolenic and linoleic acids (Fig. 12.5).

Thus, it is apparent that most of the water-stress compounds affect spore germination either by completely inhibiting it as in (A) (with the

Fig. 12.4. TLC plate spot bioassay of some water-stress compounds and phytoalexins. Each compound was tested at three levels (1.0, 0.1 and 0.01 μmol) and applied to the plate in 30 μl of solvent. 1, nonanoic acid; 2, decanoic acid; 3, undecanoic acid; 4 and 12, wyerone acid; 5, ABA; 6, proline; 7 and 14, pisatin; 8, farnesol; 9, α-linolenic acid; 10, oleic acid; 11, linoleic acid; 13, propionic acid; 15, ethanol control; 16, distilled-water control.

notable exception of ABA and proline), or by retarding rates of germination as in (B). In these respects the water-stress compounds acted in the same fashion as the phytoalexins, pisatin and wyerone acid. However, unlike the phytoalexins, which also inhibited germ-tube growth, the water-stress compounds had no effect on the growth of the fungus once the spores had germinated (except for α-linolenic and linoleic acids).

The reasons for the different responses to the water-stress compounds in the two bioassays are probably due to the different ways in which the compounds were exposed to the fungi. In the TLC spot bioassay, which is frequently used for preliminary determination of antifungal properties of compounds, the fungal spores are applied in a nutrient solution to the test compound which is deposited as a solid within a small area of silica gel coating a TLC plate. Thus concentrations of the compounds can be very high in a localised area. In contrast, liquid culture bioassays involve dissolving test compounds in a nutrient solution which is then inoculated with spores. Consequently, comparison of concentration effects on fungal spore germination and growth cannot readily be made between the two types of bioassay. Nevertheless investigators have used both methods to assay for fungal toxicity.

As shown in Fig. 12.4, the C_9, C_{10} and C_{11} saturated short chain fatty acids gave a 'discrete' type of fungal inhibition, whereas other compounds observed to be fungitoxic (e.g. the unsaturated longer-chain fatty acids and phytoalexins) gave a 'diffuse' type of inhibition. The reasons for this difference are not clear, but may be due to the saturated short-chain fatty acids being more strongly adsorbed onto the silica gel of the TLC plate than the other compounds such that their movement away from the site of application is restricted. In contrast, the other compounds might diffuse some distance away from the site of application because they are not so strongly adsorbed onto the silica gel. Consequently, when the TLC plate is sprayed with spores of *C. herbarum*, there is a clear distinction between the area of the plate containing the saturated short-chain fatty acids and the surrounding area. Other fungitoxic compounds showed a gradual decrease in the inhibition of fungal growth (presumably due to the decreasing 'concentrations' of the substances present in the silica gel) towards the edges of the 'diffuse' spots.

The TLC spot bioassay indicated that several stress compounds were toxic to the fungus *C. herbarum*, namely the C_9, C_{10} and C_{11} saturated short-chain fatty acids, the unsaturated longer-chain fatty acids, and farnesol. There are no previous studies regarding the toxicity of farnesol to fungi, but the toxicity of the unsaturated longer-chain fatty acids (oleic, linoleic and especially α-linolenic acids) to *C. herbarum* does not support reports that oleic acid (Rigler & Greathouse, 1940), the potassium salt of oleic acid (Puritch, Tan & Hopkins, 1981) and linoleic acid (Rigler & Greathouse, 1940) show little or no fungitoxic properties and can be metabolised by fungi and used as a nutritional source (Johnson, 1957; Lewis & Johnson, 1967). In agreement with the findings of this present study, however, are reports (using liquid culture and/or agar plate bioassays) that saturated short-chain fatty acids or their salts are toxic to many fungi, including *Cladosporium resinae* (Teh, 1974), *Botrytis cinerea* (Puritch *et al.*, 1981), *Paecilomyces variotii*, and members of the *Aspergillus glaucus* group (Lord, Lacey, Cayley & Manlove, 1981). Puritch *et al.* (1981) and Lord *et al.* (1981) observed that the fungitoxicity of the saturated short-chain fatty acids increased with increasing carbon chain length to a maximum effectiveness around C_9 or C_{10}, while the C_{11} and C_{12} saturated short-chain fatty acids were markedly less fungitoxic. A similar increase in fungitoxicity with increasing chain length was observed in the present study with the TLC plate spot bioassay, although in this case the C_{11} fatty acid, undecanoic acid, was the most fungitoxic. The toxicity of the C_9, C_{10} and C_{11} saturated short-chain fatty acids, the unsaturated longer-chain fatty acids and farnesol to *C. herbarum* may be due to a direct action of these compounds on the membranes of the fungal cells (these compounds also damage membranes of plant cells including guard cells – Plumbe, 1983). This view is supported by the observations of Teh (1974) that toxic saturated short-chain fatty acids inhibit respiration and cause rapid potassium efflux and protein leakage from cells of the fungus *C. resinae*.

The stimulation of fungal growth by the C_3 saturated fatty acid, propionic acid, in the TLC plate spot bioassay, suggests *C. herbarum* can metabolise the compound and use it as a nutritional source. This view is supported in part by Lord *et al.* (1981) who reported that although growth of *P. variotti* and *A. glaucus* group fungi in liquid culture was inhibited by concentrations of propionic acid around $100 \, mol \, m^{-3}$, the fungi could metabolise lower concentrations of the acid (25 to $50 \, mol \, m^{-3}$).

No previous studies have investigated the effect of proline on fungal growth, but the stimulation of growth observed in (A) suggests that, as with propionic acid, *C. herbarum* can use the amino acid as a nutritional source.

The observation in this study that ABA did not affect growth of *C. herbarum* contrasts with the report of Borecka & Pieniazek (1968) that

0.1 and 0.05 mol m^{-3} ABA stimulated spore germination of *Gloeosporium album* and *Botrytis cinerea* in liquid culture. The reason for this difference is unclear.

General discussion

The effects of ABA on stomata occurred rapidly with no damage to the guard cells while phaseollin and wyerone acid affected stomatal responses relatively slowly and usually caused severe damage to guard cells. Phaseollin and wyerone acid also caused damage to guard cell protoplasts by disrupting membranes while ABA had no obvious effect other than to cause contraction of the protoplasts (Plumbe, 1983). Other water-stress compounds, such as the short-chain fatty acids and farnesol which accumulate in water-stressed leaf tissue of some species and close stomata, do, however, act in a similar manner to phaseollin and wyerone acid by damaging guard cell membranes (Plumbe, 1983). It must be appreciated, however, that farnesol and certain of the fatty acids naturally accumulate in leaves under water stress and can modify stomatal movements without damaging guard cells (Fenton *et al.*, 1977; Willmer *et al.*, 1978). Presumably, therefore, under natural conditions lower concentrations than those used in this study (0.1 mol m^{-3}) accumulate in leaves to exert a non-damaging effect on guard cells enabling closure of stomata to occur. It is possible also that levels of phytoalexins can accumulate in leaf tissues which are not phytotoxic but are high enough to close stomata.

Compared with phaseollin and wyerone acid, pisatin and rishitin generally had negligible effects on stomatal responses although pisatin did show some inhibitory effect on the opening of closed stomata. Also, neither pisatin nor rishitin (at 0.1 mol m^{-3}) affected the viability of isolated guard cell protoplasts over a 1-h period (Plumbe, 1983). Thus, the relative effectiveness of the four phytoalexins at modifying stomatal responses and affecting guard cell viability was, wyerone acid > phaseollin > pisatin > rishitin. This order of effectiveness is very similar to the relative phytotoxicity of these phytoalexins towards other cell types as observed by Hargreaves (1980) and Lyon (1980).

Unlike the phytoalexins, the phenolics had no obvious damaging effect on guard cells. Thus, some of the phenolics, particularly caffeic and sinapic acid and their derivatives, warrant further attention and may have potential use as antitranspirants. Stomatal sensitivity to the simple phenolic compounds such as caffeic and sinapic acids, however, was not as great as that shown towards ABA, since all the simple phenolic compounds were tested at a concentration of $1.0 \, mol \, m^{-3}$ (preliminary studies had shown that the compounds had no large effect on stomatal responses at lower concentrations) whereas ABA was highly effective at $0.1 \, mol \, m^{-3}$. Indeed, ABA is known to affect stomatal responses in epidermal strips at concentrations as low as 10^{-5} to $10^{-7} \, mol \, m^{-3}$ (Ogunkanmi *et al.*, 1973). This difference in sensitivity shown by stomata to the simple phenolic compounds and ABA suggests, therefore, that the mode of action of these two types of compounds may be very different.

At the concentrations tested $(0.1 \, mol \, m^{-3})$ only the phytoalexins, wyerone acid and phaseollin, significantly inhibited germ-tube growth. Nevertheless, on the TLC spot bioassay the inhibiting effects of the fatty acids and farnesol were striking (as opposed to another classical water-stress compound ABA, which did not have antifungal properties) and it is possible that they serve a secondary purpose in water-stressed plants by providing protection from invading fungi at a time when the plant is likely to be more susceptible to attack.

References

Ayres, P. G. (1980). Stomatal behaviour in mildewed pea leaves: solute potentials of the epidermis and effects of pisatin. *Physiological Plant Pathology*, **17**, 157–65.

Bailey, J. A. (1961). Physiological and biochemical events associated with the expression of resistance to disease. In *Active Defence Mechanisms in Plants*, ed. R. K. S. Wood, pp. 39–65. London: Plenum Press.

Barrett, N. M. & Naylor, A. W. (1966). Amino acid and protein metabolism in Bermuda grass during water stress. *Plant Physiology*, **41**, 1222–30.

Borecka, H. & Pieniazek, J. (1968). Stimulatory effect of abscisic acid on spore germination of *Gloeosporium album* O Sterw. and *Botrytis cinerea* Pers. *Bulletin of the Academy of Polish Science*, **16**, 657–61.

Chen, D., Kessler, B. & Monselise, S. P. (1965). Studies on water regime and nitrogen metabolism of citrus seedlings grown under water stress. *Plant Physiology*, **39**, 379–86.

Fenton, R., Mansfield, T. A. & Wellburn, A. R. (1976). Effects of isoprenoid alcohols on oxygen exchange of isolated chloroplasts in relation to their possible physiological effects on stomata. *Journal of Experimental Botany*, **27**, 1206–14.

Fenton, R., Davies, W. J. & Mansfield, T. A. (1977). The role of farnesol as a regulator of stomatal opening in *Sorghum. Journal of Experimental Botany*, **28**, 1043–53.

Hargreaves, J. A. (1980). A possible mechanism for the phytotoxicity of the phytoalexin phaseollin. *Physiological Plant Pathology*, **16**, 351–7.

Hargreaves, J. A., Mansfield, J. W. & Rossall, S. (1977). Changes in phytoalexin concentrations in tissues of the broad bean plant (*Vicia faba* L.) following inoculation with species of *Botrytis. Physiological Plant Pathology*, **11**, 227–42

Johnson, G. T. (1957). Fatty acids as carbon sources for the growth of *Spicaria violacea*. *Mycologia*, **49**, 172–7.

Kemble, A. R. & MacPherson, H. T. (1954). Liberation of amino acids in perennial rye grass during wilting. *Biochemistry Journal*, **58**, 46–9.

Klarman, W. L. & Sanford, J. B. (1968). Isolation and purification of an antifungal principle from infected soybeans. *Life Science*, **7**, 1095–103.

Larqué-Saavedra, A. (1978). The antitranspirant effect of acetylsalicylic acid on *Phaseolus vulgaris* L. *Physiologia Plantarum*, **43**, 126–8.

Larqué-Saavedra, A. (1979). Stomatal closure in response to acetylsalicylic acid treatment. *Zeitschrift für Pflanzenphyiologie*, **93**, 371–5.

Lewis, H. L. & Johnson, G. T. (1967). Growth and oxygen-uptake responses of *Cunninghamella echinulata* on even-chain fatty acids. *Mycologia*, **59**, 878–87.

Lord, K. A., Lacey, J., Cayley, G. R. & Manlove, R. (1981). Fatty acids as substrates and inhibitors of fungi from propionic acid treated hay. *Transactions of the British Mycological Society*, **77**, 41–5.

Lyon, G. D. (1980). Evidence that the toxic effect of rishitin may be due to membrane damage. *Journal of Experimental Botany*, **31**, 957–66.

Michniewicz, M. & Rozej, B. (1974). Stimulation of transpiration rate in bean plants by ferulic acid. *Die Naturwissenschaften*, **61**, 33–4.

Ogunkanmi, A. G., Tucker, D. J. & Mansfield, T. A. (1973). An improved bio-assay for abscisic acid and other antitranspirants. *New Phytologist*, **72**, 277–82.

Ogunkanmi, A. B., Wellburn, A. R. & Mansfield, T. A. (1974). Detection and preliminary identification of endogenous antitranspirants in water-stressed *Sorghum* plants. *Planta*, **117**, 293–302.

Plumbe, A. M. (1983). Phytoalexins, water stress compounds and stomata. Ph.D. Thesis, University of Stirling.

Plumbe, A. M. & Willmer, C. M. (1985). Phytoalexins, water-stress and stomata. I. Do phytoalexins accumulate in leaves under water stress? *New Phytologist*, **101**, 269–74.

Puritch, G. S., Tan, W. C. & Hopkins, J. C. (1981). Effect of fatty acid salts on the growth of *Botrytis cinerea*. *Journal of Botany*, **59**, 491–4.

Rigler, N. S. & Greathouse, G. A. (1940). The chemistry of resistance of plants to *Phymatotrichum* foot rot. VI. Fungicidal properties of fatty acids. *American Journal of Botany*, **27**, 701–4.

Stevens, R. A. (1977). A structural and functional study of stomata. Ph.D. Thesis, Plymouth Polytechnical College.

Teh, J. S. (1974). Toxicity of short-chain fatty acids and alcohols towards *Cladosporium resinae*. *Applied Microbiology*, **28**, 840–4.

Willmer, C. M., Don, R. & Parker, W. (1978). Levels of short-chain fatty acids and of abscisic acid in water-stressed and non-stressed leaves and their effects on stomata in epidermal strips and excised leaves. *Planta*, **139**, 281–7.

Wilson, J. A. & Davies, W. J. (1979). Farnesol-like antitranspirant activity and stomatal behaviour in maize and *Sorghum* lines of differing drought tolerance. *Plant, Cell and Environment*, **2**, 49–57.

13
Movement of liquid water and the effects of fungal infection

HAMLYN G. JONES

East Malling Research Station, Maidstone, Kent, ME19 6BJ, UK

The fundamental principles of water flow in plants are outlined so as to provide a basis for understanding how fungal infection affects plant water relations and water movement within the plant. Particular emphasis is placed on a discussion of the effects of vascular wilt pathogens. However, I do not review the pathological literature in detail, but rather I provide a physiologist's view, attempting to highlight those areas where it seems to me that appropriate knowledge is limiting, and suggest some experiments that might help to remedy these deficiencies.

Basic theory

The basics of cell and tissue water relations have been outlined by Papendick (Ch. 1), so this discussion concentrates only on those aspects that are relevant to the flow of liquid water within plants.

Restricting ourselves to movement that occurs wholly within the liquid phase, both diffusion (in the sense that movement depends on the random thermal movements of individual molecules) and mass flow in response to pressure gradients can contribute to water flow. Because we are concerned with flow processes, which are necessarily not at equilibrium, the approach of classical thermodynamics is not appropriate. It is therefore necessary to turn to the techniques of irreversible thermodynamics which were first applied to water flow in plants in the early 1960s (Slatyer, 1967). The particular advantage of this approach is that it is applicable to systems where there may be several linked flows (of water, solutes, electrical charge, etc.) driven by a series of appropriate driving forces. Taking the simple cases of fluxes of water (J_w, mol m^{-2} s^{-1}) and of uncharged solute (J_s, mol m^{-2} s^{-1}) driven by the corresponding differences in chemical potential $\Delta\mu_w$ and $\Delta\mu_s$, we get the following pair of equations:

$$J_w = L_{ww}\Delta\mu_w + L_{ws}\Delta\mu_s \tag{1}$$

$$J_s = L_{sw}\Delta\mu_w + L_{ss}\Delta\mu_s \tag{2}$$

where the various coupling coefficients (L) relate a flow (first subscript) to a force (second subscript). Each driving force is the sum of several components, e.g. for water:

$$\Delta\mu_w = -V_w\pi + V_w\Delta P + mg\rho h \tag{3}$$

where V_w is the partial molar volume of water, π is the osmotic pressure, P is the hydrostatic pressure, m is the molecular weight of water, g is the acceleration due to gravity, ρ is the density of water and h is the height above the reference level. For charged solutes it is also necessary to allow for electrical fields. Dividing Eq. (3) through by V_w gives the more familiar expression in terms of water potential (ψ).

For studies of water movement it is convenient to replace J_w by a volume flux, J_v, which for a single solute is equal to $J_wV_w + J_sV_s$ (in most cases $= J_{vw}$). It is, after all, generally the total volume flux that is measured. Using Eq. (1) and (2), substituting the appropriate driving forces and rearranging terms, gives the following useful expression:

$$J_v = L_p(\Delta P - \sigma\Delta\pi) \tag{4}$$

where L_p is termed a hydraulic conductance coefficient ($\mathrm{m\,s^{-1}\,Pa^{-1}}$), and σ is a dimensionless reflection coefficient that is a measure of the permeability of the path to the solute. For flow across a semipermeable membrane, that is one that allows water to pass freely but is completely impermeable to the solute, $\sigma = 1$. When solutes can pass freely, as for example for flow in the xylem, or in a fungal hypha (see Eamus & Jennings, Ch. 2) $\sigma = 0$ and J_v is then proportional to the pressure difference across the pathway.

Xylem transport

The major long-distance pathway for water movement in plants is in the vessels (tracheids in Gymnosperms) of the xylem. The structure of vessels and their interconnections within a singular vascular bundle of a herbaceous plant, indicating the inter-vessel pits, is shown in Fig. 13.1a. The vascular wilt pathogens, especially those in the genera *Verticillium* and *Fusarium*, as well as *Ceratocystis* (which is mainly important in deciduous woody trees), are well known as specialised organisms that have succeeded in colonising the xylem vessels of a wide range

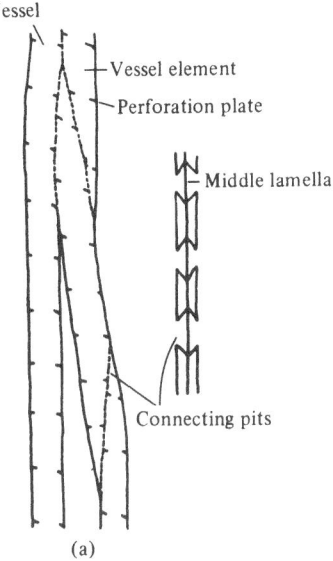

Vessel

Vessel element

Perforation plate

Middle lamella

Connecting pits

(a)

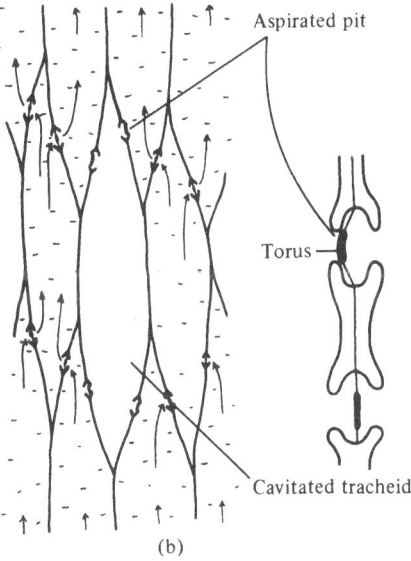

Aspirated pit

Torus

Cavitated tracheid

(b)

Fig. 13.1. (a) Diagram of the arrangement of vessels within a vascular bundle, showing the individual elements and perforation plates, as well as the vessel-to-vessel pits (inset). Note that the relative vessel diameters are represented much enlarged. (b) The structure of coniferous tracheids showing bordered pits and their action in isolating a cavitated tracheid.

Table 13.1. *Relationships between different measures of hydraulic conductance*

Conductance (L_p) (m s^{-1} Pa$^{-1)}$	$= J_v/\Delta P$	Resistance $(R_p) = 1/L_p$
Conductivity (L) (m^2 s^{-1} Pa^{-1})	$= J_v l/\Delta P$	Resistivity $(R)\ = 1/L$

of plant species (Mace, Bell & Beckman, 1981). It is now well established that the characteristic wilt symptoms result, at least partly, from effects on water flow within the vascular system (see e.g. Talboys, 1968; Ayres, 1978; Hall & MacHardy, 1981), so I propose first to outline the current state of knowledge on vascular water flow.

Within the xylem pathway flow depends only on ΔP, and reduces to:

$$J_v = L_p\Delta P = \Delta P/R_p \tag{5}$$

where R_p is the hydraulic resistance. It is worth noting here that the volume flux density, J_v, is equivalent to an average velocity. Eq. (5) is an example of the general transport equation and is often referred to as an Ohm's Law analogue. Although hydrostatic pressure is the appropriate driving force for mass flow, for tall trees it is necessary to allow for gravitational potential (Eq. 3) and replace ΔP by $\Delta P + \rho g\Delta h$. This substitution will be implicit in all that follows, but can be ignored with safety for most herbaceous plants because a 10 m elevation difference is equivalent to only 0.1 MPa. A further common simplification is to approximate ΔP by $\Delta\psi$, since it is ψ rather than xylem pressure that is measured by instruments such as psychrometers.

It is appropriate at this stage to compare the various expressions that have been used for the constant that relates flow rate to driving force, so as to permit comparisons between different authors. The main possibilities are summarised in Table 13.1. The area that is used as the basis for calculation can be that of the vessel lumina (common for L_p), or of the vascular bundles, or of whole stems, while in some cases even leaf areas have been used. Conductance measurements are most convenient for analysing flow in parallel paths, while their reciprocals (especially resistance) are more useful for analysing flow in a largely series pathway such as the transpiration stream.

Resistance networks and the control of leaf water status

The water flow pathway through plants can be represented by a more or less complex resistance network so that the overall result of an infection that affects a particular component resistance can be analysed using standard techniques of electrical network analysis (Brophy, 1972). The overall consequence of a given change in a component resistance depends on its position in the pathway (Fig. 13.2). Where there are several paths in parallel, increasing the resistance of one (even to infinity) will have an effect on the overall resistance that is small and proportional to the reciprocal of the number of parallel paths. Similarly, for a series pathway, a large increase in the resistance of a low-resistance component (e.g. the stem) has only a relatively small effect on the total, as noted below.

Typically the main flow resistances are in the roots (often about 50% of the total) and the petiole or leaf (often about 35% of the total), with the remaining 10–20% being in the stem (e.g. Boyer, 1974; Lands-

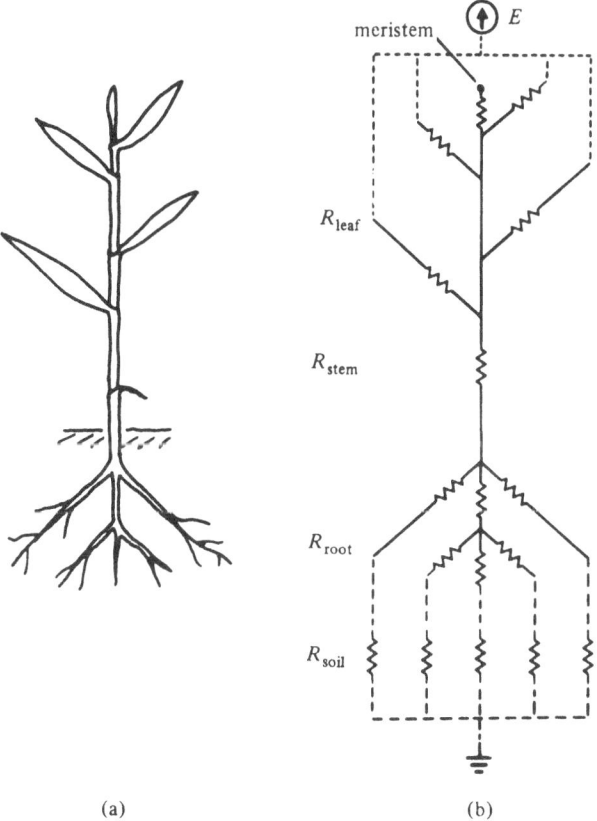

(a)   (b)

Fig. 13.2. Flow pathways in a transpiring plant (a) shown as an electrical analogue (b) (from Jones, 1983).

berg, Blanchard & Warrit, 1976; Black, 1979*a,b*). A particularly large component of the leaf and petiole resistance appears to be within the lamina (Boyer, 1974; Black, 1979*a*), possibly because of the small size of conduits there. Although the root resistance is usually a large component of the total, this is not always so (Begg & Turner, 1970). Because the stem resistance tends to be small, doubling it (which would require complete blockage of half the vessels) would only increase the total resistance by about 15%.

Much emphasis in the literature of plant pathology has been on the effects of disease on resistances, but it is important to realise that large changes in flow resistance are not necessarily reflected in large changes in leaf water status or physiological function. This is because plants have an extremely efficient feedback control of leaf water status, preventing it falling to damaging levels except in extreme circumstances (Jones, 1985). The mechanism of this control can be appreciated if Eq. (5) is rearranged and expressed in terms of leaf and soil water potentials (ψ_{leaf} and ψ_{soil}), transpiration rate (J_{wv}) and the total liquid-phase hydraulic resistance (R_{sp}, which includes the resistance at the root-soil interface):

$$\psi_{leaf} = \psi_{soil} - J_{wv}R_{sp} \tag{6}$$

Assuming adequate soil water, any increase in resistance that results from infection and which tends to lower ψ_{leaf}, can be compensated by a corresponding decrease in J_{wv}. These two quantities are largely independent since the liquid phase resistance has no direct effect on J_{wv}, which is controlled by the gas phase resistance through the stomata (Jones, 1983). It follows that changes in R_{sp} are not necessarily reflected in wilting symptoms.

The possibility of this control by the stomata raises the question of the role of transpiration. Although some liquid water movement is necessary for effective nutrient uptake and redistribution, this is far less than is brought about by normal transpiration rates (Luttge & Higinbotham, 1979). Similarly the cooling effect of transpiration is important only in extremely hot conditions (Jones, 1983), so transpiration control can compensate for fungus-induced changes in resistance with little detrimental effect. The main possible disadvantage of stomatal closure is the resulting reduction in photosynthesis (and hence productivity), but again the reduction in photosynthesis will not normally be proportional to any reduction in stomatal conductance. This is because the photosynthetic apparatus is frequently operating at CO_2 levels close to saturation.

Poiseuille's Law

Hydrodynamic theory indicates that for flow in a cylindrical tube, L_p depends on the radius (r) and length (l) of the tube, on the viscosity of the fluid (η), on the surface characteristics of the tube wall and on the flow rate. Whenever the pattern of flow is laminar, the Poiseuille–Hagen Law holds and flow per unit area of tube is given by:

$$J_v = (r^2/8\eta l)\Delta P \tag{7}$$

while flow per conduit $(Q, \text{m}^3\text{s}^{-1})$ is

$$Q = (\pi r^4/8\eta l)\Delta P \tag{8}$$

This law holds over a wide range of flows, viscosities and tube dimensions, only failing for very small tubes (less than a few nm diameter) when diffusive effects become important, or when flow is turbulent. Flow is normally laminar when the value of Reynold's number $(Re = J_v \rho r/\eta)$ is less than about 2000. The maximum value of Re would occur in large vessels with a high flow rate. Assuming extreme values for J_v of $0.8\,\text{m s}^{-1}$ (Passioura, 1972) and r of 250 µm (Zimmermann, 1983) gives an upper limit of about 150, which is well within the range for laminar flow.

Many studies on flow through xylem have indicated that the apparent resistance to flow is between about 2 and 100 times that expected from Poiseuille's Law (Giordano, Salleo, Salleo & Wanderlingh, 1978; Jeje & Zimmermann, 1979; Zimmermann, 1983), though values close to theoretical have been observed (see Dimond, 1966; Zimmermann & Brown, 1980). Only a few studies have tested this equation on single tubes or vessels (e.g Giordano *et al.*, 1978); most others have studied flow through stem segments with a range of vessel sizes, so are at best only approximate. The spacing and size of wall sculpturings and thickenings can affect the flow resistance (Jeje & Zimmermann, 1979), for example close spiral thickenings can reduce the effective vessel diameter. There is also evidence that the flow resistance can be greater than expected when the walls are porous (van Booren, 1976).

Resistance of pits and pit membranes

In addition to any flow resistance within the vessels or tracheids there must also be a resistance to flow through the intervening pits, but the extent to which this contributes to the overall resistance is not well known. In many species there are large areas of vessel-to-vessel contact (Fig. 13.1a) which have a high density of pits through which water can flow (see Zimmermann, 1983, for a detailed discussion of

Table 13.2. *Calculation of the relative contributions of resistances in vessel lumina and in intervessel walls for* Raphis excelsa *(from information in Zimmermann, 1983)*

	Lumen	Pits in secondary wall	Pit membrane
Relative cross-sectional area	1	10	100
Length of path/vessel	0.2 m	5 μm	100 μm
Pore diameter	100 μm	1 μm	25 nm
Relative resistance	100	2.5	2

xylem anatomy). Simple calculation shows that the resistance of the pit membranes between vessels is unlikely to contribute much to the overall resistance, and this component has been ignored by many workers (e.g. Dimond, 1966). Table 13.2 shows, for example, that for the palm *Raphis excelsa* the contribution of the pores and the pit membranes is only about 5% of the total, largely because of the extensive area of overlap between vessels (see Fig. 13.1a). This result may overestimate the pit resistance of many species since the size of pores in pit membranes may range up to about 1 μm (Staemm & Wagner, 1961; Van Alfen, McMillan, Turner & Hess, 1983). The small size of the pores in pit membranes renders them liable to blockage by polysaccharides and other compounds released by fungal or host enzymes. Van Alfen *et al.* (1983) have demonstrated that at least in alfalfa the mean diameter of pit pores is smaller in leaf traces (100 nm) than in stems. The blockage of these pores by macromolecules produced by wilt pathogens could explain some of the increased flow resistance that has been observed, and why the increase is often most dramatic and damaging within the petiole or the leaf itself (Duniway, 1971a), where water has to pass through normal cell walls to reach evaporation sites at the surface of mesophyll cells.

The characteristic bordered pits between tracheids in conifers (Fig. 13.1b) have much larger pores in the pit membranes (around 5 μm diameter as compared to about 5 nm in cell walls), so they also constitute only a very small resistance, as do the perforation plates between individual vessel elements. (The almost complete absence of vascular wilts in Gymnosperms, with only *Verticicladiella wagenerii* showing clear similarity to the vascular wilts of Angiosperms (Smith, 1967), is probably a result of the absence of long vessels in Gymnosperms with the consequent lack of advantage in vascular colonisation.) It is clear from this type of calculation that the dominant pathway of long-distance flow in the xylem must be within the vessels rather than within the cell wall matrix. Flow through cell walls is likely to be important only in long-distance transport as a means of bypassing breaks in the normal pathway, and possibly as a pathway for water movement in the root cortex and the leaf mesophyll.

Xylem cavitation

An often neglected, but important, factor that reduces the conductivity of xylem is the occurrence of embolisms or cavitations within the vessels (Milburn, 1979). These interrupt the continuity of the water column. Because the water flowing through the xylem in a transpiring plant can be under substantial tension it is unstable and easily fractured. The mechanism of this cavitation is not clearly established but probably requires the presence of nucleation sites within the vessel, whether these may be on particles or surfaces within the vessel, or whether they may be small air bubbles entering through defects in the cell walls. Although fungal action could degrade cell walls enough to permit entry of air, air is not likely to be drawn into healthy vessels at the tensions normally existing in plants, since typical wall pores of about 5 nm diameter would prevent air entry under tensions of at least 30 MPa (Jones, 1983). Similarly the small size of pit membrane pores localises any embolisms within a single vessel or tracheid, so that water can continue to flow through alternative vessels (Fig. 13.2).

Immediately after a cavitation event, the vessel is probably filled with water vapour under vacuum (actually at an absolute pressure equal to the vapour pressure of water – about 20 kPa) so is unlikely to refill with liquid water unless the hydrostatic pressure in the vessels rises above -0.1 MPa (referred to atmospheric). After a few hours or more, air may diffuse into a cavitated vessel rendering reversal even more difficult (Milburn, 1979).

A common response to cavitation is the production of tyloses (outgrowths from the xylem parenchyma cells that can occlude vessels completely), while other responses include the secretion of gums and gels, and the suberisation of the walls of non-functional vessels (Zimmermann, 1983). Resin impregnation of cell walls is particularly important in conifers. All these examples may represent a 'safety' response isolating the damage and preventing spread of infection, and may be particularly important where the cavitation is caused by fungal infection. The frequency of these responses to infection provides some evidence that the primary effect of infection is often on the induction of cavitations. It seems unlikely, therefore, that the tyloses directly affect water flow, as has been suggested by some (e.g. Cooper, 1981).

Unfortunately, cavitations are extremely difficult to detect, most evidence for their occurrence is circumstantial, and based, for example, on the detection of acoustic emissions (at around 500 Hz) which are thought to originate from the vibrations set up in the cell walls when they are released from tension by the cavitation process (Milburn, 1979). More recently detectors operating at ultrasonic frequencies, which are less subject to background noise, have improved detection (e.g. Tyree & Dixon, 1983). Acoustic sensors, however, cannot be used to identify the specific vessels(s) affected. Studies of xylem conductivity based on pressure-induced flow through isolated segments also cannot detect effects of *in vivo* cavitations, because the applied pressure may refill the vessels; even vacuum-induced flow could cause refilling. Conversely many sampling techniques are likely to cause cavitations during stem excision or injection, even if great care is taken to use ultrapure degassed water. This is because air remaining in intercellular spaces is likely to be sucked in when transpiring vessels are cut under water. It is, however, possible to detect groups of cavitated vessels by studying the patterns of staining obtained with a relatively non-mobile dye such as decolorised basic fuchsin (Talboys, 1955; see also Sauter (1984) who used a double-staining technique in an attempt to distinguish recently cavitated vessels containing water vapour, from those containing air). Solutions of gelatin containing indian ink particles (which are too large to penetrate pit membranes – J. A. Milburn, personal communication) can be used to identify functional vessels near the end of cut stem segments. However, it is clear that better techniques are required for the study of cavitations.

Although pathologists have not seriously considered xylem cavitation as a factor in wilt diseases, some instances of vessel collapse in infected plants have been reported (Chambers & Corden, 1963). Such collapsed vessels could contribute to increased flow resistances, but they are unlikely to be related to cavitation and are more likely to reflect an abnormal development pattern.

Effects of vascular wilts

It is now well established that vascular wilt pathogens can grow and spread within xylem vessels. What is not clear, however, is whether they grow in functioning vessels, or whether entry of the fungus causes cavitation and the subsequent failure of the vessel. This is important since any cavitation would have a large effect on water flow that would not be detected either by traditional anatomical studies or by studies on pressure-induced flow through stem segments.

Zimmermann (1983) has argued that one would expect fungal entry into a vessel under tension to cause cavitation by permitting air entry, and has presented evidence indicating the importance of cavitations in Dutch Elm disease (Newbanks, Bosch & Zimmermann, 1983). Unfortunately, the evidence for fungal growth in uncavitated vessels is largely indirect because of the difficulty of pinpointing cavitations. There are two main lines of evidence indicating that the vascular wilt pathogens are specialised enough to enter and grow in fully functional vessels. Firstly the rate of spread of the pathogen is often too rapid to be explained by mycelial growth alone: for example, in hop wilt, *Verticillium albo-atrum* can move from the base to the top of 5 m of stem in only 10 to 15 d (Talboys, 1968). It seems that movement involves microconidia being swept up in the transpiration stream (thus implying functional vessels) until stopped by pit membranes, at which point they germinate, grow into the next vessel and produce a new generation of microconidia (MacHardy & Beckman, 1981). Furthermore, Sewell & Wilson (1964) demonstrated that conidia were present in substantial numbers in the xylem sap of naturally infected hop plants. Secondly, dyes can indicate, rather imprecisely, those vessels that are conducting (Talboys, 1955; Newbanks *et al.*, 1983), and some apparently conducting vessels have been found to contain mycelium (Talboys, 1958). The conclusion that some functioning vessels are colonised by the pathogen does not, however, imply that cavitation is not an important consequence of infection.

Possible mechanisms by which vascular wilt pathogens might avoid cavitating vessels include the penetration of vessels only at night when ψ is high and xylem tension is small, or else the development of particularly tight seals round the penetration sites, though I know of no specific evidence on these points. I would argue that the success of vascular wilts is directly related to their ability to colonise vascular systems without causing embolism. This agrees with the common observation that infection of a resistant host results in extensive tylosis with the production of large amounts of gums and tissue browning (Robb, Smith & Busch, 1982), all of which are characteristic responses to cavitation even in the absence of the pathogen. In fact, differences between plants in response to infection often correspond to their wounding responses. These points suggest that only when a susceptible host is infected can the pathogen successfully penetrate the vessels without causing embolism. Further work clearly is necessary to resolve the question of the importance of cavitation in vascular wilt diseases.

The main approaches that have been used to investigate the effects of wilt pathogens on water flow have included both anatomical studies and direct measurements of flow resistances.

Anatomical studies. Several workers have attempted to estimate the effect of fungal mycelium within vessels on flow resistances by means of Poiseuille's equation, Eq. (7), though it is difficult to estimate an appropriate vessel diameter when mycelium is present. Equally it is difficult to estimate representative resistances from cross-sections of the xylem because of the longitudinal variability of fungal distribution (Chambers & Corden, 1963). These studies have been well reviewed by Talboys (1968), though a more recent attempt to map the distribution of mycelium in wilted plants has been reported (MacHardy *et al.*, 1976). To complement these studies, Waggoner & Dimond (1954) have attempted to estimate the effect of different amounts of mycelium using flow measurements through narrow glass tubes as model vessels with pieces of wire as model hyphae. Both theoretical and experimentally based studies of this type have shown that infection apparently increases resistances severalfold in any part of the pathway, but since the density of hyphae is normally greatest near the base of the plant, the effect should be greatest there (e.g. Talboys, 1968). Unfortunately it is not possible, using this type of calculation, adequately to account for the effects of gels blocking xylem vessels, or even less to account for the effects of gels, gums, suberins, and other fungal or host metabolites in blocking flow in cell walls and pit membranes. Furthermore, although the effects of tyloses on flow resistances can be estimated, it is most unlikely that tylosed vessels are functional and not previously cavitated and air-filled.

Another unresolved question in the calculation of flow resistances from Poiseuille's equation is the extent to which infection alters the viscosity of sap. Although Waggoner & Dimond (1954) could not detect any effect of infection on the viscosity of xylem sap, it is likely that the sap they analysed was extracted primarily from healthy conduits.

Measurements of flow resistances. Because of the difficulties with Poiseuille calculations, more useful estimates of the effects of vascular wilts on flow resistances can be obtained from direct physiological measurements of resistances. The main approach used by pathologists has been to force water through cut stem segments under pressure (or sometimes by applying vacuum) and to calculate the resistance from Eq. (6) (e.g. Ludwig, 1952; Duniway, 1971a; Dryden & Van Alfen, 1983). Unfortunately this approach involves damage to the conducting pathway and possible cavitation. Slightly better, perhaps, is to study flow through systems cut at only one end, such as detopped root systems (Duniway, 1971a), or leaves with varying amounts of plant still attached (Duniway, 1971b; Landsberg *et al.*, 1976), but these studies are still subject to the same problems.

To date, the most convincing study of the effects of wilt diseases on flow resistances is that of Duniway (1971b). Although the approach involved sequential cutting of the stem or leaf under water, flow resistances during transpiration were estimated from changes in evaporation rate when the water potential gradient was maintained constant (using leaf density as determined with a beta-gauge) as an indirect measure of ψ_{leaf}. Particular deficiencies with the original study include the need to cut the stem to estimate partial resistances and the fact that transpiration only occurred from one leaf, so that the relative importance of resistances in the petiole and leaf were overestimated (because of the branched nature of the pathway). Notwithstanding these problems, Duniway was able to show that Fusarium wilt of tomato led to large increases in all the component resistances, though wilting was apparent only when the petiole resistance increased significantly. Hall (Ch. 14) discusses further how fungal infection affects flow resistances and leaf-water status.

It should be better to estimate *R* from measurements on intact plants, using measurements of ψ at intermediate points to determine the magnitudes of the various component resistances. One method (Begg & Turner, 1970; Black, 1979b) for determining the stem water potential at any point is to measure (with a pressure chamber) ψ of a leaf inserted at that point that has previously been enclosed in a darkened plastic bag and allowed to equilibrate with the stem for about one hour. Alternatively stem water potentials could be measured with *in vivo* stem psychrometers (Campbell & Campbell, 1974; McBurney & Costigan, 1982), or indirectly using the beta-gauge, or from stem-diameter measurements (So, 1979).

Water movement over short distances

Water movement from cell to cell, whether it flows through the plasmalemma and the cell wall matrix or in the symplasm through the plasmodesmata, is important to plant functioning. Where the flow is through membranes it is necessary to use Eq. (4) rather than Eq. (5) to estimate J_v. Clearly any factor such as infection that affects the integrity of the membrane (and hence the value of σ) will also affect J_v.

In tissues such as the root cortex or the leaf mesophyll the relative contributions of flow through the apoplast and through the symplast are inversely proportional to their resistances. These clearly depend on the relative cross-sectional areas and pathlengths as well as on permeabilities. The importance of flow through membranes in such water movement clearly depends on all these factors, and no generalisations can be made (see Weatherley, 1983, for a recent review). In addition, movement from cell to cell entirely within the symplast involves movement through the plasmodesmata, which comprise only a small fraction of the surface area between adjacent cells.

The effects of root rots and other similar fungal diseases on the movement of water across the root cortex and endodermis are worth considering at this point, though more details may be found in the chapter by Hall (Ch. 14). The lytic enzymes that are released by rotting cause rapid tissue breakdown including wall degeneration and membrane damage leading to cell death, with water soaking frequently occurring. However, since water flow through such a tissue is largely a passive process, it often can continue, as long as some liquid continuity is retained. The water flow will be reduced with time, largely as a result of the oxidation of phenolic substances released from the cells and the general impregnation of cell walls with these and other impermeable materials.

There is one place in the normal transpiration pathway where the water must flow through a membrane, that is in order to penetrate the root endodermis, where the apoplastic pathway is blocked by the suberised casparian strip. However, water may enter the symplast anywhere in the root cortex. In this zone active ion transport is an important factor determining water movement. Any fungus that affects membrane integrity or function will therefore have an effect on water movement across the endodermis, since under conditions of low transpiration active accumulation of solutes inside this zone causes osmotic water flow and root pressure (see e.g. Luttge & Higinbotham, 1979). The dependence on flow rate of the root resistance to water uptake that is often reported has been linked to flushing of the actively accumulated solutes away from the endodermis by high mass flow rates (Fiscus, 1975; Powell, 1978).

Another way in which fungal infection can affect water and ion relations at this level is by the production of toxins that affect both ion transport at membranes and their hydraulic permeabilities, although evidence suggests that *Fusarium* infection does not affect the permeability of the endodermis to ions. For example, using detopped tomato root systems, Duniway (1971a) found no evidence for infection causing an increased permeability of the roots to salts. The effects of fungal toxins on host plant water relations are discussed by Ayres & Paul (Ch. 15), so will not be considered further here. It is also worth noting that infection often can lead to large changes in endogenous levels of the natural host plant growth regulators such as auxins and abscisic acid. The latter compound is known to increase root permeability to water and also has important effects on ion transport in the root cortex (Luttge & Higinbotham, 1979).

Acknowledgement. I am particularly grateful to Dr P. W. Talboys for extensive discussion and guidance during the preparation of this paper.

References

Ayres, P. G. (1978). Water relations of diseased plants. In *Water Deficits and Plant Growth*, vol. 5, ed. T. T. Kozlowski, pp. 1–60. New York: Academic Press.

Begg, J. E. & Turner, N. C. (1970). Water potential gradients in field tobacco. *Plant Physiology*, **46**, 343–6.

Black, C. R. (1979a). The relative magnitude of the partial resistances to transpirational water movement in sunflower (*Helianthus annuus* L.). *Journal of Experimental Botany*, **30**, 245–53.

Black C. R. (1979b). A quantitative study of the resistances to transpirational water movement in sunflower (*Helianthus annuus* L.). *Journal of Experimental Botany*, **30**, 947–53.

Boyer, J. S. (1974). Water transport in plants: mechanism of apparent changes in resistance during absorption. *Planta*, **117**, 187–207.

Brophy, J. J. (1972). *Basic Electronics for Scientists*. New York: McGraw-Hill.

Campbell, G. S. & Campbell, M. D. (1974). Evaluation of a thermocouple hygrometer for measuring leaf water potential in situ. *Agronomy Journal*, **66**, 24–7.

Chambers, H. L. & Corden, M. E. (1963). Semeiography of *Fusarium* wilt of tomato. *Phytopathology*, **53**, 1006–10.

Cooper, R. M. (1981). Pathogen-induced changes in host ultrastructure. In *Plant Disease Control*, ed. R. C. Staples, pp. 105–42. New York: Wiley.

Dimond, A. E. (1966). Pressure and flow relations in vascular bundles of the tomato plant. *Plant Physiology*, **41**, 119–31.

Dryden, P. & Van Alfen, N. K. (1983). Use of the pressure bomb for hydraulic conductance studies. *Journal of Experimental Botany*, **34**, 523–8.

Duniway, J. M. (1971a). Water relations of *Fusarium* wilt in tomato. *Physiological Plant Pathology* **1**, 537–46.

Duniway, J. M. (1971*b*). Resistance to water movement in tomato plants infected with *Fusarium*. *Nature*, **230**, 252–3.

Fiscus, E. L. (1975). The interaction between osmotic- and pressure-induced flow in plant roots. *Plant Physiology*, **55**, 917–22.

Giordano, R., Salleo, A., Salleo, S. & Wanderlingh, F. (1978). Flow in xylem vessels and Poiseuille's Law. *Canadian Journal of Botany*, **56**, 333–8.

Hall, R. & McHardy, W. E. (1981). Water relations. In *Fungal Wilt Diseases of Plants*, ed. M. E. Mace, A. A. Bell & C. H. Beckman, pp. 255–98. New York: Academic Press.

Jeje, A. A. & Zimmermann, M. H. (1979). Resistance to water flow in xylem vessels. *Journal of Experimental Botany*, **30**, 817–27.

Jones, H. G. (1983). *Plants and Microclimate*. Cambridge: Cambridge University Press.

Jones, H. G. (1985). Physiological and environmental control of evaporation from plants and implications for plant water status. In *Crop Water Requirements*, ed. A. Perrier *et al.*, Proceedings of the Paris Conference, I.N.R.A., Paris (in press).

Landsberg, J. J., Blanchard, T. W. & Warrit, B. (1976). Studies on the movement of water through apple trees. *Journal of Experimental Botany*, **27**, 579–96.

Ludwig, R. A. (1952). Studies on the physiology of hadromycotic wilting in the tomato plant. *MacDonald Agricultural College, Technical Bulletin* 20.

Luttge, U. & Higinbotham, N. (1979). *Transport in Plants*. New York: Springer-Verlag.

Mace, M. E., Bell, A. A. & Beckman, C. H. (1981). *Fungal Wilt Diseases of Plants*. New York: Academic Press.

McBurney, T. & Costigan, P. A. (1982). Measurement of stem water potential of young plants using a hygrometer attached to the stem. *Journal of Experimental Botany*, **33**, 426–31.

MacHardy, W. E. & Beckman, C. H. (1981) Vascular wilt *Fusaria*: infection and pathogenesis. In *Fusarium: Diseases, Biology and Taxonomy*, ed. P. E. Nelson, T. A. Toussoun & R. J. Cook, pp. 365–90. University Park, Pa.: Penn State University Press.

MacHardy, W. E., Busch, L. V. & Hall, R. (1976). *Verticillium* wilt of chrysanthemums: quantitative relationship between increased stomatal resistance and local vascular dysfunction preceding wilt. *Canadian Journal of Botany*, **54**, 1023–34.

Milburn, J. A. (1979). *Water Flow in Plants*. London: Longman.

Newbanks, D., Bosch, A. & Zimmermann, M. H. (1983). Evidence for xylem dysfunction by embolization in dutch elm disease. *Phytopathology*, **73**, 1060–3.

Passioura, J. B. (1972). The effect of root geometry on the yield of wheat growing in stored water. *Australian Journal of Agricultural Research*, **23**, 745–52.

Powell, D. B. B. (1978). Regulation of plant water potential by membranes of the endodermis in young roots. *Plant, Cell and Environment*, **1**, 69–76.

Robb, J., Smith, A. & Busch, L. (1982). Wilts caused by *Verticillium* species. A cytological survey of vascular alterations in leaves. *Canadian Journal of Botany*, **60**, 825–37.

Sauter, J. J. (1984). Detection of embolization of vessels by a double staining technique. *Journal of Plant Physiology*, **116**, 331–42.

Sewell, G. W. F. & Wilson, J. F. (1964). Occurrence and dispersal of *Verticillium* conidia in xylem sap of hop (*Humulus lupulus* L.). *Nature*, **204**, 901.

Slatyer, R. O. (1967). *Plant Water Relationships*. London: Academic Press.

Smith, R. S. (1967). *Verticicladiella* root disease of pines. *Phytopathology*, **57**, 935–8.

So, H. B. (1979). An analysis of the relationship between stem diameter and leaf water potentials. *Agronomy Journal*, **71**, 675–9.

Stamm, A. J. & Wagner, E. (1961). Determining the distribution of intrastructural openings in wood. *Forest Products Journal*, **11**, 141–4.

Talboys, P. W. (1955). Detection of vascular tissues available for water transport in the hop by colourless derivatives of basic dyes. *Nature*, **175**, 510.

Talboys, P. W. (1958). Association of tylosis and hyperplasia of the xylem with vascular invasion of the hop *Verticillium albo-atrum*. *Transactions of the British Mycological Society*, **41**, 249–60.

Talboys, P. W. (1968). Water deficits in vascular disease. In *Water Deficits and Plant Growth, vol. 2*, ed. T. T. Kozlowski, pp. 255–311. New York: Academic Press.

Tyree, M. T. & Dixon, M. A. (1983). Cavitation events in *Thuja occidentalis* L.? Ultrasonic acoustic emissions from the sapwood can be measured. *Plant Physiology*, **72**, 1094–9.

Van Alfen, N. K., McMillan, B. D., Turner, V. & Hess, W. (1983). The role of pit membranes in macro-molecule induced wilt of plants. *Plant Physiology*, **73**, 1020–3.

Van Booren, R. (1976). Porous pipe flow. *Physical Fluids*, **19**, 481–2.

Waggoner, P. E. & Dimond, A. E. (1954). Reduction in water flow by mycelium in vessels. *American Journal of Botany*, **41**, 637–40.

Weatherley, P. E. (1983). Water uptake and flow in roots. In *Encyclopedia of Plant Physiology*, vol. 12B, ed. O. L. Lange, P. S. Nobel, C. B. Osmond & H. Ziegler, pp. 79–109. Berlin: Springer-Verlag.

Zimmermann, M. H. (1983). *Xylem Structure and the Ascent of Sap*. Berlin: Springer-Verlag.

Zimmermann, M. H. & Brown, C. L. (1980). *Trees: Structure and Function*. Berlin: Springer-Verlag.

14
Effects of root pathogens on plant water relations

ROBERT HALL

Department of Environmental Biology, University of Guelph, Ontario N1G 2W1, Canada

Root-infecting fungi that cause visible symptoms of root disease, such as root rot, club root and root-browning, are referred to in this review as rhizopathic. Many rhizopathic fungi cause wilt and other symptoms of water stress but there are relatively few studies on the characteristics of this stress. This is in contrast to the vascular wilt diseases which, though fewer in number (Kommedahl & Windels, 1979), have been studied much more intensively (Hall & MacHardy, 1981). Rhizopathic water stress was reviewed briefly by Duniway (1973) and Ayres (1978, 1981) in general reviews of the effect of plant pathogens on plant water status.

Rhizopathic water stress is revealed in different ways (Kommedahl & Windels, 1979). Sometimes it is revealed as sudden wilting of mature plants, as in black shank of tobacco caused by *Phytophthora parasitica* var. *nicotianae*, at other times as part of seedling death, as in snapdragon (*Antirrhinum* sp.) infected with *Pythium ultimum*. Slow development of water stress, with or without wilt, seems to be very common, as in infections of woody plants with *Phytophthora cinnamomi*.

Early work on rhizopathic water stress centred on black shank of tobacco and *Phytophthora* remains the pathogen genus most studied. The following account is based largely on studies involving seven species in five genera: *Phytophthora, Fusarium, Phymatotrichum, Aphanomyces* and *Gaeumannomyces*. This very small sample may not be representative of rhizopathic fungi in general. However, the information available is remarkably consistent and has been interpreted by the authors and in this review according to modern views (Kramer, 1983) of the catenary flow model described by van den Honert (1948). This model has been used widely to explain water relations of diseased plants (Duniway, 1973;

Ayres, 1978, 1981; Hall & MacHardy, 1981). It relates flow of water (J_v) directly to a driving force, provided by differences in water potential $(\Delta\psi)$ or vapour pressure (Δc), and inversely to resistance (R). Equation 1 presents a simple form of the model.

$$\text{Flow} = J_v = \frac{\Delta\psi_{\text{root to leaf}}}{R_{\text{root to leaf}}} = \frac{\Delta c_{\text{leaf to air}}}{R_{\text{leaf to air}}} \tag{1}$$

This model assumes that flow is equal in all parts of the plant and that resistances in different parts are independent of each other and of the flow rate. These assumptions are not entirely valid for healthy plants and it may be that they are even less applicable to diseased plants. Nevertheless, the model provides a generally satisfactory way of explaining water relations in healthy and diseased plants.

I shall discuss first the effects of rhizopathic fungi on plant water status, then examine the possible causes of rhizopathic water stress and, finally, consider briefly rhizopathic water stress as it is affected by environmental water and as it may relate to the general problem of plant water stress in natural and managed ecosystems.

Water status in rhizopathic diseases
Transpiration

Most studies show that root diseases that cause wilt reduce transpiration. An early study by Schramm & Wolf (1954) reported that *Phytophthora parasitica* var. *nicotianae* reduced transpiration of tobacco. They did not relate the reduction in transpiration to the onset of wilt but noted that the difference in transpiration rates between healthy and infected plants increased with time. Later, De Roo (1969) examined transpiration of one-year-old potted rhododendron plants apparently naturally infected with *Phytophthora cinnamomi*. This fungus normally infects the roots and lower stem. Detached shoots from diseased plants, and intact plants deprived of water, transpired less and absorbed less water than healthy, well watered plants. Diseased plants further resembled dehydrated plants in that the rate of absorption of water by detached shoots was greater than the rate of transpiration.

Sterne, Kaufmann & Zentmyer (1978) examined the effects of *P. cinnamomi* on transpiration of 8- to 10-year-old avocado trees growing in the field (Fig. 14.1). In healthy trees, transpirational flux usually reached a maximum value between 1200 and 1400 hours, with a mean value at 1400 hours of $1.78\ \mu\text{g cm}^{-2}\,\text{s}^{-1}$ for leaves exposed to full sunlight (high photosynthetically active radiation, PAR, Fig. 14.1). In leaves of diseased trees, transpirational flux was much lower (less than $0.5\,\mu g\,cm^{-2}\,s^{-1}$) and, although it remained fairly constant through the day, appeared to peak between 1000 and 1200 hours, about two hours earlier than in healthy plants. Transpiration rates were lower in shaded leaves (low PAR, Fig. 14.1) and differences between infected and healthy trees were less apparent.

The tops of the diseased trees had an average disease rating of 2.5 (0 = healthy tree and 5 = dead tree). It is likely that these trees showed symptoms of wilting, leaf discoloration, die-back and reduced vegetative growth and that some individual leaves and shoots used for measurements were wilted. However, the relationship between transpiration changes and the onset of wilt was not established in these studies. The question was explored in the following studies.

Safir & Schneider (1976) determined the effect of *Aphanomyces cochlioides* on transpiration of four-week-old potted plants of sugarbeet from the weight loss per pot. Five days after inoculation, roots of susceptible and resistant plants had similar root rot ratings of 1.5 (1 = blackening near the base of the hypocotyl and 2 = more than half the hypocotyl blackened). Above-ground symptoms, including wilt, were evident in susceptible plants 3 to 5 d after inoculation and in resistant plants 7 d

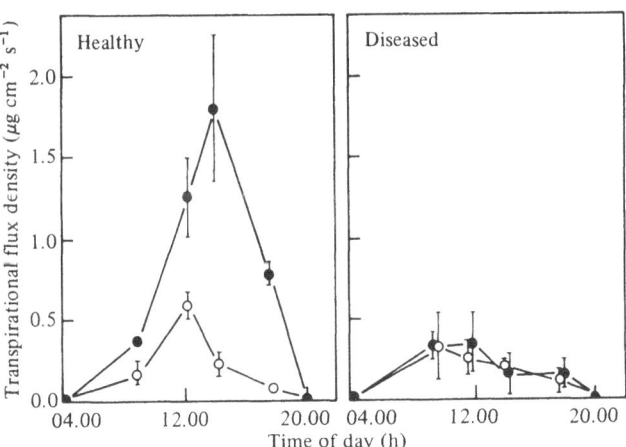

Fig. 14.1. Effect of root disease caused by *Phytophthora cinnamomi* and high (●) and low (○) levels of PAR on diurnal changes in transpirational flux density of avocado leaves. Healthy and diseased trees were well watered (soil matric potential -0.01 to $-0.03\,MPa$). High PAR (sun leaves) = 168 to 210 nmol cm^{-2} s^{-1}. Low PAR (shaded leaves) = 4 to 13 nmol cm^{-2} s^{-1}. From Sterne *et al.* (1978).

after inoculation. Five days after inoculation, transpiration ($g \, dm^{-2} h^{-1}$) was 20% lower in resistant plants and 45% lower in susceptible plants. These data indicated that a decline in transpiration preceded wilting of resistant plants. Nine days after inoculation, root rot ratings were 1.60 for resistant plants and 2.54 for susceptible plants (3 = most of the hypocotyl blackened). Transpiration was unchanged in healthy plants but in inoculated plants was 60% lower in the resistant line and 84% lower in the susceptible line.

P. cinnamomi reduced transpiration of Isopogon ceratophyllus (Dell & Wallace, 1981; Dawson & Weste, 1982). In the latter study the first symptoms of wilt appeared three months after inoculation. At this time, transpiration ($g \, g^{-1}$ shoot dry weight h^{-1}) of healthy plants showed a marked diurnal rhythm with peak values at 1400 hours of 0.44 and 0.33 in two trials. Transpiration of infected plants was constant through the day and rates were much lower (approximately 0.1 and 0.02 in two trials). These data showed that transpiration was lower in infected plants at the time wilting was first observed. Similarly, transpiration of seedlings of Eucalyptus sieberi infected with P. cinnamomi did not decline until the shoots wilted, 3 to 15 d after inoculation (Dawson & Weste, 1984).

Schneider (1984) reported that Fusarium moniliforme did not alter transpiration of corn plants even when it increased resistance to liquid water movement through the plants. Transpiration rates were maintained by a reduction of ψ_{leaf}, i.e. an increase in $\Delta\psi_{root \, to \, leaf}$. However, ψ_{leaf} did not decline sufficiently to cause wilt in these experiments. These results indicate that changes in ψ_{leaf} or resistance to water flow are earlier indicators of water stress than are changes in transpiration.

Stomatal conductance

Root infection often leads to reduced stomatal conductance (increased diffusive resistance). In sugarbeet infected with Aphanomyces cochlioides wilt occurred 3 to 5 d after inoculation in susceptible plants and 7 d after inoculation in resistant plants (Safir & Schneider, 1976). Five days after inoculation, diffusive resistance of individual leaves was 50% (resistant plants) to 167% (susceptible plants) higher in infected plants than in healthy plants. Nine days after inoculation, diffusive resistance had increased 2.5 times (resistant plants) to 3.5 times (susceptible plants) as a result of infection.

Sterne et al. (1978) reported that leaf conductance of avocado was lower in trees infected with Phytophthora cinnamomi than in healthy trees. In healthy trees, conductance of leaves exposed to full sunlight (sun leaves) reached values of 0.07 to 0.08 cm s⁻¹ at 0830 hours and remained there until 1800 hours, before declining to zero at 2000 hours. In shaded leaves, conductance reached a peak value of $0.05\,\text{cm s}^{-1}$ at 0830 hours then declined to zero at night. In diseased trees, conductance values for sun leaves and shaded leaves were similar and reached values of 0.03 to $0.04\,\text{cm s}^{-1}$ at 0900 hours before declining to zero at night.

Dawson & Weste (1982) reported that stomatal conductance of glass-house-grown plants of *Eucalyptus macrorhyncha* (field-susceptible) and *E. goniocalyx* (field-resistant) was not altered by inoculation with *P. cinnamomi*. Similarly, under field conditions, infection did not consistently reduce leaf conductance, although it did change the pattern so that sometimes conductance was higher and at other times lower in infected trees than in uninfected trees. The authors concluded that the infected plants were not water-stressed under the conditions studied and may show effects of disease only in dry periods after rain. In a subsequent study, Dawson & Weste (1984) reported that leaf conductance in seedlings of *Eucalyptus sieberi* infected with *P. cinnamomi* did not decline until plants wilted (stage 3, Fig. 14.2).

Schneider & Pendery (1983) examined the effect of *Fusarium moniliforme* on water relations of corn under field conditions. There were no visible symptoms of wilt but infected plants subjected to an early season water stress behaved like plants subjected to chronic water stress. Diffusive resistance values during the day tended to be higher in infected stressed plants than in other treatments.

Leaf water potential

Root infections commonly reduce ψ_{leaf}. Leaf water potential (as measured by psychrometric techniques), leaf xylem potential (ψ_{xylem}) (measured with a pressure chamber) and relative water content (rwc) are considered to be essentially equivalent measures of leaf water status, and among the earlier indicators of water stress. Infection of roots with *P. cinnamomi* reduced one or more of these parameters in potted plants of rhododendron (De Roo, 1969) and *Isopogon ceratophyllus* (Dell & Wallace, 1981), and in mature forest trees of *Eucalyptus obliqua* (Weste, 1980) and *E. marginata* (Shea, Shearer & Tippett, 1982).

Sterne *et al.* (1978) reported that ψ_{xylem} was greater in healthy avocado trees than in trees infected with *P. cinnamomi* (Fig. 14.3). In healthy trees, night values for sun and shade leaves averaged $-0.175\,\text{MPa}$. During the day, ψ_{xylem} declined to minimum values at 1200 to 1500 hours of -0.9 to $-1.10\,\text{MPa}$ in sun leaves and -0.7 to $-0.8\,\text{MPa}$ in shaded

leaves. In sun and shade leaves of infected trees, ψ_{xylem} averaged
-0.8 MPa at night then declined steadily after sunrise to -1.4 MPa at
1800 hours before returning to night values.

Dawson & Weste (1982) reported that *P. cinnamomi* had reduced
ψ_{leaf} and rwc of *Isopogon ceratophyllus* by the time wilt symptoms began
to appear in one plant. It appears that most of the leaves were not
wilted, indicating that ψ_{leaf} declined before wilt appeared. *I. ceratophyllus*
is highly susceptible to infection by *P. cinnamomi* (all root pieces plated
were infected) and glasshouse-grown plants show symptoms of water
stress such as closure of stomata during the day. By contrast, the fungus
did not alter ψ_{xylem} or the diurnal pattern of stomatal movement of less
susceptible glasshouse-grown species of *Eucalyptus*, even though 40%
of root pieces of one species were colonised. In a subsequent study
with young seedlings of *Eucalyptus sieberi*, Dawson & Weste (1984)
showed that ψ_{xylem} was not significantly reduced by infection of roots
with *P. cinnamomi* until wilt occurred.

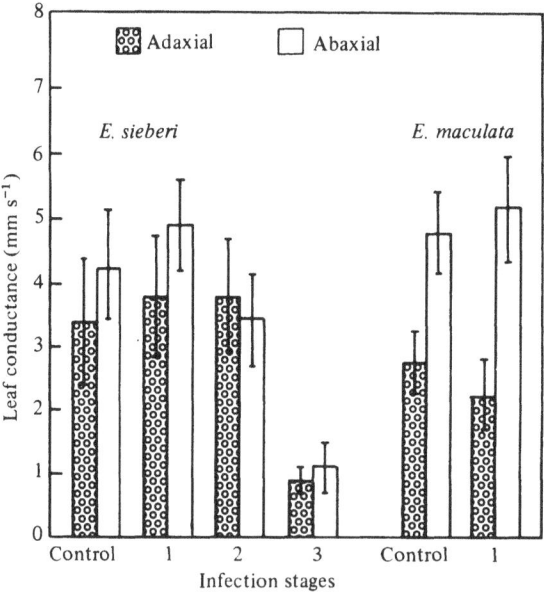

Fig. 14.2. Effect of infection with *Phytophthora cinnamomi* on leaf conductance of *Eucalyptus sieberi* (susceptible) and *E. maculata* (resistant). In
infected plants three infection stages were distinguished: in stage 1, lesions
formed in roots about 10 h after inoculation; in stage 2, lesions extended
and hydraulic conductivity in roots was reduced to 50% of controls; in
stage 3, leaves wilted and leaf conductance was further reduced. From
Dawson & Weste (1984).

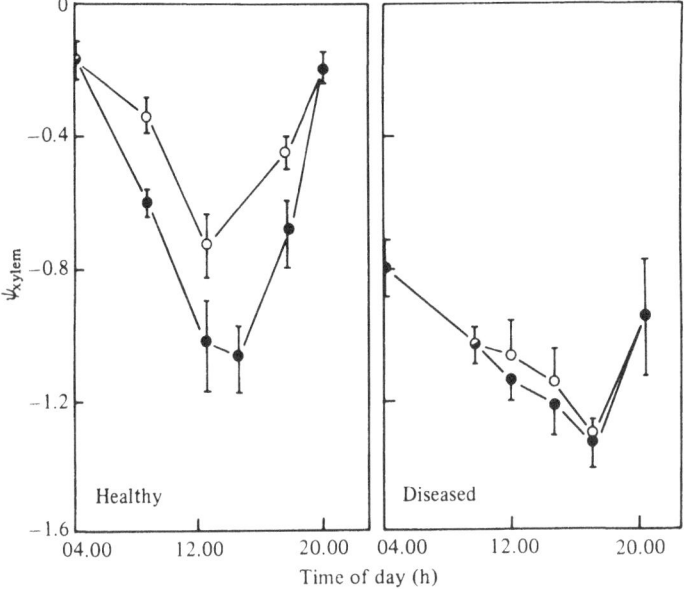

Fig. 14.3. Effects of root disease caused by *Phytophthora cinnamomi* and high (●) and low (○) levels of PAR on diurnal changes in ψ_{xylem} of avocado leaves. Healthy and diseased trees were well watered (soil matric potential −0.01 to −0.03 MPa). High PAR (sun leaves) = 168 to 210 nmol cm^{-2}s^{-1}. Low PAR (shaded leaves) = 4 to 13 nmol cm^{-2}s^{-1}. From Sterne *et al.* (1978).

Schneider & Pendery (1983) and Schneider (1984) found that *Fusarium moniliforme* reduced ψ of corn leaves only if the plants had been subjected to a previous period of drought. In the field, the reductions were as great as 0.5 MPa (Schneider & Pendery, 1983).

Leaf water relations

Maintenance of normal relations in infected plants would indicate that the selective permeability of leaf cell membranes is not altered and, thus, stress must arise through other mechanisms.

Infection of safflower plants with *Phytophthora cryptogea* did not alter the relation of ψ to the percentage of leaves which appeared wilted (Duniway, 1975). There was no effect of infection on the relation of osmotic potential (ψ_π) to ψ in safflower leaves infected with *P. cryptogea* (Duniway, 1975) (Fig. 14.4a) or in cotton infected with *Phymatotrichum omnivorum* (Olsen, Misaghi, Goldstein & Hine, 1983). Infection did not change the relation of rwc to ψ in safflower infected with *P. cryptogea*

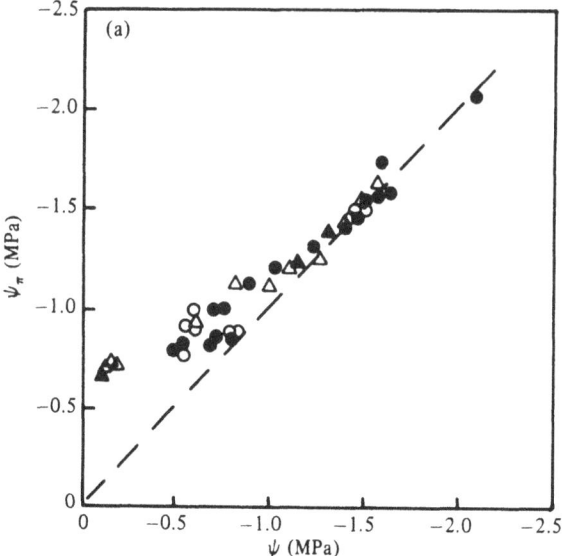

Fig. 14.4a. Relation of ψ_π to ψ in leaves of healthy and *Phytophthora crypto-gea*-infected safflower plants. (○) healthy plants developing water stress, (●) infected plants developing water stress, (△) healthy plants recovering from water stress, (▲) infected plants recovering from water stress. From Duniway (1975).

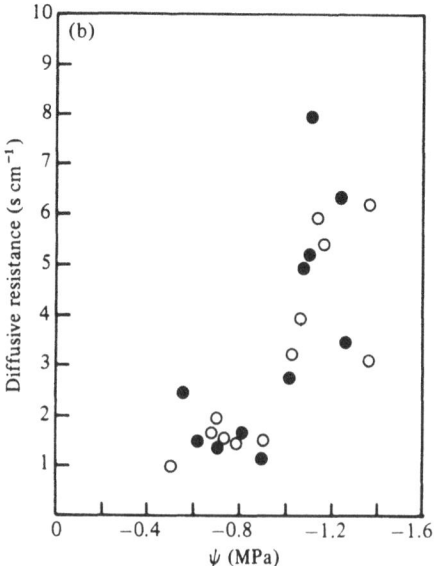

Fig. 14.4b. Relation of diffusive resistance to ψ of primary leaves in seedlings of sugarbeet inoculated with *Aphanomyces cochlioides* (●) or deprived of water (○). From Safir & Schneider (1976).

(Duniway, 1975), in cotton infected with *P. omnivorum* (Olsen *et al.*, 1983), or in corn infected with *Fusarium moniliforme* (Schneider & Pendery, 1983). The relation of diffusive resistance to rwc or ψ was unchanged in safflower infected with *P. cryptogea* (Duniway, 1975), in cotton infected with *P. omnivorum* (Olsen *et al.*, 1983), in sugarbeet infected with *Aphanomyces cochlioides* (Safir & Schneider, 1976) (Fig. 14.4*b*) and in corn infected with *F. moniliforme* (Schneider & Pendery, 1983).

In healthy plants it is common to find that ψ_{leaf} declines as transpiration rate increases (Kramer, 1983). Sterne *et al.* (1978) observed this in avocado (Fig. 14.5). However, for a given rate of transpiration, values for ψ_{xylem} were lower in water-stressed trees than in well watered trees and showed hysteresis. For a given transpiration rate, potentials were higher between 0400 hours and 1400 hours when transpiration rates were increasing than between 1400 hours and 0400 hours when transpiration rates were decreasing. Trees infected with *Phytophthora cinnamomi* behaved like those deprived of water. They showed hysteresis in the

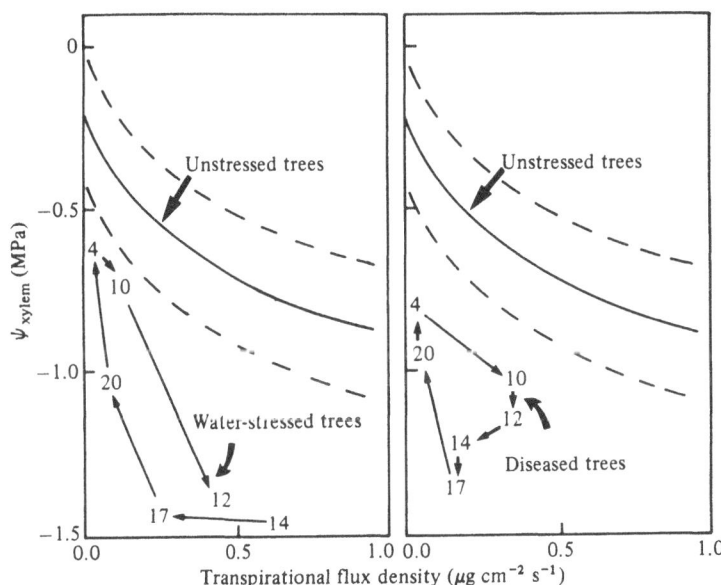

Fig. 14.5. Relation of ψ_{xylem} to transpirational flux density in leaves of unstressed avocado trees (irrigated every 7 to 8 d), of water-stressed trees (without irrigation for 30 d) and of diseased trees infected with *Phytophthora cinnamomi*. The solid line for healthy unstressed trees was plotted from a regression equation developed from 110 individual data points; dashed lines indicate confidence limits $P = 0.05$. For water-stressed and diseased trees, numbers indicate the time of day (e.g. 4 = 0400 hours). From Sterne *et al.* (1978).

relation of ψ_{leaf} to transpiration rate and had lower ψ_{leaf} for a given transpiration rate than healthy, unstressed trees.

Schneider (1984) examined the effect of *Fusarium moniliforme* on corn plants which were either continuously well watered or well watered then deprived of water for a short period then well watered again for one week before measurements were taken. In healthy plants, both well watered and stressed, and in infected well watered plants, ψ_{leaf} declined steadily as transpiration rate increased. In infected plants which had been subjected to a previous drought period ψ_{leaf} was lower than in other treatments and showed a different pattern of response to changes in transpiration rate. As transpiration rate increased, ψ_{leaf} fell abruptly, then rose abruptly, then declined more steadily. The sudden rise in ψ_{leaf} was interpreted as evidence for removal of resistance to water flow in the xylem.

Recovery of leaf water potential or turgor

A simple but informative experiment is to detach a wilted shoot or leaf from an infected plant under degassed water and leave the cut end immersed. Recovery of turgor indicates that normal osmotic mechanisms are operating in the leaves. A reduced rate of recovery, compared to that of healthy droughted tissues, indicates that resistance to flow has increased; such resistance must be located below the lowest cut that permits normal rates of recovery of leaf turgor.

Recovery of turgor in detached shoots or leaves has been reported for several rhizopathic diseases. However, the conditions under which recovery occurred depended on the host–parasite combination. In tobacco infected with *Phytophthora parasitica* var. *nicotianae* local lesions develop in the roots and basal portions of the stem. Wilted shoots cut just above the lesion recovered turgor within 2 h (Powers, 1952). Stems detached below the lesion did not recover turgor. Turgor was restored when water was introduced into the wilted stem of an intact plant above a lesion. Recovery occurred in safflower infected with *Phytophthora cryptogea* when cut leaf bases, but not cut stem bases or intact plants, were immersed in water (Duniway, 1975). In cotton infected with *Phymatotrichum omnivorum*, complete recovery was reported when leaf disks were immersed in water (Olsen *et al.*, 1983). Recovery occurred in corn infected with *Fusarium moniliforme* when cut stem bases were immersed in water (Schneider, 1984), and in rhododendron infected with *Phytophthora cinnamomi* when detached shoots were immersed in water (De Roo, 1969). In all these studies it was concluded that infection of roots or lower stems did not impair the osmotic mechanism of leaves but deprived leaves of water by increasing resistance to water flow through roots or stems.

Absorption of water

There is very little direct information on the effect of rhizopathic infection on the movement of water into roots. The effect of *Gaeumanno-myces graminis* var. *tritici* on water uptake in wheat has been examined by Asher (1972) and Kararah (1976). Asher (1972) reported that shoots of infected plants contained a lower percentage of water than shoots of uninfected plants and concluded that infection reduced the ability of roots to absorb water. Kararah (1976) enclosed an inoculated seminal root in a micropotometer and all other roots of the same plant in a water-saturated atmosphere. All leaves were removed except one. Eight and 16 d after inoculation water uptake through inoculated roots was reduced to 40% and 2%, respectively, of uninoculated roots. The initial reduction of water uptake occurred before the xylem was invaded. High restriction of water uptake was associated with colonisation of the xylem and by day 16 most of the xylem elements were obstructed by hyphae or other plugging materials. However, it is not clear from these studies whether radial or axial movement of water through roots was reduced by infection.

Hydraulic resistance

The observations reviewed so far, which show that infected plants behave like plants deprived of water, have usually been inter-preted as evidence that infection restricts movement of water to leaves. More direct evidence that root infection increases hydraulic resistance (decreases conductance) in roots or stems is provided by several studies.

Schramm & Wolf (1954) applied a suction of 20 kPa to cut stumps of tobacco plants. Very little or no liquid was pulled through plants infected with *Phytophthora parasitica* var. *nicotianae* whereas large (un-specified) amounts were pulled through healthy root systems. Using the same host–parasite system, Powers (1952, 1954) observed that shoots regained turgor if water was introduced to the stem above the lesion, that bridge grafts around stem lesions delayed wilting, and that aqueous solutions of eosin dye moved through roots to the stem lesion but not beyond. He concluded that wilting of infected plants was caused by an obstruction to the flow of water through the vascular elements of the roots or lower stem.

Rates of recovery of rwc and ψ of leaves from water-stressed healthy safflower plants and plants infected with *Phytophthora cryptogea* were similar when cut bases of leaves were immersed in water (Fig. 14.6) (Duniway, 1975). However, when roots of intact plants were immersed in water, the initial rate of recovery of infected plants was one-eighth that of healthy plants. When the roots were removed and the cut stem base was immersed in water, the initial rate of recovery of infected plants was one-sixteenth that of healthy plants. Most infected plants

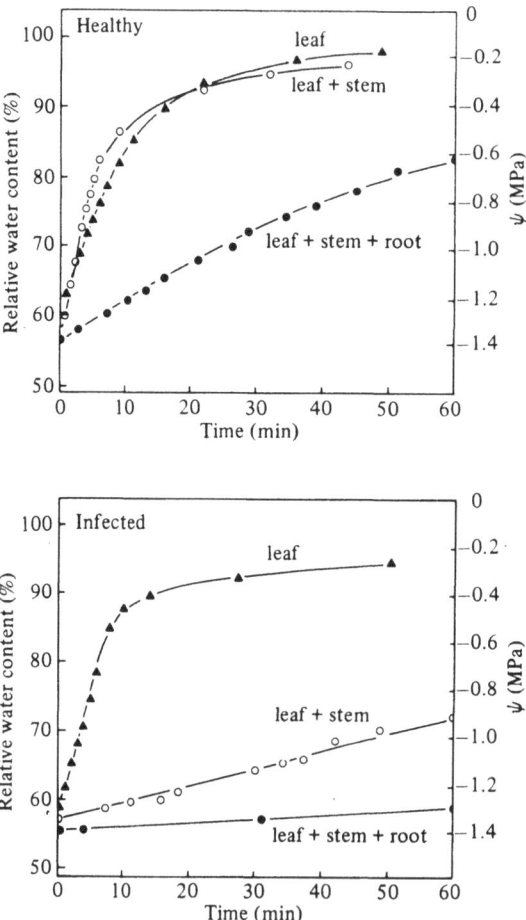

Fig. 14.6. Relative water contents of leaves from healthy and *Phytophthora cryptogea*-infected safflower plants during recovery from water stress. Recovery was initiated at time zero by supplying water to the roots (●), to the base of the stem (○) or to the base of the leaf (▲). The right-hand axis gives the equivalent ψ. From Duniway (1975).

Table 14.1. *Effect of infection of safflower with* Phytophthora cryptogea *on resistance to liquid flow in leaf, stem and root under applied pressure*

| Plant part | Resistance ($\times 10^5$ MPa s cm^{-1}) | | |
| | Healthy | Infected | |
		5 d	7 d
Leaf	0.27**	0.265**	0.265**
Stem	0.05 to 0.08†	2.07*	6.0*
Root	0.27†	2.07*	31.0*
Total	0.62††	4.4††	37.2††

Adapted from Duniway (1977: **from text; †calculated from text; *from Figure 1B; ††from Table 3).

did not recover from wilting when the cut stem base was immersed in water. It was concluded that *P. cryptogea* increased resistance to water flow in roots and stems of safflower, but quantitative estimates of the resistance were not possible.

Table 14.1 presents representative values for resistances obtained directly, or calculated, from a subsequent report (Duniway, 1977). Wilting was first observed 5 d after inoculation when resistances had increased 7-fold in roots, 25- to 40-fold in stems and 5- to 7-fold in the total plant. By day 7 root resistance had increased 155-fold, stem resistance had increased 60-fold and total plant resistance had increased 40- to 60-fold (Fig. 14.7).

Schneider & Pendery (1983) calculated resistance to water flow between roots and leaves of corn plants from observed rates of water uptake from soil and plant water potentials observed in the field. Compared to uninfected plants or unstressed infected plants, resistance was approximately doubled in plants that were infected with *Fusarium moniliforme* and had been previously water stressed. Similar results were found by Schneider (1984) when corn was grown in pots where roots but not stems were infected and neither roots nor leaves showed symptoms of infection. Healthy well watered plants, healthy plants previously subjected to drought and infected well watered plants showed similar levels of resistance to water flow between root and leaf. However, resistance in infected plants previously subjected to drought was increased 30 to 150%.

To determine the location of the resistance, pots were immersed in water and stems were severed at the first node. When the cut ends of shoots were immersed in water, ψ_{leaf} rose from -1.1 MPa in infected stressed plants and about -0.75 MPa in other treatments to about -0.1 MPa in all treatments after 1 h. The similarity in rates of recovery and final ψ_{leaf} indicated there were no differences among treatments in resistance of shoots to water movement. It was concluded that the higher resistance to liquid water movement in infected stressed plants was located in the roots.

Some evidence was obtained about the nature of the resistance. In plants exposed to vapour pressure deficits of 1.1 or 1.6 kPa the increased resistance in infected stressed plants persisted over the experimental period of 8 h. However, in plants exposed to vapour pressure deficits of 2.3 or 3.1 kPa, resistance remained high until the plant had transpired about 180 g of water then rapidly declined to the level of the other treatments, indicating to the authors the novel concept of a transient resistance. It was suggested that the immediate cause of the increased resistance developed during the night and was then removed or rendered inoperative after a certain volume of water had been transpired during the day. Gels, as described by Van der Molen, Beckman & Rodehorst (1977), might behave in this fashion. However, even this transient resis-

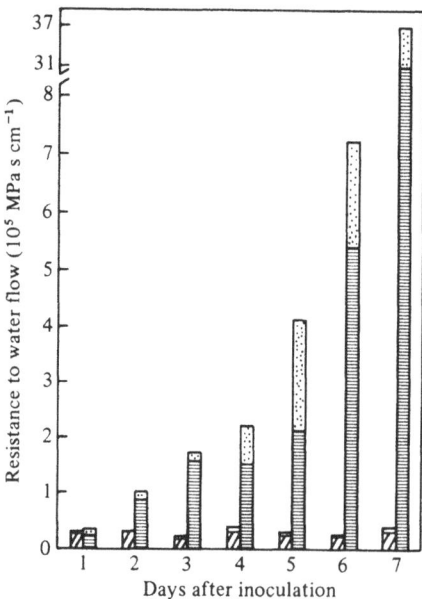

Fig. 14.7. Influence of infection of safflower by *Phytophthora cryptogea* on root and stem resistance to water flow. Infected plants wilted at day 5. From Duniway (1977).

tance would place a stress on the plant so that growth and productivity would be reduced over the long term.

Nine days after inoculation of susceptible sugarbeet with *Aphanomyces cochlioides* resistance to water movement through the whole plant, as calculated from transpiration rates and $\Delta\psi_{soil\ to\ leaf}$, was ×11 higher in infected plants than in controls (Safir & Schneider, 1976). However, at this time, symptom expression in the shoots was well advanced and it is not clear from these data that increased resistance caused the symptoms of water stress.

Olsen *et al.* (1983) examined the effect of *Phymatotrichum omnivorum* on resistances to water flow through roots, lower stems and petioles of field-grown cotton plants. At the initial stages of wilting, infection did not alter resistance in petioles but increased average resistance ×24 in the lower stem and ×113 in the root. In infected plants, average root resistance was reported as ×27 the average lower-stem resistance. However, the authors noted that root resistances were even higher than the reported values since these averages did not include infinite resistances (no measurable flow through roots at a pressure of 0.1 MPa) detected in four of the root systems studied. Whether the increased resistance related to radial or longitudinal flow of water and the reason for increased resistance in root and lower stem were not determined.

Infection of *Eucalyptus sieberi* by *Phytophthora cinnamomi* led to an ordered series of events in which decreased hydraulic conductivity in roots preceded water stress in leaves (Dawson & Weste, 1984). In stage 1, lesions formed about 10 h after inoculation. In stage 2, lesions extended and hydraulic conductivity of roots was reduced to 50% of controls. In stage 3, root hydraulic conductivity was reduced to 90% of controls and water stress occurred in leaves, as shown by wilt and by reductions in transpiration rate, rwc, leaf conductance (Fig. 14.2) and ψ_{xylem}. The decline in root hydraulic conductivity began 2 to 14 d after inoculation and always preceded symptoms of water stress in the shoot.

Causes of rhizopathic water stress
Colonisation of the vascular system

Rhizopathic fungi may invade the stele, cortex, epidermis or endodermis. They show great variability in selectivity for tissues and in the amount of damage they cause (Kommedahl & Windels, 1979; Hornby & Fitt, 1981). Some colonise and damage xylem extensively, while others damage the cortex but are slow to damage the stele. Plant death caused by root-infecting fungi typically follows extensive damage to the xylem, phloem or meristematic tissues. Infections of other tissues such as the cortex, periderm or epidermis are less harmful. Most studies show that rhizopathic fungi cause water stress only if they colonise or damage the xylem. This is not surprising in view of our understanding of water transport in healthy plants and of mechanisms of hadromycotic wilt. The following examples illustrate the point.

Thielaviopsis basicola primarily causes seedling root rot and stunting in *Gossypium hirsutum* and *G. barbadense* without vascular infection and without wilt. However, it can invade the vascular tissue of the crown of *G. barbadense* in mid to late season and cause sudden wilt (Watkins, 1981). *Pythium myriotylum* typically causes root rot in peanut. The root system is reduced in size and decayed. The shoots are stunted and wilt readily during drought. Wilted plants may recover turgor and outgrow the disease if conditions are suitable. However, the fungus occasionally causes rapid wilt associated with limited root rot but extensive discoloration of the vascular system (Porter, Smith & Rodriquez-Kabana, 1984). *Phytophthora cinnamomi* caused wilt in four species where hyphae grew extensively in the xylem but did not cause wilt in two species where hyphae made limited growth in xylem (Weste & Cahill, 1982).

Rhizopathic and hadromycotic wilts may be viewed as members of a spectrum in which there is variation in the amount of cortical and vascular damage. This can be illustrated by four Fusarium wilts of potato (Hooker, 1981). All four species invade xylem and cause wilt. *Fusarium oxysporum* f. sp. *tuberosi* is a vascular wilt fungus which causes vascular discoloration and wilt but no cortical rot. In contrast, *F. eumartii*, *F. avenaceum* and *F. solani* are rhizopathic fungi and cause root rot, stem rot, vascular discoloration and wilt. Some cortical rot occurs and is more evident with *F. solani* than the other two species.

Some root rots develop slowly and may cause mild water stress but rarely wilt. *Stagonospora meliloti* invades roots and crowns of alfalfa and orange-red deposits develop in the cortex and xylem. However, there are few foliar symptoms until eventually roots become necrotic and die. Slow development of water stress appears to be related to slow development of damage in the xylem (Graham, Stuteville, Frosheiser & Erwin, 1979). Similarly, *Fusarium solani* f. sp. *phaseoli* causes root rot and slow wilting in adult peanut plants and few hyphae are observed in vessels (Porter *et al.*, 1984).

Fitt & Hornby (1978) compared the effects of seven fungi on water and solute transport in wheat. *Aureobasidium bolleyi* and *Gaeumanno-*

myces graminis var. *tritici* damaged the phloem and reduced shoot growth, shoot water content and solute uptake, while *Phialophora radicicola*, *Pythium scleroteichum* and *Wojnowicia graminis* did not penetrate root steles and did not reduce shoot growth or solute uptake. Kararah (1976) also reported that uptake of water can be reduced in wheat infected with *Gaeumannomyces graminis* var. *tritici* when the phloem is destroyed but before the xylem is extensively invaded. Ultimately, root rot and damage to phloem and xylem cause complete cessation of water and solute absorption (Manners & Myers, 1981).

Vascular invasion may not lead to wilt. For example, *Rhizoctonia solani* can invade the vascular system but wilt is not commonly reported as a symptom of infection by this fungus (Bateman, 1970; Kommedahl & Windels, 1979; Hooker, 1981). *Sclerotium bataticola* invades the vascular system of sunflower but does not cause wilt (Kommedahl & Windels, 1979).

Increased hydraulic resistance in roots or stem

Numerous studies have shown that 5-fold or greater increases in hydraulic resistance occur in roots or lower stems of infected plants and precede or coincide with symptoms of water stress in leaves (Powers, 1952, 1954; Schramm & Wolf, 1954; Duniway, 1975, 1977; Safir & Schneider, 1976; Schneider & Pendery, 1983; Olsen *et al.*, 1983; Schneider, 1984; Dawson & Weste, 1984). From Eq. (1) this increased resistance would be expected to reduce the flow of water to leaves unless compensating changes occurred in $\Delta \psi$ within the plant. The evidence indicates that the changes that do occur in leaves, such as an increase in diffusive resistance, a reduction in stomatal aperture and possibly a decrease in ψ_π, are unable to compensate for the increased resistance to flow through roots or stems. As a result, the amount of water flowing to leaves is less than the amount lost by transpiration, the water content of leaf cells declines, ψ_p (turgor) and ψ_{leaf} decline and leaves may eventually wilt. In the studies that identified increased hydraulic resistance as the origin of water stress, no evidence was found that leaves lost turgor as a result of membrane damage.

Mobile chemicals

Root-rotting fungi destroy roots through the activity of a wide range of mobile chemicals which may be classified according to their activity (e.g. enzymes, toxins) or chemical composition (e.g. proteins, polysaccharides). These chemicals break down the structure of cells and it would not be surprising, therefore, if they caused obstruction of the xylem or were transported to leaves and caused water stress by toxic effects. However, the evidence that rhizopathic water stress is caused by the production of mobile chemicals in infected plants is largely indirect.

Mobile chemicals active in the vascular system of infected plants may cause an increase in hydraulic resistance. Wolf (1933) reported that tobacco plants wilted when placed in culture medium that had supported growth of the black shank pathogen *Phytophthora parasitica* var. *nicotianae* but not when placed in fresh medium or aqueous extracts from diseased plants. In a continuation of this study, Wolf & Wolf (1954) found that leaves that had wilted in healthy sap or charcoal-treated infected sap recovered turgor when immersed in water, whereas those that had wilted in untreated infected sap did not. They suggested this was evidence that a wilt toxin was produced in infected plants. These authors and Powers (1952) recognised that mobile chemicals might cause the increased resistance to flow observed in infected tobacco plants. Duniway (1977) noted that resistance was increased in stems of safflower one to two internodes above the extent of colonisation by *Phytophthora cryptogea* and suggested this 'action at a distance' could involve mobile chemicals that caused obstruction to water flow. Cotty & Misaghi (1984) reported wilt-inducing factors in cotton infected with *Phymatotrichum omnivorum* and in cultures of the fungus.

Keen, Wang, Bartnicki-Garcia & Zentmyer (1975) reported that β-glucans from mycelium of *Phytophthora cinnamomi*, *P. palmivora* and *P. megasperma* var. *sojae* produced wilting symptoms in soybean, cacao, tomato and *Persea indica*. The effects were not host-specific. In addition, the same symptoms were produced by mycolaminarin phosphate, algal laminarin, pustulan and carboxymethyl cellulose. Woodward, Keane & Stone (1980) showed that β-glucans isolated from culture filtrates of *Phytophthora cinnamomi*, *P. cryptogea* and *P. nicotianae* caused wilting of *Eucalyptus sieberi* and *E. cypellocarpa*. Again the effects bore no relation to the pathogenicity of the fungus to the plant. Polysaccharide materials have been implicated in vascular wilts caused by fungi (Keen, Long & Erwin, 1972; Nachmias, Buchner & Krikun, 1982). They may cause obstruction to flow in the xylem or (see Jones, Ch. 13) may increase viscosity of xylem sap. Enzymes and other chemicals produced by the pathogen or the plant may also cause vascular obstructions. The literature on vascular wilts cannot be reviewed here but contains a wealth of material useful to studies on the mechanism of rhizopathic wilts.

Toxins are important in many rhizopathic diseases (Kommedahl & Windels, 1979). Examples include oxalic acid produced by *Aspergillus niger* in peanut and by *Penicillium oxalicum* in maize, and the host-specific toxins of *Helminthosporium victoriae* and *Periconia circinata*. These toxins produce changes in membrane permeability that are commonly followed by death of root cells, reduced transpiration and foliar wilt. Keen *et al.* (1975) and Woodward *et al.* (1980) suggested that the β-glucans from *Phytophthora* species might cause wilt by a direct toxic action on leaf cells. Fitt & Hornby (1978) noted that water content and growth of shoots were reduced one week after wheat was planted in sand infested with *Cochliobolus sativus* or *Fusarium culmorum*, that is several weeks before root steles were colonised. Culture filtrates of *C. sativus* reduced plant growth. Leaf cell membranes could be affected by toxins from rhizopathic fungi. As discussed previously, relationships between parameters of water status such as ψ_{leaf}, ψ_π and rwc indicate such effects do not occur in safflower infected with *Phytophthora cryptogea*, in cotton infected with *Phymatotrichum omnivorum*, in sugarbeet infected with *Aphanomyces cochlioides* or in corn infected with *Fusarium moniliforme*.

Reduction in size of the root system

Larger root systems give plants access to greater amounts of soil moisture (Passioura, 1983). Therefore a reduction in the size of the root system is likely to hasten plant water stress as the soil dries by reducing the surface area available for water uptake. In addition, reduction in the rate of growth of roots may lead to more rapid drying of the soil around the root, resulting in decreased hydraulic conductance of the soil and lower ψ_{soil}. Both these factors would restrict water uptake by the plant. Sterne *et al.* (1978) suggested that *Phytophthora cinnamomi* caused wilt in avocado by these mechanisms.

Several studies show that water stress can occur where there is limited infection of roots or reduction of root mass. Olsen *et al.* (1983) found no differences in root dry weights of healthy and *Phymatotrichum omnivorum*-infected cotton plants when infected plants began to wilt. Similarly, on the first day after inoculation of safflower with *Phytophthora cryptogea*, healthy and infected plants had the same fresh weight of roots but infection decreased flow rates through roots by about 30% (Duniway, 1977). Moreover, fresh weight of roots on infected plants did not change between one and seven days after inoculation but in this time wilt developed and flow of water through roots declined from 2 to 0.2 μl s^{-1}. In *Eucalyptus sieberi* infected with *Phytophthora cinna-*

momi water transport through roots was reduced by 91% when less than one-sixth of the root system was infected (Dawson & Weste, 1984).

These studies do not eliminate the possibility that extensive decay of roots contributes to water stress. Whether loss of roots through decay causes water stress may depend on the effect of root rot on the root/shoot ratio and on the conditions for plant growth. If the ratio is lower than normal, water stress is more likely to develop, especially under hot, dry conditions. However, if the plant has adapted to root rot so that the root/shoot ratio is not changed appreciably then a small root system *per se* is less likely to cause wilt.

Hormonal changes

Cytokinins are believed to be produced in roots and have been reported to increase transpiration and stomatal aperture of graminaceous plants (Blackman & Davies, 1983). Therefore a reduction in root mass could reduce the flow of cytokinins to leaves with attendant reduction in transpiration. Increased resistance to water flow in the xylem in which they are transported (Bidwell, 1979) could reduce the supply of cytokinins to leaves. Cytokinin levels in xylem exudate and leaves may be reduced by water stress in uninfected plants (Itai & Vaadia, 1971). Levels of cytokinin activity may also be reduced by infection with vascular wilt fungi (Misaghi *et al.*, 1972; Patrick, Hall & Fletcher, 1977) but only after the initial symptoms of wilt. It seems unlikely that the rhizopathic water stresses examined to date are caused by alterations in cytokinin supply to leaves. Any changes that may occur in cytokinin levels in leaves or xylem fluid are more likely to be the result than the cause of pathogen-induced water stress in leaves. Nevertheless, observed reductions in cytokinin activity would have a profound effect on the metabolic activity of leaves and could contribute to their senescence and death.

Altered metabolism in roots

Kramer (1933) noted that water was absorbed more readily through dead than through living roots and proposed that living cells in roots resist water flow. Therefore, disruption of root cell membranes by pathogens would be expected to reduce rather than increase radial resistance to water flow. (It is unlikely that membrane damage would cause water to leak from roots into soil. It is more likely that transpirational pull would override any effects of altered permeability or the possibility that the ψ_π of cells was higher than ψ_{soil}). Rhizopathic fungi could reduce water absorption through a variety of effects on living cells in roots. Kramer (1983) refers to a number of phenomena in healthy roots that indicate how root infection might decrease water uptake. For example, root pathogens may reduce water absorption by reducing the rate of root elongation. The reason for this is that maturation of endodermal tissues occurs much nearer the tip in slowly elongating roots; thus the region of low resistance to water uptake is shortened.

Summary of likely mechanisms of rhizopathic water stress

Present evidence indicates that increased hydraulic resistance in roots and lower stem plays a major role in the development of water stress, although some rhizopathic diseases (e.g. Victoria blight of oats) may cause water stress by impairing osmotic function in leaf cells, and loss of roots and death or reduced metabolic function may contribute to stress. A common sequence of events in rhizopathic diseases characterised by plant water stress may be as follows. Hyphae penetrate radially through the cortex of roots and enter the xylem. Damage to root cells is indicated by an increase in loss of electrolytes. Death of cortical and possibly other cells of the root leads to visible root rot symptoms. Reduction in root mass by infection is not sufficient to cause the observed reduction in water flow through the plant. Hyphae proliferate in the xylem and cause discoloration of vascular tissue in roots and possibly in stems. Resistance to water movement through roots and sometimes through the stem is increased. This reduces the water supply to the leaves, leading first to reductions in ψ and rwc, and then to reduced stomatal conductance, stomatal aperture and transpiration. If these changes persist for prolonged periods they cause metabolic alterations that lead to reduced growth and productivity. Water relations in water-stressed leaves of infected plants are comparable to those in uninfected plants deprived of water. Osmotic mechanisms in leaf cells are unimpaired at this stage and water stress in infected plants is not caused by damage to leaf cell membranes. When the resistance in roots and lower stem becomes five or more times the resistance in healthy plants, leaves wilt. This may happen within days or years after infection. Wilt may happen suddenly or develop slowly, or may not occur at all even though leaves are water-stressed. Severe water stress is followed by debilitation or death of leaves or entire plants.

Significance of root pathogens to plant water stress

Root diseases are numerous and widespread. The US Department of Agriculture Handbook 165 (Anonymous, 1960) lists 3300 root diseases caused by 235 species in 81 genera of fungi. Of this number of diseases, 64% are root rots and 7% are vascular wilts (Kommedahl & Windels, 1979). On a world scale the contributions of root-rotting fungi to plant water stress must therefore be substantial. Many of these diseases characteristically cause wilting and it is likely that many additional root diseases place a water stress on plants without producing wilt. The magnitude and agricultural and ecological significance of this water stress are not known. A mild water stress imposed by a pathogen may mimic an environmental water stress and stimulate osmotic adjustments in the plant (Morgan, 1984) or increase the tolerance of the plant to drought or extremes of temperature (Grierson, Soule & Kawada, 1982). If infection occurs early in the season a reduction in the size of the root system and adjustments in the shoot/root ratio and ψ_π of leaves may prevent or delay water stress later in the season. If infection occurs late in the season the plant may not be able to adjust and may show water stress. In general, however, the continuous irritation by the pathogen, whenever infection occurs, is more likely to be harmful than beneficial.

The large capacity of plants to produce roots would equip them to replace roots destroyed or inactivated by pathogens. However, the existence of an apparently excessive capacity for root growth suggests that there are significant environmental constraints on root survival. Root pathogens may be important among these constraints. The possible importance of minor root diseases (Salt, 1979), and of feeder root necrosis (Wilhelm, 1965), to plant water status deserves more study. The general ability of root pathogens to restrict growth by reducing absorption of water and minerals is reflected in the suggestion by Bald (1969) that root diseases could be measured by their effects on leaf area. Reducing root diseases through resistance or other means will undoubtedly make a significant contribution to reducing plant water stress.

References

Anonymous. (1960). *Index of Plant Diseases in the United States.* Agricultural Handbook United States Department of Agriculture Number 165, 531 pp.

Asher, M. J. C. (1972). Effect of *Ophiobolus graminis* infection on growth of wheat and barley. *Annals of Applied Biology*, **70**, 215–23.

Ayres, P. G. (1978). Water relations of diseased plants. In *Water Deficits and Plant Growth*, vol. 5, *Water and Plant Disease*, ed. T. T. Kozlowski, pp. 1–60. New York: Academic Press.

Ayres, P. G. (1981). Effects of disease on plant water relations. In *Effects of Disease*

on the Physiology of the Growing Plant, ed. P. G. Ayres, pp. 131–48.
Cambridge: Cambridge University Press.

Bald, J. G. (1969). Estimation of leaf area and lesion sizes for studies on soil-borne pathogens. *Phytopathology*, **59**, 1606–12.

Bateman, D. F. (1970). Pathogenesis and disease. In *Rhizoctonia solani, Biology and Pathology*, ed. J. R. Parmeter, pp. 161–71. Berkeley: University of California Press.

Bidwell, R. G. S. (1979). *Plant Physiology*, 2nd edn. New York: Macmillan.

Blackman, P. G. & Davies, W. J. (1983). The effects of cytokinins and ABA on stomatal behaviour in maize and *Commelina. Journal of Experimental Botany*, **34**, 1619–26.

Cotty, P. J. & Misaghi, I. J. (1984). Wilt-inducing factors from cotton infected with *Phymatotrichum omnivorum* (Shear) Dugger. *Phytopathology*, **74**, 817 (abstr.).

Dawson, P. & Weste, G. (1982). Changes in water relations associated with infection by *Phytophthora cinnamomi. Australian Journal of Botany*, **30**, 393–400.

Dawson, P. & Weste, G. (1984). Impact of root infection by *Phytophthora cinnamomi* on the water relations of two *Eucalyptus* species that differ in susceptibility. *Phytopathology*, **74**, 486–90.

Dell, B. & Wallace, I. M. (1981). Recovery of *Phytophthora cinnamomi* from naturally infected jarrah roots. *Australasian Plant Pathology*, **10**, 1–2.

De Roo, H. C. (1969). Sap stress and water uptake in detached shoots of wilt-diseased and normal rhododendrons. *HortScience*, **4**, 51–2.

Duniway, J. M. (1973). Pathogen-induced changes in host water relations. *Phytopathology*, **63**, 458–66.

Duniway, J. M. (1975). Water relations in safflower during wilting induced by Phytophthora root rot. *Phytopathology*, **65**, 886–91.

Duniway, J. M. (1977). Changes in resistance to water transport in safflower during the development of Phytophthora root rot. *Phytopathology*, **67**, 331–7.

Fitt, B. D. L. & Hornby, D. (1978). Effects of root-infecting fungi on wheat transport processes and growth. *Physiological Plant Pathology*, **13**, 335–46.

Graham, J. H., Stuteville, D. L., Frosheiser, F. I. & Erwin, D. C. (1979). *A Compendium of Alfalfa Diseases*. St Paul: The American Phytopathological Society.

Grierson, W., Soule, J. & Kawada, K. (1982). Beneficial aspects of physiological stress. *Horticultural Reviews*, **4**, 247–71.

Hagedorn, D. J., ed. (1984). *Compendium of Pea Diseases*. St Paul: The American Phytopathological Society.

Hall, R. & MacHardy, W. E. (1981). Water relations. In *Fungal Wilt Diseases of Plants*, ed. M. E. Mace, A. A. Bell & C. H. Beckman, pp. 255–98. New York: Academic Press.

Hooker, W. J., ed. (1981). *Compendium of Potato Diseases*. St Paul: The American Phytopathological Society.

Hornby, D. & Fitt, B. D. L. (1981). Effects of root-infecting fungi on structure and function of cereal roots. In *Effects of Disease on the Physiology of the Growing Plant*, ed. P. G. Ayres, pp. 101–30. Cambridge: Cambridge University Press.

Itai, C. & Vaadia, Y. (1971). Cytokinin activity in water-stressed shoots. *Plant Physiology*, **47**, 87–90.

Kararah, M. A. (1976). Host–parasite relationships in the take-all disease of wheat caused by *Gaeumannomyces graminis* var. *tritici*. Ph.D. Thesis, University of Southampton, UK.

Keen, N. T., Wang, M. C., Bartnicki-Garcia, S. & Zentmyer, G. A. (1975). Phytotoxicity of mycolaminarins – β-1,3-glucans from *Phytophthora* spp. *Physiological Plant Pathology*, **7**, 91–7.

Keen, N. T., Long, M. & Erwin, D. C. (1972). Possible involvement of a pathogen-produced protein–lipopolysaccharide complex in *Verticillium* wilt of cotton. *Physiological Plant Pathology*, **2**, 317–31.

Kommedahl, T. & Windels, C. E. (1979). Fungi: pathogen or host dominance in disease. In *Ecology of Root Pathogens*, ed. S. V. Krupa & Y. R. Dommergues, pp. 1–103. Oxford: Elsevier Scientific Publishing Company.

Kramer, P. J. (1933). The intake of water through dead root systems and its relation to the problem of absorption by transpiring plants. *American Journal of Botany*, **20**, 481–92.

Kramer, P. J. (1983). *Water Relations of Plants*. New York: Academic Press.

Manners, J. G. & Myers, A. (1981). Effects on host growth and physiology. In *Biology and Control of Take-all*, ed. M. J. C. Asher & P. J. Shipton, pp. 237–48. New York: Academic Press.

Misaghi, I., DeVay, J. E. & Kosuge, T. (1972). Changes in cytokinin activity associated with the development of *Verticillium* wilt and water stress in cotton plants. *Physiological Plant Pathology*, **2**, 187–96.

Morgan, J. M. (1984). Osmoregulation and water stress in higher plants. *Annual Review of Plant Physiology*, **35**, 299–319.

Nachmias, A., Buchner, V. & Krikun, J. (1982). Comparison of protein–lipopolysaccharide complexes produced by pathogenic and non-pathogenic strains of *Verticillium dahliae* Kleb. from potato. *Physiological Plant Pathology*, **20**, 213–21.

Olsen, M. W., Misaghi, I. J., Goldstein, D. & Hine, R. B. (1983). Water relations in cotton plants infected with *Phymatotrichum*. *Phytopathology*, **73**, 213–16.

Passioura, J. B. (1983). Roots and drought resistance. *Agricultural Water Management*, **7**, 265–80.

Patrick, T. W., Hall, R. & Fletcher, R. A. (1977). Cytokinin levels in healthy and *Verticillium*-infected tomato plants. *Canadian Journal of Botany*, **55**, 377–82.

Porter, D. M., Smith, D. H. & Rodriquez-Kabana, R. (1984). *Compendium of Peanut Diseases*. St Paul: The American Phytopathological Society.

Powers, H. R. Jr (1952). Water movement in tobacco plants affected by black shank. *Plant Disease Reporter*, **36**, 127.

Powers, H. R. Jr (1954). The mechanism of wilting in tobacco plants affected by black shank. *Phytopathology*, **44**, 513–21.

Safir, G. R. & Schneider, C. L. (1976). Diffusive resistances of two sugarbeet cultivars in relation to their black root disease reaction. *Phytopathology*, **66**, 277–80.

Salt, G. A. (1979). The increasing interest in 'minor pathogens'. In *Soil-Borne Pathogens*, ed. B. Schippers & W. Gams, pp. 289–312. New York: Academic Press.

Schneider, R. W. (1984). Transient changes in hydraulic resistance caused in corn roots by *Fusarium moniliforme*. *Phytopathology*, **74**, 1230–3.

Schneider, R. W. & Pendery, W. E. (1983). Stalk rot of corn: mechanism of predisposition by an early season water stress. *Phytopathology*, **73**, 863–71.

Schramm, R. J. & Wolf, F. T. (1954). The transpiration of black shank-infected tobacco. *Journal of the Elisha Mitchell Scientific Society*, **70**, 255–61.

Shea. S. R., Shearer, B. & Tippett, J. (1982). Recovery of *Phytophthora cinnamomi* Rands from vertical roots of jarrah (*Eucalyptus marginata* Sm.). *Australasian Plant Pathology*, **11**, 25–8.

Sterne, R. E., Kaufmann, M. R. & Zentmyer, G. A. (1978). Effect of Phytophthora root rot on water relations of avocado: interpretation with a water transport model. *Phytopathology*, **68**, 595–602.

van den Honert, T. H. (1948). Water transport as a catenary process. *Discussions of the Faraday Society*, **3**, 146–53.

Van der Molen, G. E., Beckman, C. H. & Rodehorst, E. (1977). Vascular gelation: a general response phenomenon following infection. *Physiological Plant Pathology*, **11**, 95–100.

Watkins, G. M., ed. (1981). *Compendium of Cotton Diseases*. St Paul: The American Phytopathological Society.

Weste, G. (1980). Vegetation changes as a result of invasion of forest on krasnozem by *Phytophthora cinnamomi*. *Australian Journal of Botany*, **28**, 139–50.

Weste, G. & Cahill, D. (1982). Changes in root tissue associated with infection by *Phytophthora cinnamomi*. *Phytopathologische Zeitschrift*, **103**, 97–108.

Wilhelm, S. (1965). *Pythium ultimum* and the soil fumigation growth response. *Phytopathology*, **55**, 1016–20.

Wolf, F. T. (1933). The pathology of tobacco black shank. *Phytopathology*, **23**, 605–12.

Wolf, F. T. & Wolf, F. A. (1954). Toxicity as a factor in tobacco black shank. *Journal of the Elisha Mitchell Scientific Society*, **70**, 244–55.

Woodward, J. R., Keane, P. J. & Stone, B. A. (1980). Structures and properties of wilt-inducing polysaccharides from *Phytophthora* species. *Physiological Plant Pathology*, **16**, 439–54.

15
Foliar pathogens alter the water relations of their hosts with consequences for both host and pathogen

P. G. AYRES and N. D. PAUL

Department of Biological Sciences, University of Lancaster, Bailrigg, Lancaster LA1 4YQ, UK

The gradient of water potential ($\Delta\psi$) between soil and leaves is rarely as great as 3.0 MPa and is often considerably less; $\Delta\psi$ between mesophyll cells and the atmosphere is often 90 MPa (air of 50% rh at 20 °C). Thus, the greatest force driving water movement through the plant occurs at the leaf–air interface. The difference of potential energy causes water to evaporate from cell surfaces and there is then a net diffusional flux of water vapour away from the leaf, primarily via stomatal pores. Plants manage to keep their water relations in balance if, over a period of one or a few daily cycles, they are able to take up from the soil approximately the same amount of water that they lose through evapotranspiration. In most circumstances they are enabled to do this because there is a great resistance to water movement at the leaf surface. This resistance (r_{cut}) resides in the hydrophobic cuticle which covers the outer surface of epidermal cells of leaves and also the stem in herbaceous plants (the stems of woody plants have outer cells wholly impregnated with hydrophobic materials) (Table 15.1). It comprises the largest component of total leaf resistance (r_{leaf}), which itself is equal to approximately half the sum of those resistances to flow of liquid water in the plant described by Jones in Chapter 13. The cuticle is normally broken only by the stomatal pores that allow diffusion of CO_2 to sites of photosynthesis inside the leaf. These pores are necessary because the cuticle is a barrier to movement of CO_2 as well as H_2O. However, by intricately regulated movements of guard cells surrounding the stomatal pores, the diameters of these pathways can be reduced to such an extent that the stomatal

Table 15.1. *Representative values of resistance to the diffusion of water vapour (after Nobel, 1983)*

	Resistance, $s\,cm^{-1}$
Boundary layer	0.13–1.3
Stomata open	
mesophytes	0.31–6.0
trees	10.0
Cuticle	
mesophytes	25–100
trees	50–200
Intercellular air spaces	0.1–0.5

resistance (r_{st}) to the escape of water vapour becomes almost as high as that offered in parallel by the cuticle.

Driving forces and resistances are crucial here because as foliar pathogens, and pathogens that invade other aerial organs, enter the plant they either breach the cuticle or pass through stomatal pores, in either case preventing normal function of the latter. Later, they may produce spore-bearing structures that erupt from the leaf, tearing open the epidermis and reducing the leaf's resistance to water loss (Owera, Farrar & Whitbread, 1981). R_{leaf} includes component resistances attributable to the boundary layer (r_a) and to intercellular air spaces (r_{int}).

$$R_{leaf} = r_a + r_{int} + \frac{r_{st}r_{cut}}{r_{st} + r_{cut}}$$

It seems probable that sporulating mycelium on the surface of a leaf may alter r_a and that r_{int} may be altered by intercellular mycelium which builds into densely aggregated structures, such as aeciosori of rusts, but no measurements are reported in the literature. Ignorance of such factors is regrettable and can be only partly justified by the observation that r_a and r_{int} make only small and insignificant contributions to R_{leaf} in healthy plants.

In each stage of the infection cycle foliar pathogens alter resistances to the escape of water so potentially upsetting the balance in the plant between water uptake and transpiration. The degree of disturbance will depend *inter alia* on the percentage of shoot area affected and the position of the infection site(s). As an example of the first, Gordon & Duniway

(1982*a*) found that steeply declining ψ_{soil} caused only a slow decline in ψ_{leaf} of sugar beet infected by powdery mildew (*Erysiphe polygoni*) when only one leaf was infected, but a very rapid decline when a majority of the seven or eight leaves on the plant were infected. As an example of the second, P. Tissera (unpublished results from this laboratory) has found that ψ of uninfected distal halves of leaves of faba bean (*Vicia faba*) infected in the proximal half with rust (*Uromyces viciae-fabae*) is lower than that of the distal halves of completely uninfected leaves. It may be expected that tissues which, as in the last example, are 'downstream' to an infection site, i.e. whose connection with the roots is interrupted by the infection site, will be most susceptible to a change in water relations. Perhaps the most important tissues in the plant in this context are those where active cell division and expansion are occurring. In mesophytes, including most crop species, lowering ψ by as little as 0.1 to 0.2 MPa can inhibit cell growth and leaf expansion (Bradford & Hsaio, 1982).

Water relations of infected plants

It is arguable that turgor potential (ψ_p) is more critical than ψ for cell expansion and some other water relations-related processes (Bradford & Hsaio, 1982). The following discussion centres on ψ, however, because ψ_p has not often been quantified in studies of diseased plants, in spite of the fact that it is partly governed by the osmotic potential (ψ_π) of the tissue and the latter can be affected by foliar pathogens in several ways. Both necrotrophic and biotrophic pathogens inhibit photosynthesis at the site of infection (Buchanan, Hutcheson, Magyarosy & Montalbini, 1982), so they may reduce the amount of organic solutes *per se* and also the supply of energy for the active accumulation of inorganic and organic solutes by cells. (Obligate biotrophs also inhibit the export of photoassimilates out of infected leaves: Whipps & Lewis, 1982.) What is particularly significant where ψ_π and ψ_p are concerned is that while biotrophs typically cause no gross impairment of host membrane permeability, e.g. *Puccinia hordei* (brown rust of barley) (Ahmad, Farrar & Whitbread, 1985), necrotrophic pathogens often impair semipermeability, sometimes at a distance in advance of the mycelium, causing a loss of solutes from host cells. The contrasting patterns of behaviour are discussed separately.

Our knowledge of the water relations of plants infected by foliar pathogens comes almost entirely from rust and powdery mildew infections and, thus, falls in the first category below.

Table 15.2. *Dimensions of (a) open sori caused by some representative rust fungi and (b) maximum stomatal pores in species known to display wide opening*

Plant	Rust	Dimensions of sorus, mm
(a) Faba bean (*Vicia faba*)	*Uromyces viciae-fabae* uredinium	0.5–1.1 long × 0.5–0.6 wide
Birch	*Melampsoridium betulinum*[a] uredinium	0.5 × 0.5, but with a peridium
Groundsel	*Puccinia lagenophorae* aecium	0.2–0.3, approx. circular (excluding peridium)
Leek	*Puccinia allii* uredinium	0.6 × 0.2
Wheat	*Puccinia recondita*[a] uredinium	1–2, approx. circular
Wheat	*Puccinia striiformis*[a] uredinium	0.5–1.0 × 0.3–0.4
		Dimensions of stomatal pore, μm
(b) Faba bean (*Vicia faba*)		14 wide × 25 long
Commelina communis		30 wide × 30 long

[a] From Wilson & Henderson (1966).

Minimal damage to host cell membranes

Although both rusts and mildew inhibit stomatal opening in their hosts (Ayres, 1981a), rusts pose the more serious threat to their host because during sporulation they rupture the plant's epidermis in order to release spores. The openings they cause in the leaf surface are enormous compared with the widest stomatal pore (Table 15.2), and increased transpiration results (see table 1 in Ayres, 1978, for full references). Powdery mildews grow mainly on the surface of their hosts, that is, in a comparatively dry environment. Their mycelium appears to have a relatively high resistance to loss of water vapour (Gordon & Duniway, 1982b). Also, a hydrophobic neckband deposited in the fine hypha connecting superficial mycelium with intracellular haustorium (Bushnell & Gay, 1978) probably stops water escaping from the leaf via an apoplastic (cell wall) pathway of potentially low resistance.

Our investigations of a small annual weed, groundsel (*Senecio vulgaris*), infected by *Puccinia lagenophorae* showed that the increased transpiration following rupture of aecia was sufficient to reduce ψ in infected leaves of well watered plants (Paul & Ayres, 1984). When groundsel was subjected to progressive drought stress, caused by withholding water from 2 to 10 d after inoculation, ψ fell similarly in healthy and rusted plants but, when watering recommenced on day 10 to restore further daily losses, ψ of healthy plants stopped falling while that of rusted plants continued downwards and leaves wilted (Fig. 15.1). Depression of ψ was associated with a lowered rate of expansion of leaf area (Fig. 15.2a) and a reduction in the rate of net photosynthesis per plant (Fig. 2b).

Mild water stress typically causes partial stomatal closure in healthy plants. This restricts diffusion of water vapour out of the leaf more than the diffusion of CO_2 into the leaf, so increasing water-use-efficiency (W.U.E. = CO_2 fixed per unit of water transpired). In our experiments, healthy groundsel showed the expected increase in W.U.E. in response to the cessation of watering, but rust-infected groundsel was unable to respond in the same way (Fig. 15.3).

The experiments with groundsel lasted only 15 d after inoculation so there were only small differences between the final sizes of healthy and rusted plants. Differences in growth between healthy and infected plants in general become more important as the length of any particular study

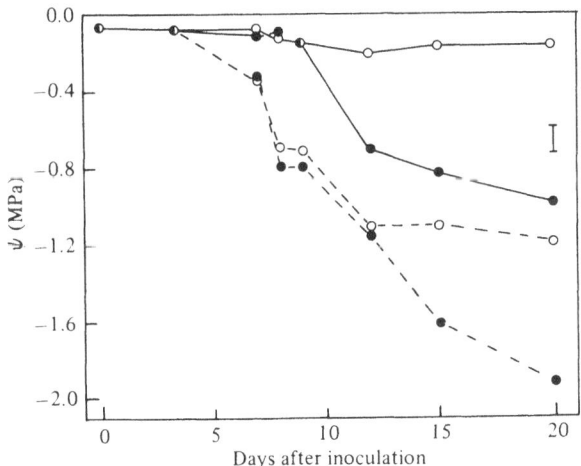

Fig. 15.1. ψ of groundsel (*Senecio vulgaris*) either well watered daily (———) or deprived of water from 2 to 10 d after inoculation and then watered only to replace daily transpirational losses (-----). Plants infected by rust (*Puccinia lagenophorae*), closed symbols; Healthy controls, open symbols. Bar indicates L.S.D. at $P = 0.05$. (From Paul & Ayres, 1984.)

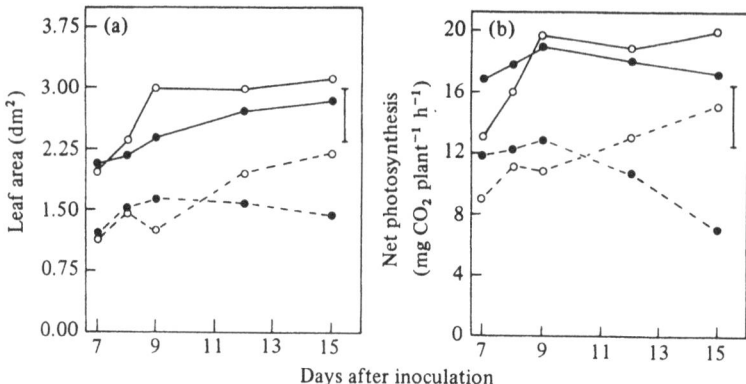

Fig. 15.2. (a) Leaf area and (b) net photosynthesis per plant in groundsel (*Senecio vulgaris*), healthy (open symbols) or infected by rust (*Puccinia lagenophorae*) (closed symbols). Watering regime and symbols as in Fig. 15.1. Net photosynthesis measured by infrared gas analysis. (From Paul & Ayres, 1984.)

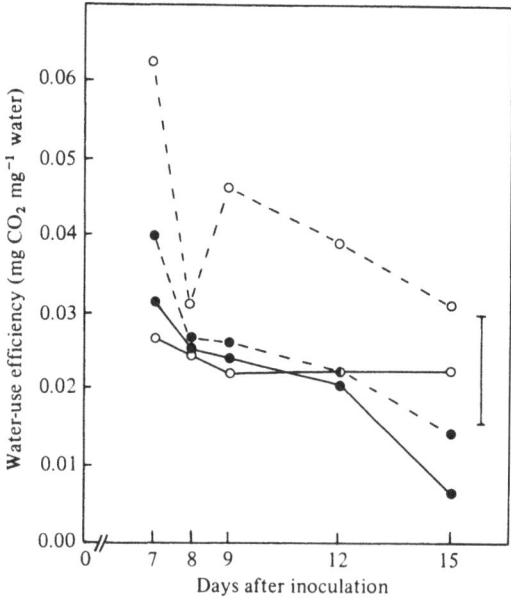

Fig. 15.3. Water-use efficiency of groundsel (*Senecio vulgaris*), healthy (open symbols) or infected by rust (*Puccinia lagenophorae*) (closed symbols). Watering regime and symbols as in Fig. 15.1. (From Paul & Ayres, 1984.)

increases. Thus, whereas transpiration per unit area of leaf area or dry matter may be increased by infection, transpiration per plant will probably be reduced. The earlier that infection occurs, the more the total requirement for water will be reduced. This last proposition was well illustrated half a century ago by Murphy (1935) who studied the water consumption and yield of oats grown at either 85% or 50% soil moisture content and inoculated with crown rust (*Puccinia coronata*) at different stages of development (Fig. 15.4). Rust infection made the plants less efficient in the use of water, i.e. raised the value of water used divided by total yield, just as it reduced W.U.E. in groundsel.

When referring to infections in the field, Murphy (1935) made the important observations that 'heavy epiphytotics of crown rust are usually preceded by a period of wet weather with attendant high humidity and soil moisture. These conditions, while highly favourable to the rust, also favour top growth and root development suffers. Then as rust appears and increases, the ratio of roots to tops is still further decreased by the effect of the fungus'. It should be remembered that (a) foliar

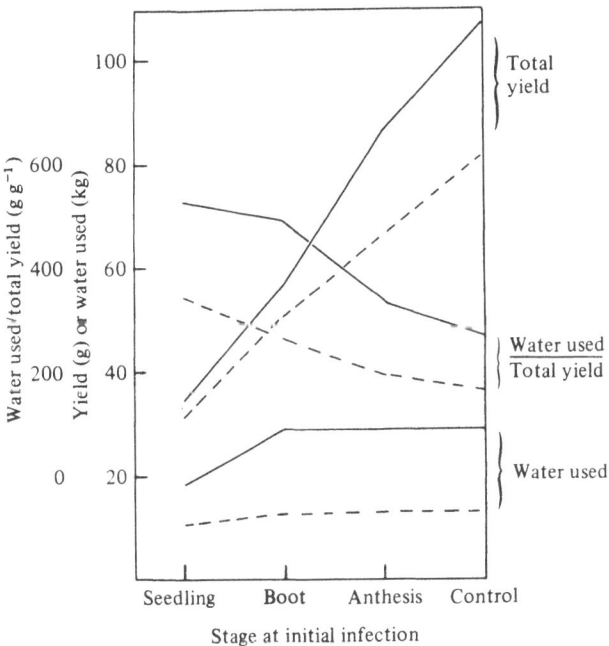

Fig. 15.4. Yield and water-usage of oats (*Avena sativa*) grown at 85% (———) or 50% (-----) soil moisture content and infected with crown rust (*Puccinia coronata*) at seedling, boot or anthesis stage. (Modified from Murphy, 1935.)

pathogens are most likely to cause water stress in the field when dry conditions follow those very specific moist or wet conditions that most pathogens require while in transit between hosts (see Chs 4, 5, 8), and (b) the water relations of infected plants are critically affected by root growth in relation to the availability of soil water (see Ch. 14). (Relationships between root growth and water and nutrient fluxes in infected plants are described by Ayres, 1985.)

Foliar infections commonly reduce the absolute size (dry weight) of the root system (Ayres, 1985). In our studies of groundsel, we have found that rust infection increases the ratio of leaf area to unit dry weight of root, i.e. each unit of root has to absorb and transport more water than controls in order to maintain the same supply of water to the leaves. However, we do not yet know what effects *P. lagenophorae*, or other rusts, have on the morphology of their hosts' root system. Experience of powdery mildew diseases suggests that effects will probably depend on the type of plant in question, in particular whether it is dicotyledonous

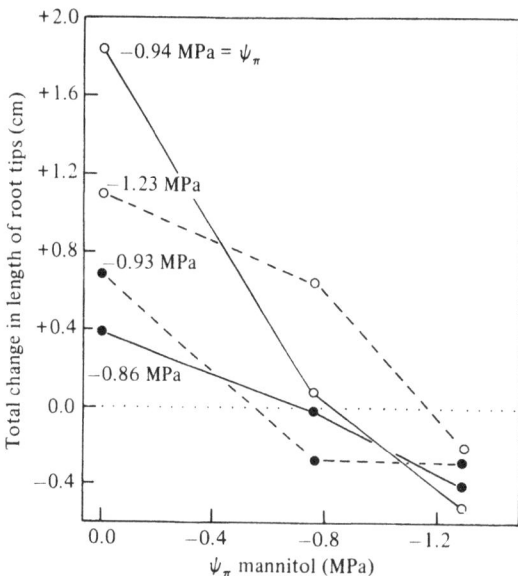

Fig. 15.5. Total extension of 12 root tips of pea (*Pisum sativum*) after 6 h in mannitol solutions of different ψ_π. Plants were healthy (open symbols), or infected by powdery mildew (*Erysiphe pisi*) (closed symbols). Root tips, 0.5 cm long, were excised from peas grown in Hoagland's solution (−0.05 MPa) continuously (———) or stressed by transfer to Hoagland's solution plus polyethylene glycol 4000 (−0.30 MPa) 4 d after inoculation (-----). (From Ayres, 1981*b*.)

with a dominant, leading tap root or monocotyledonous with several fibrous roots of similar length and hydraulic conductivity.

Extrapolation from mildewed to rusted plants may be justifiable where roots are concerned, since both infections reduce photosynthesis and the supply of assimilates to the roots (Whipps & Lewis, 1982). Both may interfere with that continued growth of roots into deep moist soil which is a vital part of the plant's response to water stress. When peas were grown and infected in solution culture, mildew (*Erysiphe pisi*) inhibited solute accumulation in root tips with the result that, when the tissues were put in media of low ψ_π, turgor was not maintained and extension growth stopped (Ayres, 1981*b*) (Fig. 15.5). Thus, when peas were grown in 60-cm-long tubes, packed with soil that was allowed to dry from the surface downwards, roots of mildewed plants showed no growth (they lost dry weight) in the upper, drier regions of the soil profile. In contrast, growth of the leading tap root into the lowest, wettest region of the profile was not inhibited by mildew (Fig. 15.6).

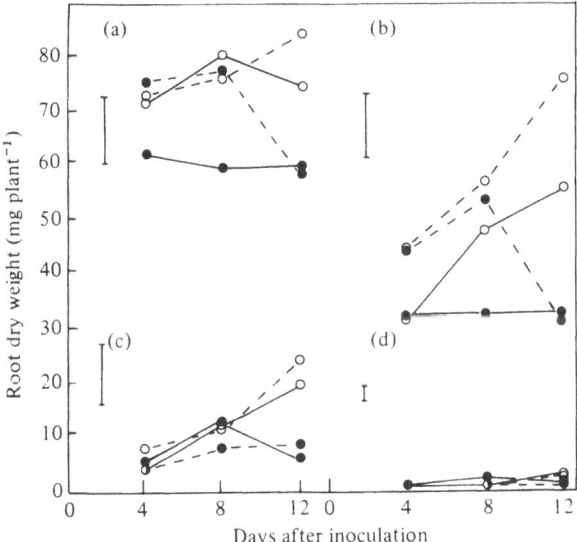

Fig. 15.6a–d. Root growth of pea (*Pisum sativum*) at different depths of soil after inoculation with powdery mildew (*Erysiphe pisi*). Soil was watered every day (———) or after inoculation on day 4 and day 8 only (-----). Healthy, open symbols; mildewed, closed symbols. Bars indicate least significant differences at $P = 0.05$. Soil depth: (a) 0–15 cm; (b) 15–30 cm; (c) 30–45 cm; (d) 45–60 cm. ψ_{soil} in unwatered soil, approx. −0.6; −0.3; −0.05; < −0.05 MPa respectively. (From Ayres, 1981*b*.)

Extrapolation from mildewed to rusted plants must be made with great caution where shoot–water relations are concerned because the demands of infected plants for water may be so much less in the former group. For example, in mildewed pea, transpiration in the light, the period when most water is lost from the plant, is reduced under all conditions except those of severe water stress because stomata become immobilised in an almost-closed position (Ayres, 1981*a*). In some plants, such as pea, the demand for water may be further reduced under conditions of severe stress because infected leaves desiccate prematurely and die. As a result of such changes, ψ of shoot apices of mildewed pea (Ayres, 1981*b*) or of healthy, upper leaves of mildewed barley (Williams & Ayres, 1981) may be higher than in comparable tissues of healthy plants under similar water stress.

Plant water deficits can arise not only as a result of summer drought, they can arise also in winter when soil water is frozen and unavailable to plants whose shoots are exposed during the day to temperatures above zero. Such conditions are frequently encountered by evergreen perennials, autumn-sown crops and annuals, such as groundsel, that overwinter vegetatively. Little is known about the impact of pathogens under winter conditions but we have found rust-infected groundsel plants to be more vulnerable to damage than are uninfected controls.

We made an experimental study of the water relations of healthy and rust-infected groundsel kept with roots in soil at $-6\,°C$, $-2\,°C$ or ambient temperature in an unheated, partly glazed glasshouse over winter; in a separate experiment we examined the recovery of plants put in a controlled-environment room after a period at $-2\,°C$. Reductions of ψ and ψ_p, which were greater in rusted than in healthy plants, occurred at soil temperatures of $-2\,°C$ and $-6\,°C$ (Paul & Ayres, 1985). Rusted plants in particular showed some wilting. When recovery after a period of cooling was studied, ψ rose in rusted plants but not as much as it did in healthy controls (Fig. 15.7). Turgor returned to pre-freezing values in rusted plants only because ψ_π remained lower than in controls. Rusted plants were slower than healthy plants to commence growth during the recovery period; in particular, leaf area did not increase as rapidly in rusted as in healthy plants (Fig. 15.8). It is not known why rusted plants were unable to remobilise solutes after low soil temperatures were relieved but it is suggested that this failure, and (or) the lowered ψ, were major factors contributing to the slower expansion of leaf area and slower increase in dry weight in rusted than in healthy plants after thawing of the soil.

Where infection initiates water stress, or exacerbates stress arising from either summer drought or soil freezing, that stress may be a contributory cause of delayed leaf expansion, as in the above case. This is very harmful for the continued growth of the plant since it has been clearly shown for a number of arable crops that dry matter production is approximately linearly related to intercepted solar radiation, which itself depends upon leaf area (Milford *et al.*, 1980; Monteith, 1981). Small, diseased plants will be at a progressive disadvantage both above and below ground as they compete with healthy neighbours.

Severe damage to membranes

It was stated earlier that fungal pathogens can drastically affect host membranes, so altering solute and water relations. The causes of such alterations are several and often not clearly defined. Thus, injury

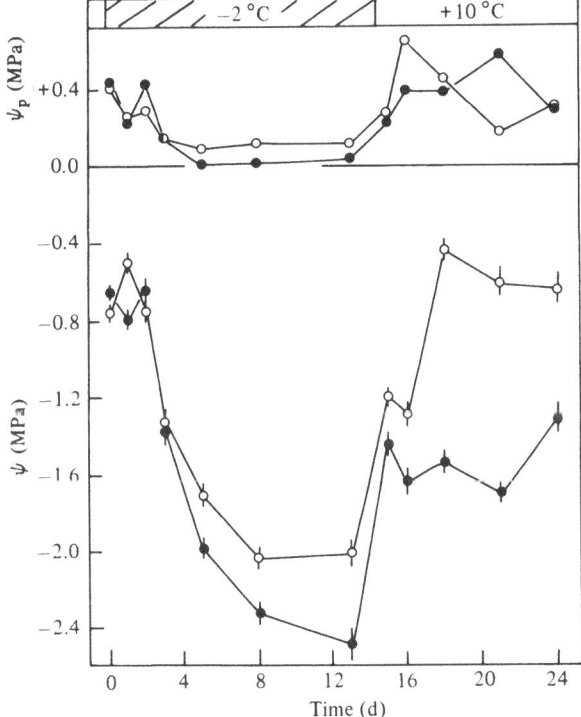

Fig. 15.7. ψ and ψ_p in leaves of groundsel (*Senecio vulgaris*), healthy (open symbols) or infected by rust (*Puccinia lagenophorae*) (closed symbols), during a period of low soil temperature ($-2\,°C$), and after return to normal soil temperatures. (From Paul & Ayres, 1985.)

to the plasma membrane may be the cause or effect of injuries to other membranes, such as those of organelles (Ayres, 1984). Although there is little good evidence to show that fungi use enzymes in a direct attack on the plasma membrane (Moreau & Rawa, 1984), injury to the plasma membrane may result from degradation of cell walls by extracellular polysaccharidases produced by fungi (Byrde, 1982). The plant may produce membrane-damaging substances as part of its response to infection; for example, the rapid desiccation of leaves of powdery mildew-infected peas subjected to water stress was attributed by Ayres (1980) to membrane damage caused by pisatin, the phytoalexin synthesised by the host. Effects of phytoalexins on stomata are discussed by Willmer & Plumbe (Ch. 12).

Some phytopathogenic fungi liberate toxins that alter membrane permeability or transport processes. The properties of these many and various substances are reviewed elsewhere (Daly, 1981; Daly & Knoche,

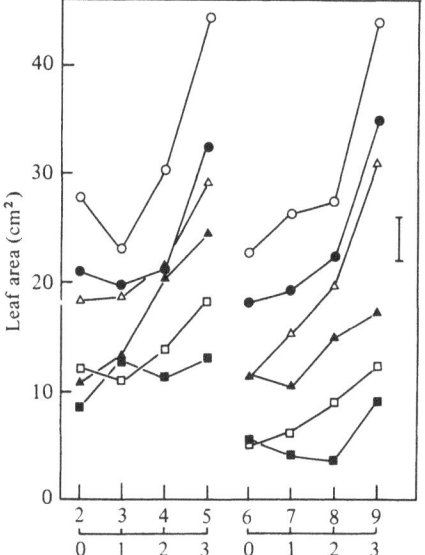

Weeks since start of low soil temperature (above)
Weeks since end of low soil temperature (below)

Fig. 15.8. Leaf area of groundsel (*Senecio vulgaris*), healthy (open symbols) or infected by rust (*Puccinia lagenophorae*) (closed symbols), during a 3-week recovery period following either 2 or 6 weeks when soil was held at $-6\,^{\circ}$C ($\square$ ■), $-2\,^{\circ}$C ($\triangle$ ▲) or ambient winter temperatures ($\bigcirc$ ●) in an unheated glasshouse. Bar indicates L.S.D. at $P = 0.05$. (From Paul & Ayres, 1985.)

1982), but a few are worth individual mention here. Cercosporin, produced by members of the genus *Cercospora*, generates both singlet oxygen and superoxide during irradiation, and these free radicals cause increases in the ratio of saturated to unsaturated fatty acids in membranes (Daub & Briggs, 1983). This results in an increase in the gel phase of the membrane, fluidity is lost, permeability increases and cell death may follow. This non host-specific toxin is interesting because its mode of action is comparatively well understood and, more significantly, it may have general relevance since the changes it causes are similar to those occurring in naturally ageing tissue.

Generally, altered permeability and solute loss caused by pathogens are irreversible; as a result, cells dehydrate and die even when there is an adequate supply of water. Such changes are typically localised around the infection site. However, a few pathogens liberate toxins that may be transported by diffusion, or by mass flow in the transpiration stream, considerable distances away from the infection site. Helminthosporoside, produced by *Helminthosporium sacchari*, the causal organism of Eyespot of sugar cane, gives rise to 'runners' (reddish-brown areas of cell necrosis that may extend 30 cm up the leaf from the Eyespot lesion, where the fungus is confined) in clones that are most susceptible to infection (Strobel, 1976; Scheffer & Livingston, 1980). Fusicoccin, the non host-specific toxin produced by *Fusicoccum amygdali*, the causal organism of peach and almond canker, is celebrated because of its ability to stimulate proton fluxes across cell membranes. Stomatal opening is promoted in both light and dark, leading to uncontrolled transpiration and severe water deficits *in vivo* (Turner & Graniti, 1976).

Development of the pathogen

Host death due to extreme water stress, whether localised or at a distance from the infection site, will inhibit growth of biotrophic fungi but, conversely, it may facilitate the growth of necrotrophic fungi. However, this last statement must be qualified. Harrison (1980, 1983) has studied the role of partially characterised toxins produced by *Botrytis fabae* in relation to the growth of lesions on leaves of field bean (*Vicia faba*) at different humidities. He found that under conditions of low humidity the concentration of toxic compounds in an infected leaf becomes high enough to kill healthy cells surrounding infected tissue. The dead tissue then dries out preventing further fungal growth and spread of the lesion. Toxin activity *in vitro* was enhanced in media of low ψ_π. In water-saturated air, however, the toxic compounds diffuse through the leaf becoming too dilute to kill uninfected tissue. Tissue does not become desiccated and the fungus continues to spread. Many questions remain to be answered about lesion development in this host/ fungus combination but Harrison's study reminds us that we are sadly ignorant of the influence of plant-water relations on the development of lesions caused by necrotrophic fungi. On one hand, how much does the dissemination of enzymes, toxins and other fungal products depend on mass flow of water through the apoplast? On the other hand, as Hancock & Huisman (1981) have asked, how do changes in the host's cell walls in response to infection, e.g. lignification, restrict the flow of water to the pathogen?

Although cell death is inimicable to the growth of biotrophic fungi, it should be remembered that water deficits in tissues infected by fungi such as rusts and mildews rarely reach lethal levels. Therefore, it is important to consider the effect that moderate water deficits may have on fungal development. Water deficits arising from both drought and salinity stress (the latter lowers the osmotic component of ψ_{soil} and makes water uptake by the plant more difficult) tend to inhibit the development of biotrophic fungi. The reasons for this may be many and various. Some stress-related factors may affect the fungus as it attempts to invade the host. For example it has sometimes been argued that reduction in stomatal aperture prevents rust fungi from gaining access to the leaf through the stomatal pore. Also, thickening of cuticle and epidermal cell walls that accompanies stress may strengthen the physical barrier a leaf offers to invasive growth (Royle, 1976). In a detailed study of the development of adult plant resistance (APR) to powdery mildew in barley, which is enhanced by plant growth under dry conditions, Ayres & Woolacott (1980) concluded that APR could not be attributed simply to increased cell wall or cuticle thickness. We suggested that specific substances that are harmful to the fungus accumulate in either cell wall or cuticle as tissue ages, and that this accumulation is promoted by water stress. Effects of water stress on predisposition to disease are reviewed by Schoeneweiss (Ch. 9).

When water deficits arise in tissues, the ψ_{π} of cells tends to decrease either because solutes are concentrated as water is withdrawn or because solutes are synthesised as part of an osmoregulatory response. Mycelium growing within this tissue will probably have to lower its own ψ_{π} in order to maintain a gradient of ψ between itself and the host. Accumulation of extra solutes by the mycelium will be at the expense of raw materials and energy that would otherwise have directly supported growth. This may be another factor responsible for that inhibition of fungal development caused by plant water stress.

If vegetative development of the fungus is restricted then so too may be its reproductive development. Growth of mycelium of *E. pisi* on the surface of pea leaves, and production of conidia by that mycelium, were reduced in parallel by drought-induced water stress (Ayres, 1977) (Fig. 15.9). Similar stress delayed the onset of conidial production and reduced the numbers of conidia produced in barley leaves infected by *E. graminis hordei*; the effects were greatest on the uppermost leaves where, as mentioned before, drought enhances APR to the establishment of the mildew (Woolacott & Ayres, 1984). Conidia originating from mycelium of drought-stressed leaves were better able to infect a new generation of drought-stressed plants than were conidia originating from well watered plants (Table 15.3). That the host plant can affect in this manner the infectivity of spores produced from it was previously unrecognised, but this phenomenon may have important epidemiological consequences since the multiplication rate of a pathogen depends upon both the number and the *quality* of the spores it produces in unit time (Zadoks & Schein, 1979). Recent work in this laboratory shows that salinity-

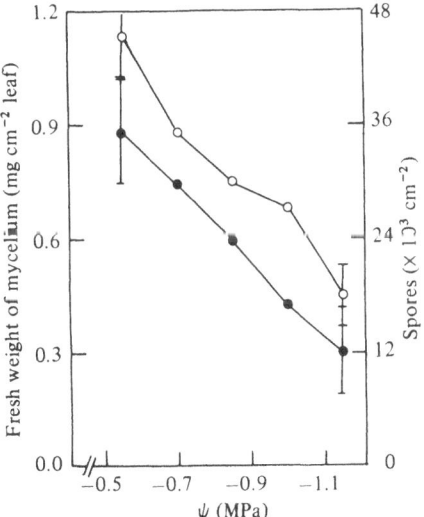

Fig. 15.9. Mycelial growth (○) and spore production (●) on leaves of field pea (*Pisum sativum*) 7 d after inoculation with powdery mildew (*Erysiphe pisi*). Watering was stopped at different times after inoculation to produce a range of water stress and results are presented for groups of plants, each with a span of ψ of 0.15 MPa. (From Ayres, 1977.)

Table 15.3. *Infectivity of conidia of powdery mildew (Erysiphe graminis hordei) originating from barley plants grown at two soil-water levels, inoculated on to plants grown under the same two soil conditions; and water relations of the inoculated plants (from Ayres & Woolacott, 1980)*

			Inoculated plants grown in wet soil (recipients)		Inoculated plants grown in dry soil (recipients)		Least significant difference between treatment means $P = 0.05$
			Conidia from plants grown in wet soil (donors)	Conidia from plants grown in dry soil (donors)	Conidia from plants grown in wet soil (donors)	Conidia from plants grown in dry soil (donors)	
Development of mildew 64 h after inoculation	% conidia giving rise to one or more:	Germ tube	38.2[+] (38.3)[‡]	36.8 (38.1)	31.8 (34.2)	44.4 (41.7)	4.2[‡]
		Appressorium	31.3 (34.0)	31.8 (34.2)	25.0 (30.7)	35.0 (36.0)	4.6
		Elongating secondary hypha	16.6 (23.9)	15.6 (23.1)	1.3 (5.4)	4.9 (11.9)	3.6
Water relations (MPa)	End of dark period	Water potential	-0.25[§]		-1.00		
		Osmotic potential	-1.35		-2.15		
		Turgor potential	1.10		1.15		
	Middle of light period	Water potential	-0.55		-1.70		
		Osmotic potential	-1.10		-2.20		
		Turgor potential	0.55		0.50		

[+] Mean of 10 replicates.
[‡] Angular transformation of percentage.
[§] Mean of three replicates sampled on the final day of incubation.

induced water stress has similar effects in the combination *E. pisi*/pea (Wyness & Ayres, 1985).

We do not know whether comparable changes occur in other bio-trophic fungi or what factors are responsible for the altered infectivity of conidia. It would be helpful to have values for the water and solute relations of conidia of different species and their parent mycelia. Psychro-metric measurements showed conidia of *E. graminis hordei* had $\psi_\pi = -4.6 \pm 0.4$ MPa when they derived from drought-stressed plants whereas conidia had $\psi_\pi = -3.1 \pm 1.2$ MPa when they derived from well watered controls (Ayres & Woolacott, 1980). Unfortunately, measurements of these parameters are few because they are difficult, requiring large numbers of spores, and because absolute values are questionable in view of the long time needed for equilibration before stable values can be obtained. Properties of the mycelium cannot be measured *in situ*, yet its removal from the leaf almost certainly affects the values recorded. Thus, the dependence of fungal water relations on host water relations must be largely a matter for speculation at present. Nevertheless, the facts remain that infectivity is affected by the water relations of the parent plant and that in influencing those water relations the fungus can to an extent determine its own fate.

References

Ahmad, I., Farrar, J. F. & Whitbread, R. (1985). Membrane integrity in leaves of barley infected by brown rust; an examination using tracer efflux and *in vivo* chlorophyll fluorescence. *New Phytologist*, **99**, 107–15.

Ayres, P. G. (1977). Effects of leaf water potential on sporulation of *Erysiphe pisi* (pea mildew). *Transactions of the British Mycological Society*, **68**, 97–100.

Ayres, P. G. (1978). Water relations of diseased plants. In *Water Deficits & Plant Growth*, vol. **5**, ed. T. T. Kozlowski, pp. 1–60. London: Academic Press.

Ayres, P. G. (1980). Stomatal behaviour in mildewed pea leaves: solute potentials of the epidermis and effects of pisatin. *Physiological Plant Pathology*, **17**, 157–65.

Ayres, P. G. (1981*a*). Responses of stomata to pathogenic microorganisms. In *Stomatal Physiology*, ed. P. G. Jarvis & T. A. Mansfield, pp. 205–21. Cambridge: Cambridge University Press.

Ayres, P. G. (1981*b*). Root growth and solute accumulation in pea in response to water stress and powdery mildew. *Physiological Plant Pathology*, **19**, 169–80.

Ayres, P. G. (1984). Interactions between environmental stress injury and biotic disease physiology. *Annual Review of Phytopathology*, **22**, 53–75.

Ayres, P. G. (1985). Effects of infection on root growth and function; consequences for plant nutrient and water relations. In *Plant Diseases: Infection, Damage and Loss*, ed. J. G. Jellis & R. K. S. Wood, pp. 105–17. London: Blackwell Scientific.

Ayres, P. G. & Woolacott, B. (1980). Effect of soil water on the development of adult plant resistance to powdery mildew in barley. *Annals of Applied Biology*, **94**, 255–63.

Bradford, K. J. & Hsiao, T. C. (1982). Physiological responses to moderate water stress. In *Encyclopedia of Plant Physiology*, vol. **12B**, *Water Relations and Carbon Assimilation*, ed. O. L. Lange, P. S. Nobel, C. B. Osmond & H. Zeigler, pp. 263–234. Berlin: Springer-Verlag.

Buchanan, B. B., Hutcheson, S. W., Magyarosy, A. C. & Montalbini, P. (1982). Photosynthesis in healthy and diseased plants. In *Effects of Disease on the Physiology of the Growing Plant*, ed. P. G. Ayres, pp. 13–28. Cambridge: Cambridge University Press.

Bushnell, W. R. & Gay, J. (1978). Accumulation of solutes in relation to the structure and function of haustoria in powdery mildews. In *The Powdery Mildews*, ed. D. M. Spencer, pp. 183–235. London: Academic Press.

Byrde, R. J. W. (1982). Fungal 'pectinases': from ribosome to plant cell wall. *Transactions of the British Mycological Society*, **79**, 1–14.

Daly, J. M. (1981). Mechanisms of action. In *Toxins in Plant Disease*, ed. R. B. Durbin, pp. 331–94. New York: Academic Press.

Daly, J. M. & Knoche, H. W. (1982). The chemistry and biology of pathotoxins exhibiting host-selectivity. *Advances in Plant Pathology*, **1**, 83–138.

Daub, M. E. & Briggs, S. P. (1983). Changes in tobacco cell membrane composition and structure. *Plant Physiology*, **71**, 763–6.

Gordon, T. R. & Duniway, J. M. (1982*a*). Stomatal behaviour and water relations in sugar beet leaves infected by *Erysiphe polygoni*. *Phytopathology*, **72**, 723–6.

Gordon, T. R. & Duniway, J. M. (1982*b*). Photosynthesis in powdery mildewed beet leaves. *Phytopathology*, **72**, 718–23.

Hancock, J. G. & Huisman, O. C. (1981). Nutrient movement in host–pathogen systems. *Annual Review of Phytopathology*, **19**, 309–31.

Harrison, J. G. (1980). The production of toxins by *Botrytis fabae* in relation to growth of lesions on field bean leaves at different humidities. *Annals of Applied Biology*, **95**, 63–71.

Harrison, J. G. (1983). Growth of lesions caused by *Botrytis fabae* on bean leaves in relation to foliar bacteria, non-enzymic phytotoxins, pectic enzymes and osmotica. *Annals of Botany*, **52**, 823–38.

Milford, G. F. J., Biscoe, P. V., Jaggard, K. W., Scott, R. K. & Draycott, A. P. (1980). Physiological potential for increasing yields of sugar beet. In *Opportunities for Increasing Crop Yields*, ed. R. G. Hurd, P. V. Biscoe & C. Dennis, pp. 71–83. London: Pitman.

Monteith, J. L. (1981). Does light limit crop production? In *Physiological Processes Limiting Crop Production*, ed. C. B. Johnson, pp. 23–38. London: Butterworths.

Moreau, R. A. & Rawa, D. (1984). Phospholipase activity in cultures of *Phytophthora infestans* and in infected potato leaves. *Physiological Plant Pathology*, **24**, 187–99.

Murphy, H. C. (1935). Effect of crown rust infection on yield and water requirement of oats. *Journal of Agricultural Research*, **50**, 387–411.

Nobel, P. S. (1983). *Biophysical Plant Physiology and Ecology*. San Francisco: Freeman.

Owera, S. A. P., Farrar, J. F. & Whitbread, R. (1981). Growth and photosynthesis in barley infected with brown rust. *Physiological Plant Pathology*, **18**, 79–90.

Paul, N. D. & Ayres, P. G. (1984). Effects of rust and post-infection drought on photosynthesis, growth and water relations in groundsel. *Plant Pathology*, **33**, 561–70.

Paul, N. D. & Ayres, P. G. (1985). Water relations and growth of rust-infected groundsel (*Puccinia lagenophorae* Cooke; *Senecio vulgaris* L.) during and after exposure to freezing soil temperatures. *Physiological Plant Pathology*, **27**, 185–96.

Royle, D. J. (1976). Structural features of resistance to plant diseases. In *Biochemical Aspects of Plant–Parasite Relationships*, ed. J. Friend & D. R. Threlfall, pp. 161–93. Phytochemical Society Symposium Series No. 13, London.

Scheffer, R. S. & Livingston, R. S. (1980). Sensitivity of sugar cane clones to toxin from *Helminthosporium sacchari* as determined by electrolyte leakage. *Phytopathology*, **70**, 400–4.

Strobel, G. A. (1976). Phytotoxins as tools in studying plant disease resistance. *Trends in Biochemical Science*, **1**, 247–9.

Turner, N. C. & Graniti, A. (1976). Stomatal response of two almond varieties to fusicoccin. *Physiological Plant Pathology*, **9**, 175–82.

Whipps, J. M. & Lewis, D. H. (1982). Patterns of translocation, storage and interconversion of carbohydrates. In *Effects of Disease on the Physiology of the Growing Plant*, ed. P. G. Ayres, pp. 47–83. Cambridge: Cambridge University Press.

Williams, G. M. & Ayres, P. G. (1981). Effects of powdery mildew and water stress on CO_2 exchanges in uninfected leaves of barley plants. *Plant Physiology*, **68**, 527–30.

Woolacott, B. & Ayres, P. G. (1984). Effects of plant age and water stress on production of conidia by *Erysiphe graminis* f.sp. *hordei* examined by non-destructive sampling. *Transactions of the British Mycological Society*, **82**, 449–54.

Wilson, M. & Henderson, D. M. (1966). *British Rust Fungi*, Cambridge: Cambridge University Press.

Wyness, L. E. & Ayres, P. G. (1985). Water or salt stress increases the infectivity of *Erysiphe pisi* conidia taken from stressed plants. *Transactions of the British Mycological Society* (in press).

Zadoks, J. C. & Schein, R. D. (1979). *Epidemiology and Plant Disease Management*, Oxford: Oxford University Press.

16
Water relations of mycorrhizal fungi and their host plants

D. J. READ and R. BOYD

Department of Botany, University of Sheffield, Sheffield, S10 2TN, UK

It is now widely recognised that most plant roots growing in natural environments develop symbiotic associations with fungi to form mycorrhizas. Studies of the physiology of the mycorrhizal associations have established that their significance to the plant is largely in the improvement of exploitation of mineral nutrient resources of the soil, while the fungi obtain access to a more or less continuous supply of carbon from the autotroph (Harley & Smith, 1983). Relatively little attention has been paid to the water relations of mycorrhizas. Reid (1979) has reviewed earlier work on the subject.

The hyphal elements forming mycorrhizas extend outwards from the root, which acts as a food base for the fungus, and provide an intimate system of physical communications between the soil, which is the ultimate source of water for both partners, and the plant. The structure of the external mycelial system is distinct in the two major types of mycorrhiza that are recognised, the vesicular-arbuscular (VA) mycorrhizas and ectomycorrhizas, and since this structural difference is likely to be of importance from the point of view of water relations it is appropriate to consider the two types separately.

Water relations of VA mycorrhizal systems

VA mycorrhizas are formed by zygomycetous fungi which produce intracellular infections in many herbaceous and woody plants. The external mycelial phase of the association characteristically takes the form of an anastomosing network of individual hyphae of diameters in the range 5 to 10 μm which penetrate the roots at intervals along its length.

The presence of infection has relatively little effect upon the gross morphology of the host root system, growth and root hair production being largely unaffected. It has, however, been observed (Baylis, 1975) that host plants with the relatively primitive unbranched 'magnolioid' root system supporting few root hairs are likely to respond more strongly to infection than are more advanced species, such as members of the Gramineae, which have diffuse root systems.

Since VA mycorrhizal fungi cannot be grown *in vitro* there is little experimental information concerning their responses to changing sub-strate water potential (ψ). It has been shown, however, that spores of *Glomus epigaeus* germinate most readily at soil water contents around field capacity while their germination is greatly inhibited below a ψ of -0.1 MPa (Daniel & Trappe, 1980). Analysis of infection intensities in soils maintained at different ψ provides an indirect indication that growth of the fungus is inhibited at low ψ_{soil}. Thus Kruscheva (1955) observed that as soil moisture content increased from 30 to 60% the number of wheat plants becoming infected increased from 70 to 97%. Using a range of ψ_{soil} from -0.10 to -1.40 MPa, Reid & Bowen (1979) showed that infection of the legume *Medicago truncatula* was most intense at the ψ_{soil} of -0.19 MPa and decreased at both higher and lower potentials.

There is now a body of evidence to show that VA infection influences the water relations of plants though there is some disagreement concern-ing the physiological basis of the observed effects. It has been demon-strated that in mycorrhizal (M) and non-mycorrhizal (NM) plants of comparable size, VA infection leads to increases of mean transpiration rates (J_{wv} = flux of water vapour) (Levy & Krikun, 1980; Allen, Smith, Moore & Christensen, 1981; Levy, Dodd & Krikun, 1983). These workers suggest that the increases arise from increases of leaf conductance and could find no evidence of increased hydraulic conductivity (L) in their mycorrhizal plants. They therefore conclude that the influence of infec-tion is primarily upon leaf physiology and suggest that changes in plant hormone balance are involved. In contrast, other workers (Safir, Boyer & Gerdemann, 1971, 1972; Hardie & Leyton, 1981; Nelsen & Safir, 1982*a,b*; Graham & Syvertsen, 1984) have observed large increases of L in mycorrhizal plants and the response appears to be associated with the improved phosphorus status of infected plants. The studies of Safir *et al.* (1971) showed that at low available P levels mycorrhizal infection could lead to a 60% increase of L in soya bean, and an even greater enhancement of L was later observed in onion (Nelsen & Safir, 1982*a*) in which differences between M and NM plants were as high as 400%.

Nelsen & Safir (1982*b*) showed that differences of *L* could be eliminated if sufficient P was added to NM plants to raise their P status to that of the infected plants. They point out that the failure of Levy & Krikun (1980) to obtain effects upon *L* may be attributable to the fact that both the M and NM citrus plants in the latter's study were grown with near-optimal P supply. It was suggested by Nelsen & Safir (1982*b*) that the low conductivities of non-mycorrhizal plants may be attributable to impairment of membrane function, presumably at the endodermis, which reduces permeability and so inhibits water transport. Phosphorus uptake from soil is markedly enhanced by VA infection and, since there is a linear relationship between the diffusion coefficient of P in a soil and its water content, the nutritional advantages accruing to infected plants are likely to be maximised under drought conditions. According to this interpretation of events the increase of *L* in infected plants is simply attributable to increases of internal permeability associated with enhanced P uptake and the effects of infection are entirely indirect.

A problem is that responses said by some workers to be primarily based upon improved P nutrition can sometimes be equally well explained in terms of possible effects upon water uptake. Thus, for example, Graham & Syvertsen (1984) added P to both M and NM treatments in order to obtain plants of similar size. This is desirable since differences in root size have large non-linear effects upon *L* (Fiscus, 1977). Addition of P caused a reduction of *L* in infected roots to below that of uninfected plants leading the authors to conclude that changes of *L* are dependent upon P nutrition. The response could, however, have been due to an inhibition, by the added P, of the external VA mycelium, an effect recently reported by Bethlenfalvay, Brown & Pacovsky (1982). If this was the case, reduced *L* of M plants could have arisen as a result of a lowered capacity to obtain water by way of the mycelium.

Is it not possible that increases of uptake and transport of water arise solely from increased exploitation of water resources by the VA fungal hyphae? Sanders & Tinker (1973) concluded that VA hyphae were unlikely to provide pathways through which significant quantities of water could be transported because in order to explain the increased water inflow observed in their mycorrhizal onion plants, which had 6 entry points per cm of infected root, a bulk flow velocity of $5\,\mathrm{cm\,min^{-1}}$ was required. This view was repeated following experiments with clover plants (Cooper & Tinker, 1981) in which infection was calculated to lead to only trivial increases of water flow. In these experiments there were only 1 to 4 hyphal connections between the source of water and the plants. Observations such as these have led some reviewers (see for example Fitter, 1985) to conclude that absorption of water by hyphae cannot account for increased water uptake into mycorrhizal plants. Some of the more recent experimental studies do, however, suggest that the increased J_{wv} seen in mycorrhizal plants is attributable directly to uptake by hyphae. Allen (1982) considered that a measured increase of J_{wv} of $60 \times 10^{-7} \, \text{cm}^3 \, \text{cm}^{-2} \, \text{s}^{-1}$ in mycorrhizal compared to non-mycorrhizal plants of *Bouteloua gracilis* was due to absorption by hyphae and he calculated that to supply this water a flux of $2.8 \times 10^{-8} \, \text{cm}^3 \, \text{s}^{-1}$ was required through the hyphae at each of the 6 entry points per cm of root length. This is equivalent to $100 \, \mu\text{m}^3 \, \text{h}^{-1}$ and is not unlike figures reported for fluxes of water through hyphae of *Phycomyces* by Cowan, Thain & Lewis (1972) which reach $131 \, \mu\text{m}^3 \, \text{h}^{-1}$. The fluxes through VA hyphae might be expected to be lower than those through *Phycomyces* because their diameters, and hence cross-sectional areas, are somewhat smaller. However, considerable compensation for this may be derived from the fact that the gradient of ψ along which water will be moving in the mycorrhizal situation is normally much greater due to the transpiration pull. There is experimental evidence showing an increased flux of nutrients to VA-infected plants under conditions of high cost transpiration (Cooper & Tinker, 1981). This result can only be explained in terms of bulk flow of solution along such a gradient.

Taking the results of all these studies together, it seems most likely that the increased flux of water frequently observed in VA-infected plants arises as a result of a combination of indirect and direct effects of the presence of mycorrhizal mycelium. Hardie & Leyton (1981) found an increase of both leaf conductance and root conductivity in mycorrhizal clover plants. They showed that infected plants had the ability to sustain lower ψ_{leaf} and suggested that this in turn enabled their roots and hyphae to extract water more effectively from drying soil. Increases of L in mycorrhizal plants were of the order of $25 \times 10^{-5} \, \text{mg} \, \text{cm}^{-1} \, \text{s}^{-1} \, \text{MPa}^{-1}$. Assuming that a water flux through each entry point of $100 \, \mu\text{m}^3 \, \text{h}^{-1}$ is possible, such an inflow could be achieved by having only about 10 hyphal entry points per cm of root length. Here again, however, infection was associated with enhancement of tissue P levels so that discrimination between hyphal water uptake and nutritional effects was still impossible. Comparisons based upon differences of L and J_{wv} between M and NM plants will normally be confounded in this way and a more direct approach to the problem is clearly needed. Kay Hardie (in preparation) has recently employed a method in which measurements of J_{wv} are made on individual mycorrhizal plants before and after severing their hyphae at the entry points. This treatment led to a maximum fall of equilibrium J_{wv} following hyphal removal of $9.9 \times 10^{-5} \, mg \, cm^{-2} \, s^{-1}$, and, assuming a potential inflow rate of $100 \, \mu m^3 \, h^{-1}$ as before, such a flux requires only 4 hyphal entry points per cm of root length. There is every likelihood in this situation that direct hyphal uptake of water is making a significant contribution to J_{wv}

It is evident that in these considerations the number of hyphal entry points per unit of root length is a feature of considerable importance. Most of the experiments described above were carried out in circumstances providing only 2 to 10 entry points per cm. This is an extremely low number and must lead to an underestimation of the potential for water inflow into plants. Mosse (1959) recorded between 26 and 211 entry points per cm in pot- and field-grown strawberry plants. Studies currently being carried out in our laboratory by C. Birch indicate that in undisturbed limestone grassland, the number of entry points is in the range 50 to 100 per cm of root length in seedlings of *Medicago lupulina*, *Plantago lanceolata*, *Arrhenatherum elatius* and *Festuca ovina*. Such numbers would greatly increase the potential inflow rate. The implication again is that direct uptake of water through hyphae will be of significance especially in shallow soils which are susceptible to seasonal drying.

We are now also more aware of the extent of the extra radical mycelial network which links the root surface with the soil. Tisdall & Oades (1979) measured up to 50 m of hyphal length per gram of grassland soil. Similar values have been obtained by Abbot & Robson (1985) who, in addition, found hyphal lengths of between 2.5 and 14 m per cm of infected root. If we assume a hyphal diameter of 5 μm and a root diameter of 500 μm then 1 m of hyphal length has a surface area equivalent to 1 cm of root length. It is thus evident that hyphal surface areas can be considerably greater than those of the roots with which the hyphae are associated. A perhaps more vital attribute of the external mycelium is that it extends outwards from the root surface and so it can exploit water resources beyond the depletion zones caused by uptake into the root. Unfortunately we have little accurate information of the distances over which such growth can occur but figures of the order of 1 to 10 cm seem likely (Rhodes & Gerdemann, 1975) and this would be adequate to provide extension beyond the 1-mm-thick depletion zone which would be expected around a root of 0.5 mm diameter.

Confirmation that direct hyphal water transport is likely to be of importance to slow-growing perennials of semi-arid habitats is provided by studies of the steppe grass *Agropyron smithii*. Mycorrhizal plants, grown in the field or in pots have lower stomatal resistances and higher turgor potentials than NM controls when both are subjected to identical conditions of reduced ψ_{soil} (Allen, 1984). An important feature of Allen's results (personal communication) is that these effects can be achieved without any changes being observed either in the P status or root size of infected compared with uninfected plants.

The latest research thus suggests that direct uptake of water by VA hyphae may make a far more significant contribution to the water economy of infected plants than has hitherto been accepted. Much more experimental work is required in this difficult area. Refinements of the hyphal excision method employed by Hardie will present an opportunity for advances in understanding. In addition, a better knowledge of the distribution of entry points and of extra-radical hyphae in different habitats will be needed before we are able to make reasonable extrapolations from laboratory experiments to field situations.

Water relations of ectomycorrhizal systems
Effects of water stress on growth of ectomycorrhizal fungi in pure cultures

Most studies of the effects of water stress on ectomycorrhizal fungi have employed osmotic substrates and thus have examined the effects of ψ_π rather than ψ_m. Shemakhanova (1962) examined the growth of *Boletus* spp. in solution cultures containing NaCl or KNO_3 as osmotica and observed good mycelial growth at -1.0 MPa. In liquid-culture studies using polyethylene glycol 4000, Mexal & Reid (1977) showed that optimum growth of *Suillus luteus* and *Thelephora terrestris* was obtained at -0.5 MPa while *Cenococcum graniforme* (*C. geophilum*), a fungus encountered frequently in dry conditions, had an optimum yield at -1.5 MPa. Saleh-Rastin (1976) found optimum growth of this fungus at -0.4 MPa in NaCl solutions. In soil adjusted osmotically with KCl to provide a range of ψ, Theodorou (1978) obtained maximum yields of the ectomycorrhizal fungi *Corticium bicolor*, *Rhizopogon luteolus* and *S. luteus* between -0.3 and -0.6 MPa, though growth was possible down to -4.0 MPa providing an exogenous supply of glucose was available. Theodorou suggests that two ecologically distinct groups of mycorrhizal fungi may be recognised by their different patterns of response to water stress. *S. luteus* and *Lactarius deliciosus*, for example, were able to grow at ψ below -10 MPa while others such as *Suillus granulatus* and species of *Rhizopogon* failed to grow at ψ below −3.5 MPa.

There is considerable indirect evidence that growth of ectomycorrhizal fungi is inhibited by excessively high soil water contents. Thus ectomycorrhizas normally fail to develop both in fully saturated soil (Shuja, Gilani & Khan, 1971; Theodorou, 1978) and in wet microsites (Lorio, Howe & Martin, 1972). The progress of infection after establishment may, however, be relatively insensitive to high water contents (Filer & Broadfoot, 1968). It is likely that reduced aeration in wet soil is more important than saturation itself, and that the external supply of oxygen through the plant is sufficient to supply the requirements of the fungus (Armstrong & Read, 1972; Read & Armstrong, 1972).

Effects of ectomycorrhizal infection on plant growth and survival under adverse water relations

In those woody species that are susceptible to ectomycorrhizal infection only primary roots generally have the potential for unlimited growth, the growth of first- and second-order laterals being restricted to such an extent that they are often referred to as 'short roots'. It has generally been considered that inhibition of extension of short roots is a direct result of mycorrhizal infection (Hatch, 1937; Clowes, 1951) but recently Warren-Wilson & Harley (1983) have shown that growth cessation in beech is a normal feature even of uninfected roots. Whatever the cause of this pattern of root development the result is that root densities of ectomycorrhizal woody species such as pine are frequently much lower than those of herbaceous species (Kramer & Bullock, 1966; Newman, 1969). In effect their root form is comparable with that of the 'magnolioid' roots which Baylis (1975) showed to be so responsive to VA mycorrhizal infection. The ectomycorrhizal root system is distinct from that of the VA plant in that most of the unsuberised surface area of the roots is ensheathed in a pseudoparenchymatous hyphal mantle. Connection with the soil is then provided either by the sheath itself, as in those species such as beech where the mantle is frequently smooth (Harley, 1978), or by an external mycelial system which, in pine, can extend over distances in excess of 1 m from the root system into the soil (Schramm, 1966). If this mycelium had an absorptive function it could clearly be of great significance for the water economy of the plants but until recently evidence in support of such a view has been largely circumstantial. Thus, several studies have suggested that ectomycorrhizal infection can provide protection from the effects of moderate drought.

Cromer (1935) believed that infection increased the drought resistance of *Pinus radiata* by protecting the roots from shrinkage and providing increased uptake of water from soil at low water potentials. In studies of *P. ponderosa*, Goss (1960) showed that mycorrhizal infection improved the survival of plants after short periods of exposure to drought but had little effect over longer periods. Mycorrhizas formed by *C. geophilum* on both seedlings and 2-year-old saplings of *Tilia cordata* have been found to survive ψ_{soil} as low as $-5.5\,MPa$ (Pigott, 1982).

Under conditions of extreme drought in the arid steppes, Zerova (1955) and Samtsevich (1955) found that mycorrhizal infection afforded little protection to oak seedlings, though infection could improve their vigour at intermediate levels of soil moisture. Dixon, Pallardy, Garrett & Cox (1983) have studied the water balance of bare-root and container-grown seedlings of *Quercus velutina* which had been inoculated with *Pisolithus tinctorius* or left uninoculated. They found that container-grown plants with extensive mycorrhiza development had a significantly improved water balance following transplantation. Pre-dawn ψ_{shoot} values were significantly higher in the mycorrhizal plants during mild drought. These results are at variance with those of Sands & Theodorou (1978) who found greater resistance to liquid flow in the soil–plant pathway when seedlings of *P. radiata* had been inoculated in sand with *R. luteolus*. However, these authors cautioned that the increased resistance was largely restricted to the soil and observed that mycelial strands had not developed in their experimental pots. Experiments such as these provide only indirect evidence for involvement of ectomycorrhizas in plant water relations and as in the case of VA mycorrhizas a more direct experimental approach is needed.

It has been realised increasingly that the external mycelial system could be of enormous significance if it was capable of absorption and transport of water but analysis of the structure and function of the system has been hampered by the absence of non-destructive experimental techniques. Duddridge, Malibari & Read (1980) described transparent Perspex chambers in which the development of the ectomycorrhizal mycelium could be followed. Transmission electron microscopy of the mycelial strands that developed in these chambers revealed they were differentiated structures in which central 'vessel' hyphae of large diameter $(10–20\,\mu m)$, and lacking cytoplasm, were surrounded by tightly packed cytoplasmic hyphae of narrower dimensions $(5–10\,\mu m)$ (Fig. 16.1c). They showed, by exposing distal parts of the undamaged mycelium to tritiated water, that the mycelial fans were capable of absorbing water and that

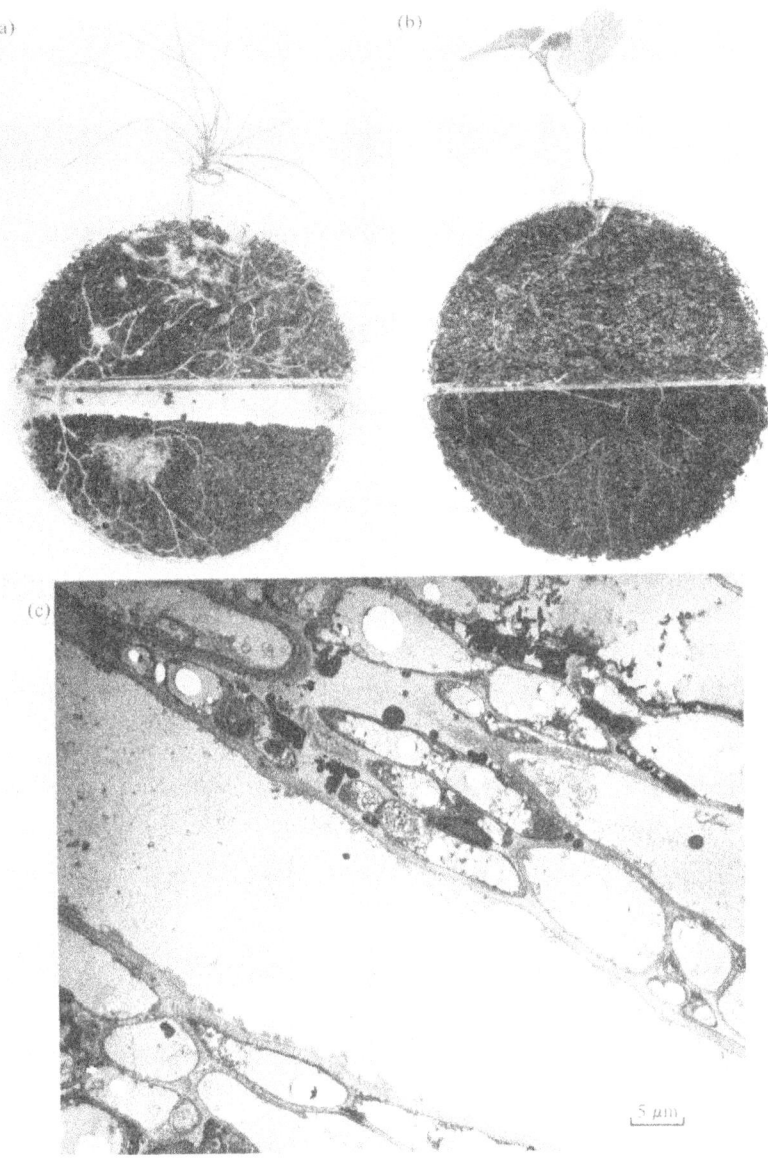

Fig. 16.1. (a) Mycorrhizal seedling of *Pinus sylvestris* growing in divided dish with single mycelial strand of *Suillus bovinus* growing from the dry to the moist peat across a plastic divide with a gap both edges of which act as diffusion barriers (natural size). (b) Mycorrhizal seedling of *Betula pendula* with mycelial strands of *Paxillus involutus* growing across the plastic divide from dry to moist peat (natural size). (c) Transmission electron micrograph of longitudinal section through a mycorrhizal mycelial strand of *Suillus bovinus*. A central vessel hypha lacking in cytoplasmic content is surrounded by a sheath of relatively isodiametric cytoplasmic hyphae. Photograph by J. A. Duddridge.

water passed through the strands to attached pine seedlings over distances in excess of 20 cm in one hour. This demonstrates that the external mycelium can function as a physiological extension of the root system to which it is attached, and Duddridge *et al.* suggested that the vessel hyphae were likely to be the major conduits through which water was transported towards the seedlings.

The critical matter here, as in the case of VA mycorrhizas, is to determine whether sufficient water can be transported to the plant to be of physiological significance. Brownlee, Duddridge, Malibari & Read (1983) grew ectomycorrhizal pine in divided Petri dishes so that the mycorrhizal fungus *Suillus bovinus* grew from dry peat, to which the roots were confined, across a diffusion barrier into moist peat (Fig. 16.1a). Severing of strands crossing the barrier led to decreases in ψ_{leaf} of up to -7 MPa within 2 h. Seedlings treated in this way died within one week of strand severance. More recently we have investigated the effects of strand severance upon transpiration and photosynthesis (Ps) of birch seedlings grown in similar split-dish systems (Fig. 16.1b). Using linked water vapour and CO_2 infrared gas analysers (ADC, Hoddesden, Herts), J_{wv} and Ps were monitored continuously before, during and after severance of strands which had grown from roots in dry peat to moist peat. Cutting of strands led to immediate decline of J_{wv} and Ps (Fig. 16.2). This indicates a direct effect of severance upon both pro-

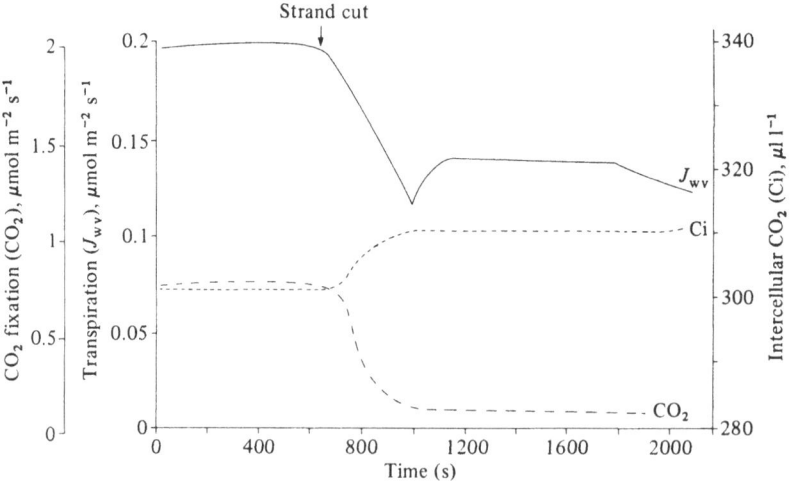

Fig. 16.2. Transpiration (J_{wv}), photosynthesis (CO_2) and internal leaf CO_2 concentration (Ci) of mycorrhizal *Betula pendula*, before and after cutting mycelial strands of *Paxillus involutus* growing from dry to moist peat.

cesses, presumably brought about by loss of guard cell turgor and stomatal closure. If, as seems reasonable, it is assumed that the bulk of the pre-severance J_{wv} was supplied by water passing through the strands from the moist peat then, by calculating the surface areas of the cut strands and the reduction of J_{wv} associated with cutting, it is possible to estimate the fluxes occurring in the strands. Values in the range of 1 to 8×10^{-3} cm^3 cm^{-2} s^{-1} have been obtained (equivalent to a rate of flow through a vessel hypha, 20 μm in diameter, of 11 to 88 μm^3 h^{-1}). These are similar to those observed in VA mycorrhizas and are somewhat lower than might be predicted from calculations of the hydraulic conductivity of vessel hyphae; however the transpiration rates of these birch seedlings were rather low. While Sands, Fiscus & Reid (1982) suggest that the ectomycorrhizal sheath does not reduce hydraulic conductance of roots, it is probable that the combined resistances of the root and sheath will also lead to a reduction of the flux through the strand system.

The experiments carried out to date using direct methods of analysis suggest that in ecto- as in VA mycorrhizas the external mycelium forms a functional extension of the root system which is capable of absorbing and transporting physiologically significant quantities of water.

Measurements of hyphal length in the split-dish systems yield values in the range from 10 to 80 m per cm of root length. These are even higher than those reported above for VA mycorrhizas. Findlay & Read (1986) have shown that ectomycorrhizal mycelia of strand-forming fungi in genera such as *Paxillus* and *Suillus* grow as a fan from infected roots, the hyphal front extending at a rate of 2 to 4 mm per day. While lower values of hyphal length would be expected in older regions of the mycelial system, the front probably forms the major water-absorbing zone and measurements in this region therefore provide a reasonable indication of the increase of effective root length obtained as a result of infection by these fungi.

Mechanism and pathway of water transport in mycorrhizal systems

Water can flow through mycelium as a result of either localised evaporation or solute accumulation, both leading to a gradient of ψ_p (see Ch. 2). In the former case, ψ_p is lowered in one part of the mycelium with respect to the remainder, while in the latter ψ_p is raised. Jennings (1984) concluded that in the mycelium of the wood-rotting fungus *Serpula lacrimans*, movement occurs as a bulk flow of solution, caused because

ψ_p, elevated at the food source by the active uptake of glucose, is higher than in non-absorbing regions. He suggests that surface droplets frequently observed at the tip of fungal hyphae are formed as a result of the positive pressure and that their production enables a high internal solute concentration to be maintained. The major difference encountered in the mycorrhizal situation is that the fungal mycelium is attached to a large evapotranspirative sink, the plant; this means that while water movement is still likely to occur by a process of bulk flow it will normally take place along a gradient of ψ_p whose direction is established by the plant. Circumstantial evidence in support of this is derived from the observation that droplet formation rarely if ever occurs in mycorrhizal mycelial systems. It is evident, however, that if bulk flow occurred continuously in the direction of the plant, hyphal growth would not be possible. Since hyphal growth does occur, and at rates not dissimilar to those observed by Jennings in *Serpula* mycelium, we must conclude that bidirectional movement of substances takes place either by means of spatial separation of function within the mycelial system or as a result of temporal separation. These possibilities can be discussed separately.

Spatial separation of transport function

Schütte (1956) described movement of dyes through some hyphae of fungal fruit bodies while others were not involved in translocation. This separation of function was not associated with morphological modification. A similar separation of function was observed in mycelial fans of *Serpula* (Brownlee & Jennings, 1982) and in this case there were morphological differences between hyphae which suggested specialisation of function. In ectomycorrhizal mycelial strands differentiation into vessel and non-vessel hyphae provides the potential for separation of functions but it is not clear whether such separation could be maintained under the large gradients of ψ imposed by the transpiration pull. If it were possible, water movement towards the plant would be expected to occur through the vessel hyphae, which have relatively large hydraulic conductivities, while carbon fluxes would occur in the reverse direction along the sheathing cytoplasmic hyphae of the strand. Even if this pattern of bidirectional flow were possible in ectomycorrhizal systems, there remains the problem of interpreting the possible events in the single hyphae of VA mycorrhizas, since bulk flow in the direction of the plant would be expected to override any translocation for which the driving force was a concentration gradient. Since in VA, as in ectomycorrhizal mycelium, evidence for a net flux of carbon against the direction of bulk flow is provided in the form of hyphal growth, we must assume that an alternative mechanism is operating. One possible explanation of the apparent conflict is that different transport processes are separated in time rather than space.

Temporal separation of translocating functions

While transpiration-driven bulk flow is likely to predominate within the plant and its attached mycelial system during the day, stomatal closure at night leads to the elimination of internal ψ gradients. Under these circumstances gradients of solute concentration would begin to determine the pattern of translocation. Relatively high solute concentrations in hyphae at or near the root could then lead to the generation of a gradient of ψ_p analogous to that described in *Serpula* and a net movement towards hyphal tips. Such a sequence of events is amenable to experimental analysis since it should be revealed in the form of a diurnal pattern of hyphal extension growth. It is most likely that this pattern would be seen in VA mycelia because spatial separation of function is unlikely to occur in these systems.

In the more morphologically differentiated ectomycorrhizal mycelium it is probable that, as in the plant, both spatial and temporal separation of function may take place. Spatial separation would allow transport to occur simultaneously along vessel hyphae towards the plant and along sheathing hyphae towards the hyphal tips during the day, a situation analogous to that seen in the xylem and phloem of the plant. A prerequisite for the occurrence of such bidirectional flow is the presence of barriers which are capable, both in terms of their physical strength and impermeability, of maintaining a separation between the two types of flow. The thickenings seen in the walls of vessel hyphae and some of the sheathing hyphae may be important in this context. At night, gradients of ψ_p would be in the same direction in vessel and sheathing hyphae, i.e. high ψ_p near the root, low ψ_p at hyphal tips. The absence of droplet formation in mycorrhizal mycelia may be due to the fact that this type of flow occurs over periods of time which are too short to generate the requisite excess pressure. There is clearly a need for experimental analysis of these processes in both VA and ectomycorrhizal systems. Root chambers provide useful structures for the measurement of growth rhythms but autoradiographic analysis of functional differentiation within the hyphal network is also urgently required.

Conclusions

It has long been accepted that the increases of plant yield associated with mycorrhizal infection arise through enhancement of mineral nutrient absorption. Since these nutrients are absorbed in solution there is an implicit acceptance of the ability of these structures to absorb water, but it has generally been considered unlikely that mycorrhizas contribute significantly to the water economy of the whole plant (Sanders & Tinker, 1973; Fitter, 1985). However, increasing awareness of the extent of the external mycelium in soil, and recent evidence that this mycelium can both absorb water and provide channels for its transport over long distances, now suggest that the mycorrhizal role in plant water relations may be far more significant than has hitherto been accepted.

The benefits of infection are likely to arise, as they do in the case of mineral nutrition, primarily as a result of supplementation of resource supply under conditions of stress. Thus, in any situation where, for geographical or seasonal reasons, water supply to the roots is restricted, the external mycelium will become increasingly beneficial as the hydraulic conductivity of unsaturated soil begins to limit uptake at the root surface. The large absorptive surface of the hyphae will in effect provide increases in the conductivity of the whole root system. Mycorrhizas may thus be of benefit to the water balance of the host largely through their capacity to provide the minimum requirements for survival during stress rather than through their ability to sustain high flow rates. In these circumstances they may also benefit the plant in terms of nutrient uptake. The two types of benefit are not, as implied in some studies, mutually exclusive. Fortunately we now have the experimental techniques to enable us to discriminate between nutritional effects and those attributable to enhancement of water uptake, and great advances in our understanding of the role of mycorrhizas in plant water relations can be expected in the near future.

Acknowledgements. We wish to thank Dr Kay Hardie for valuable comments on initial drafts of this paper, and Drs K. Hardie and M. Allen for permission to examine data which they are preparing for publication.

References

Abbot, L. K. & Robson, A. D. (1985). Formation of external hyphae in soil by four species of vesicular-arbuscular mycorrhizal fungi. *New Phytologist*, **99**, 245–56.
Allen, E. (1984). VA mycorrhizal and colonizing annuals: implications for growth,

competition and succession. In *VA Mycorrhizae and Reclamation of Arid and Semi-arid Lands*, ed. S. E. Williams & M. F. Allen, pp. 42–52, University of Wyoming Agricultural Experiment Station Scientific Report No. SA1261.

Allen, M. F. (1982). Influence of vesicular-arbuscular mycorrhiza on water movement through *Bouteloua gracilis* (H.B.K.) Lag ex Steud. *New Phytologist*, **91**, 191–6.

Allen, M. F., Smith, W. K., Moore, T. S. & Christensen, M. (1981). Comparative water relations and photosynthesis of mycorrhizal and non-mycorrhizal *Bouteloua gracilis*. *New Phytologist*, **88**, 683–93.

Armstrong, W. & Read, D. J. (1972). Some observations on oxygen transport in conifer seedlings. *New Phytologist*, **71**, 55–62.

Baylis, G. T. S. (1975). The magnolioid mycorrhiza and mycotrophy in root systems derived from it. In *Endomycorrhizas*, ed. F. E. Sanders, B. Mosse & P. B. Tinker, pp. 373–89. London: Academic Press.

Bethlenfalvay, G. J., Brown, M. S. & Pacovsky, R. S. (1982). Relationships between host and endophyte. *New Phytologist*, **90**, 537–43.

Brownlee, C. & Jennings, D. H. (1982). The pathway of translocation in *Serpula lacrimans*. *Transactions of the British Mycological Society*, **79**, 401–76.

Brownlee, C., Duddridge, J. A., Malibari, A. & Read, D. J. (1983). The structure and function of mycelial systems of ectomycorrhizal roots with special reference to their role in forming inter-plant connections and providing pathways for assimilate and water transport. *Plant & Soil*, **71**, 433–43.

Clowes, F. A. L. (1951). The structure of mycorrhizal roots of *Fagus sylvatica*. *New Phytologist*, **50**, 1–16.

Cooper, K. M. & Tinker, P. B. (1981). Translocation and transfer of nutrients in vesicular-arbuscular mycorrhizas. IV. Effect of environmental variables on movement of phosphorus. *New Phytologist*, **88**, 327–39.

Cowan, M. C., Thain, J. F. & Lewis, B. G. (1972). Uptake of potassium by the developing sporangiophores of *Phycomyces blakesleeanus*. *Transactions of the British Mycological Society*, **58**, 113–26.

Cromer, D. A. N. (1935). The significance of the mycorrhiza of *Pinus radiata*. *Bulletin of Forest Bureau of Australia*, **16**, 1–19.

Daniels, B. A. & Trappe, J. M. (1980). Factors affecting spore germination of the vesicular-arbuscular mycorrhizal fungus *Glomus epigaeus*. *Mycologia*, **72**, 457–71.

Dixon, R. K., Pallardy, S. G., Garrett, H. E. & Cox, G. S. (1983). Comparative water relations of container-grown and bare-root ectomycorrhizal and non-mycorrhizal *Quercus velutina* seedlings. *Canadian Journal of Botany*, **61**, 1559–65.

Duddridge, J. A., Malibari, A. & Read, D. J. (1980). Structure and function of mycorrhizal rhizomorphs with special reference to their role in water transport. *Nature*, **287**, 834–6.

Finlay, R. D. & Read, D. J. (1986). The structure and function of the vegetative mycelium of ectomycorrhizal plants. I. Translocation of ^{14}C-labelled carbon between ectomycorrhizal plants interconnected by a common mycelium. *New Phytologist*, (in press).

Fiscus, E. (1977). Determination of hydraulic and osmotic properties of soybean root systems. *Plant Physiology*, **59**, 1013–20.

Filer, T. H. & Broadfoot, W. M. (1968). Sweetgum mycorrhizae and soil microflora in shallow-water impoundment. *Phytopathology*, **58**, 1080.

Fitter, A. H. (1985). Functioning of vesicular-arbuscular mycorrhizas under field conditions. *New Phytologist*, **99**, 257–65.

Goss, R. W. (1960). Mycorrhiza of ponderosa pine in Nebraska grassland soils. *University of Nebraska. Agricultural Experiment Station Research Bulletin* 192, 47 pp.

Graham, J. H. & Syvertsen, J. P. (1984). Influence of vesicular-arbuscular mycorrhiza on the hydraulic conductivity of roots of two citrus rootstocks. *New Phytologist*, **97**, 277–85.

Hardie, K. & Leyton, L. (1981). The influence of vesicular-arbuscular mycorrhiza on growth and water relations of red clover. I. In phosphate deficient soil. *New Phytologist*, **89**, 599–608.

Harley, J. L. (1978). Ectomycorrhizas as nutrient absorbing organs. *Proceedings of the Royal Society of London*, **B203**, 1–21.

Harley, J. L. & Smith, S. E. (1983). *Mycorrhizal Symbiosis*. London: Academic Press.

Hatch, A. B. (1937). The physical basis of mycotrophy in the genus *Pinus*. *Black Rock Forest Bulletin*, **6**, 168 pp.

Jennings, D. H. (1984). Water flow through mycelia. In *The Ecology and Physiology of the Fungal Mycelium*, ed. D. H. Jennings & A. D. M. Rayner, pp. 143–64. Cambridge: Cambridge University Press.

Kramer, P. J. & Bullock, H. C. (1966). Seasonal variations in the proportions of suberised and unsuberised roots of trees in relation to the absorption of water. *American Journal of Botany*, **53**, 200–4.

Krushcheva, E. P. (1955). Mycorrhizas of agricultural plants. In *Mycotrophy in Plants*, ed. A. Imshenetskii (translated from Russian by Israel Programme for Scientific translation). Washington D.C.: USDA & NSF. 1967.

Levy, Y. & Krikun, J. (1980). Effect of vesicular-arbuscular mycorrhiza on *Citrus jambhiri* water relations. *New Phytologist*, **85**, 25–31.

Levy, Y., Dodd, J. & Krikun, J. (1983). Effect of irrigation, water salinity and rootstock on the vertical distribution of vesicular-arbuscular mycorrhiza in citrus roots. *New Phytologist*, **95**, 397–403.

Lorio, P. L., Howe, V. K. & Martin, C. N. (1972). Loblolly pine rooting varies with micro-relief on wet sites. *Ecology*, **53**, 1134–40.

Mexal, J. & Reid, C. P. P. (1973). The growth of selected mycorrhizal fungi in response to induced water stress. *Canadian Journal of Botany*, **51**, 1579–88.

Mosse, B. (1959). Observations on the extramatrical mycelium of a vesicular-arbuscular endophyte. *Transactions of the British Mycological Society*, **42**, 439–48.

Nelsen, C. E. & Safir, G. R. (1982a). Increased drought tolerance of mycorrhizal onion plants caused by improved phosphorus nutrition. *Planta*, **154**, 407–13.

Nelsen, C. E. & Safir, G. R. (1982b). The water relations of well-watered mycorrhizal and non-mycorrhizal onion plants. *Journal of the American Society for Horticultural Science*, **107**, 271–4.

Newman, E. I. (1969). Resistance to water flow in soil and plants. *Journal of Applied Ecology*, **6**, 1–12.

Pigott, C. D. (1982). Survival of mycorrhiza formed by *Cenococcum geophilum* Fr. in dry soils. *New Phytologist*, **92**, 513–17.

Read, D. J. & Armstrong, W. (1972). A relationship between oxygen transport and the formation of the ectotrophic mycorrhizal sheath in conifer seedlings. *New Phytologist*, **71**, 49–53.

Reid, C. P. P. (1979). Mycorrhizae and water stress. In *Root Physiology and Symbiosis*, ed. A. Riedacker & M. J. Gagnaire-Michard, pp. 392–408. IUFRO Symposium, Nancy, France.

Reid, C. P. P. & Bowen, G. D. (1978). Effect of soil moisture on VA mycorrhiza formation and root development in *Medicago*. In *The Root-Soil Interface*, ed. J. L. Harley & R. Scott Russell, pp. 211–19. London: Academic Press.

Rhodes, L. H. & Gerdemann, J. W. (1975). Phosphate uptake zones of mycorrhizal and non-mycorrhizal onions. *New Phytologist*, 75, 555–61.

Safir, G. R., Boyer, J. S. & Gerdemann, J. W. (1971). Mycorrhizal enhancement of water transport in soybean. *Science*, 172, 581–3.

Safir, G. R., Boyer, J. S. & Gerdemann, J. W. (1972). Nutrient status and mycorrhizal enhancement of water transport in soybean. *Plant Physiology*, 49, 700–3.

Saleh-Rastin, N. (1976). Salt tolerance of the mycorrhizal fungus *Cenococcum gramiforme* (Sow.) Ferd. *European Journal of Forest Pathology*, 6, 184–7.

Samtsevich, S. A. (1955). The importance of ectendotrophic mycorrhizas in tree nutrition. In *Mycotrophy in Plants*, ed. A. Imshenetskii (translated from Russian by Israel Programme for Scientific Translations). Washington D.C.: USDA & NSF 1967.

Sanders, F. E. & Tinker, P. B. (1973). Phosphate flow into mycorrhizal roots. *Pesticide Science*, 4, 385–95.

Sands, R. & Theodorou, C. (1978). Water uptake by mycorrhizal roots of radiata pine seedlings. *Australian Journal of Plant Physiology*, 5, 3019.

Sands, R., Fiscus, E. L. & Reid, C. P. P. (1982). Hydraulic properties of pine and bean roots with varying degrees of suberization. Vascular differentiation and mycorrhizal infection. *Australian Journal of Plant Physiology*, 9, 559–69.

Schramm, J. E. (1966). Plant colonization studies on black wastes from anthracite mining in Pennsylvania. *Transactions of the American Philosophical Society*, 56, 1–194.

Schütte, K. H. (1956). Translocation in the fungi. *New Phytologist*, 55, 164–83.

Shemakhanova, N. M. (1967). *Mycotrophy of Woody Plants*, Israel Program for Scientific Translations, 329 pp. Washington D.C.: USDA & NSF.

Shuja, N., Gilani, U. & Khan, A. G. (1971). Mycorrhizal associations in some angiosperm trees around New University Campus, Lahore. *Pakistan Journal of Forestry*, 21, 367–73.

Theodorou, C. (1978). Soil moisture and the mycorrhizal association of *Pinus radiata* D. Don. *Soil Biology and Biochemistry*, 10, 33–7.

Tisdall, J. M. & Oades, J. M. (1979). Stabilization of soil aggregates by the root systems of ryegrass. *Australian Journal of Soil Research*, 17, 429–41.

Warren-Wilson, J. & Harley, J. L. (1983). The development of mycorrhiza on seedlings of *Fagus sylvatica* L. *New Phytologist*, 95, 673–95.

Zerova, M. Y. (1955). Mykorrhiza formation in forest trees of the Ukrainian SSR. In *Mycotrophy in Plants*, ed. A. Imshenetskii (translated from Russian by Israel Programme for Scientific Translations). Washington D.C.: USDA & NSF 1967

17

Water availability, the distribution of fungi and their adaptation to the environment

PAUL DOWDING

Environmental Sciences Unit, Trinity College, Dublin 2, Eire

Water is the medium in which and through which all of the intra- and extra-cellular chemical reactions and solute transfers necessary for life take place. In addition, fungi require substantial amounts of water for the volume increase responsible for extension growth. Some fungi also require water in order to eject reproductive propagules or to get their spores airborne (Ingold, 1971; see also Ch. 4). Most require liquid water surrounding their spores before germination will take place. Clark (1965) has noted that fungi in soil, unlike plants, do not compete for water as their water requirements are so low relative to that present in their immediate environment. However, where fungi are intimately associated with plants they will have to compete with the plant cells for water (Cook & Papendick, 1970). In spite of these conventional requirements for water some fungal communities inhabit niches with high osmotic or matric stress or with widely fluctuating water regimes, and in such conditions it is likely that fungi do compete for water. Some fungi modify their niche in such a way as to change its water potential or retentivity, whilst others are adapted to germinate, grow and/or survive under conditions of low water potential.

The increasing volume of literature now appearing is evidence of an upsurge of interest in this subject. Much of the work reported has been on pure cultures in constant conditions, and mainly in osmotically altered liquid or semisolid media, where liquid water occupies most of the space. There are many environments or niches where fluctuations in water status are a frequent, if not regular, feature of life. In such situations water in any form does not occupy more than half the space, solutes are mostly intracellular, and most of the water stress is matric rather than osmotic. Thus, experimental findings may well not apply to field situations.

Discussion of the effects of water on fungal communities cannot be separated from that of the effects of temperature. Fluctuations in both variables are closely and dynamically associated in most of the niches occupied by terrestrial fungi. In addition, it has been shown that the ability of particular fungi to tolerate low water potential is increased at high temperatures and vice versa (Bruehl & Manadhar, 1972; Wearing & Burgess, 1979).

Sources of water in the environment

Rain

Rainfall is the most obvious input of water to niches occupied by fungi in terrestrial habitats, but it is also an important vehicle for dissolved substances and for particulates. When it first hits the surface of plant material or of soil it usually contains some dissolved minerals, carbonic acid and is saturated with oxygen, but does not ordinarily contain dissolved organic substrates. In addition, increasing amounts of anthropogenic mineral species, such as sulphate and nitrate ions, are occurring in rain in much of the Northern hemisphere (Ottar, 1978; Charlson & Rodhe, 1982). There will be an exchange of soluble materials between the rain and the leaf with its surface microflora. Microbial cells on the plant surface may become incorporated into the surface water film and will move with it. Excess water will drip or run off the surface onto which it first fell. This run-off water is nearly always enriched in soluble organic substances and with microbial cells when compared to unintercepted rain falling at the same time (Eaton, Likens & Bormann, 1973; Hart & Parent, 1974; Henderson, Harris, Todd & Grizzard, 1977). It also falls in larger drops (Geiger, 1965), which fall onto other leaves and onto the soil or organic layers overlying the soil (Ovington, 1954).

The combined amount of secondary rain and stem-flow is always less than the amount of rain originally intercepted by the canopy because all the plant surfaces retain a water film which subsequently evaporates and some water becomes airborne as microdroplets arising from each individual splash event (see Ch. 8 for further discussion). The residual water film is very important for epiphytic microorganisms. Kohnke (1975) reported that a 'good stand' of maize held 0.5 mm after each rainfall event greater than 1 mm. If one assumes a leaf area index of 5 then the film thickness on individual leaves would be between 100 and 50 μm depending on how much of the undersurface was wetted. The microdroplets arising from each splash will carry microbial cells from both the parent drop and from the water film into which it fell, and because of their high surface/volume ratio the microdroplets evaporate quickly leaving their contents, if any, truly airborne. Spores can also be splashed from litter or soil by the action of primary or secondary raindrops (see Ch. 5 for further discussion). Ashton & Macauley (1972) observed that soil splash was an important and obvious source of inoculum of the leaf-spot disease of *Eucalyptus regnans* seedlings, especially where seedlings were growing on soil that was only partly covered with leaf litter.

Once the rain has percolated through the live and dead plant material above the soil surface it enters the pore spaces in the soil. At this stage it is still carrying much oxygen as well as organic solutes of plant origin (Nihlgaard, 1970). It may be at a very different temperature to the surface layers of the soil or litter (Geiger, 1965). The sudden cooling of the surface of soil or litter by a convective shower during a warm sunny day must affect the activity of organisms in that niche. The cooling effect would be prolonged by the evaporation of surface and absorbed water.

Mists

As the size of water drops in the air decreases their rate of descent decreases and their residence time increases. Mists are defined by droplet sizes of 60 to 200 μm (Lapple, 1961), occur only in conditions of very high rh, and are important sources of water at particular altitudinal zones on several mountain ranges (e.g. the Andes and the Sierra Nevadas). In West Germany over half the precipitation at heights above 800 m was in the form of mists and fog (Geiger, 1965). The development of epiphytic lichens, particularly those with a fruticose growth form, is very noticeable in these areas. Dowding (1969) demonstrated that artificially generated mists from a rapidly spinning disc were able to pick up spores of sticky-spored fungi, such as *Ceratocystis* imperfect states and *Trichoderma viride*, but not those of *Cladosporium herbarum*.

Fogs

Fogs are defined by a droplet size, 2 to 60 μm, smaller than mists (Lapple, 1961). They are often characteristic of topographical features such as hollows or valley bottoms. Occasionally fogs will form as the result of the rapid evaporation of water from a rain-wetted forest canopy which is receiving positive radiation from the sky. Fogs are important microbiologically as the water drops initially form round airborne condensation nuclei, which in rural areas are mostly fungal spores and pollens liberated earlier the same day. The water film will start the hydration process and may become enriched with solutes leaked from the spore or pollen grain within. The stable atmospheric conditions in which fogs almost always form mean that the water-coated particles gradually settle out onto plant surfaces.

Dew

The same atmospheric conditions that lead to the formation of fogs lead in the evening and early night to dew formation on plant surfaces which have cooled below the dewpoint temperature of the air. Air tends to have a higher dewpoint temperature at the surface of and within a plant canopy than above the plant canopy, because of transpiration and reduced mixing within the stand (Geiger 1965; Oke, 1978). Dew may be augmented by upward distillation from the ground surface, by guttation from leaves, and occasionally may be heavy enough to drip off leaves as secondary rainfall. The most exposed leaves attract the most condensation (Fig. 17.1; Geiger, 1965) and amounts may be substantial. Made (1956) found that in a meadow during the three months

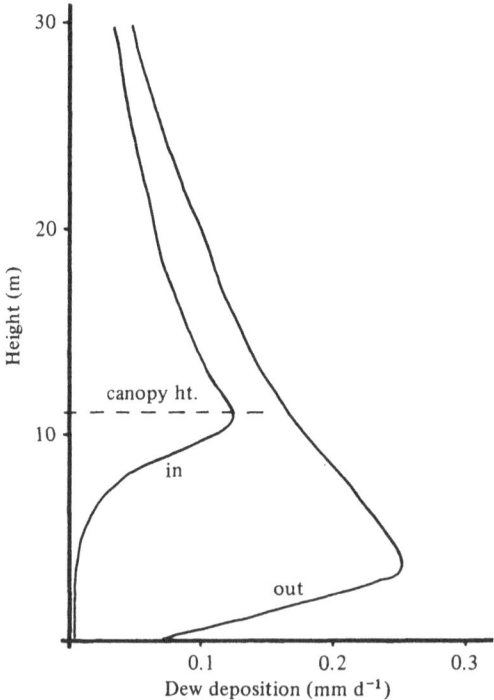

Fig. 17.1. The daily amount of dew deposition at different heights within and outside an *Abies* stand during the summer (after Geiger, 1965).

May to July an average of 2 mm of dew fell each dewy night and that the total duration of the water film was nearly 500 h in the 90 d of the study. By contrast, under a nearby beech tree the duration of dew was only 47 h in the 90 d (Fig. 17.2). The amount of dew varied with height inside and outside tree canopies but was always much less within the canopy than above or outside it. Kohnke (1975) reported that maize in the southern USA can accumulate up to 13 mm of dew in one month and soya beans 33 mm per month. For phylloplane organisms and for germinating spores of plant pathogens the nightly water film created by dew could easily be a more important water source than rain, particularly in the summer months in those areas which receive the greater part of their rain as convectional showers and thunderstorms.

Groundwater

Plant debris on the surface of the soil, and more particularly living and dead plant material in the soil itself, can be affected by the substitution of surrounding air by water. This water can be derived either from excessive rainfall which temporarily exceeds the drainage capacity

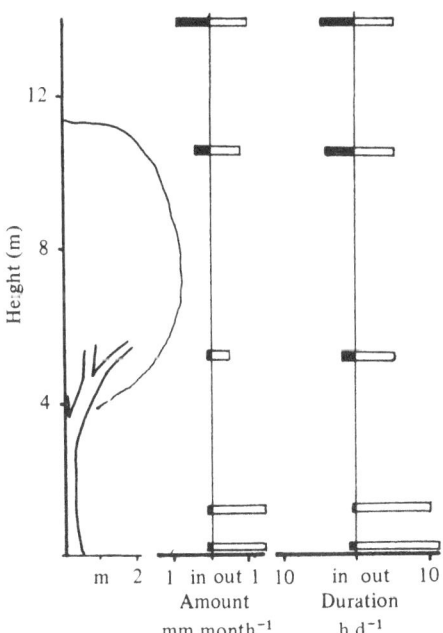

Fig. 17.2. The amount and duration of dew at different heights in and beside an isolated beech tree from June to August (after Made, 1956).

of the site or from rather longer-term changes in the level of the water table. In both cases the water will bring dissolved gases and solutes to the fungal communities in a particular situation, and microbial propagules. The increased liquid diffusion space created by the substitution of air by water in spaces adjacent to fungal niches may lead to enhanced losses of solutes by leaching (Kerr, 1964). The diffusion of gases and particularly of CO_2 and O_2 will be several orders of magnitude slower in a waterfilled soil than in an aerated soil and may, if temperatures are high enough, cause partial or complete anaerobiosis (see also Boddy, Ch. 21). In practice, soils are not uniform and pockets of anaerobic soil will exist in a largely aerobic soil and vice-versa. Water itself has a much higher specific heat than fungal substrata so that a wet substratum cools and warms more slowly than a dry one whether or not there is also heat loss/gain because of the transfer of water. The temperature of the substratum–water complex determines the solubility of O_2 and of CO_2 and in substrata where high water content has occluded the normally airfilled pore spaces the amount and rate of supply of dissolved gases can be of critical importance to aerobic fungi. High temperatures in water-saturated situations mean not only that less O_2 is available in solution but also that its rate of consumption will be high.

In this chapter only terrestrial situations will be considered. Fungal communities in permanently waterlogged soils or sediments and those on plants and plant material that is permanently submerged in still or moving water, be it fresh or saline, do not experience the magnitude or frequency of changes in water potential that are experienced by fungal communities in terrestrial environments.

Wet/dry cycles

The duration of water films on plant surfaces and in dead plant material is as important a regulator of the type of fungi present, and of their activity, as is temperature in above-ground niches (Magan, Cayley & Lacey, 1984). Unfortunately the water status and temperature of a niche are not independent and it is therefore necessary to describe the thermal environment of these niches in terms which allow simultaneous appreciation of changes in water status and in temperature. In most parts of the world there are considerable diurnal changes in both parameters, which in the case of temperature at or near the soil surface may exceed the annual amplitude of the average daily temperature. It is necessary to consider the changes in energy balance throughout diurnal, annual and phenological cycles to understand the changes in temperature and in water potential in any niche exposed to the sky (Geiger, 1965) and even in the surface layers of the soil (Rose, 1968).

Under sunny conditions and net positive energy flux, all exposed fungal niches will lose any surface water rapidly, unless the wind speed is very low and the vapour pressure of water in the air is high. Once the surface water has been lost transpiration can begin. Evaporation of dew or rainwater films from both the external surface and internal surfaces of the living leaf reduce the temperature of the leaf and will therefore reduce the gradient in water potential between the leaf and the air immediately surrounding it. The combined effect of transpiration and of reduced wind speed within a plant stand would be to increase humidity towards the base of the stand during the day (Geiger, 1965; Oke, 1978). Air movement at night lessens both the radiative cooling of the leaf surfaces and the accumulation of water-saturated air near the ground which are necessary for dew formation. Cloud cover at night will lessen the amount of radiative cooling, and during the day will reduce the radiative heating of plant and other surfaces. Dew may form at the start of a night and disappear during the night because of occlusion and an increase in wind speed (e.g. during the passage of a frontal system).

If the dew does not disappear before dawn, it will do so shortly after. It is worth noting here that after the sun has risen convective winds begin, especially in the absence of advective winds. The same principles apply to drying after a rainfall event. The uppermost leaves in a plant stand experience greater fluctuations in temperature and in water status than leaves lower in the canopy and changes in both parameters happen more frequently and more rapidly at the top of the canopy than within the canopy. The leaves within the canopy experience a higher mean water potential than the leaves at the top of the canopy but may be less frequently wetted. However, once the lower leaves are wet the water film will persist longer than at the top of the canopy.

Dead leaves which remain attached to the stem (as in *Calluna vulgaris*) or which remain standing (as in tussock-forming grasses e.g. *Molinia caerulea*) no longer have an internal water supply to keep them cool and hydrated when externally derived water films have evaporated. Consequently, when dry they tend to heat up more under a positive energy gradient than live leaves and can only lose heat by radiation and by conduction to the surrounding air. Organisms which persist in this niche after leaf death, or which subsequently colonise it, have to withstand greater diurnal variations in both temperature and water potential than do organisms on the surface of living leaves. If the dead leaves persist for longer than a few months they lose their cuticle, and with it the phylloplane yeasts, and then behave rather like blotting paper with regard to water and to their energy balance (allowing for the slightly lower albedo of dead plant material when compared to that of the white paper).

Dead leaves which separate from the plant only suffer the same fluxes in temperature and water status as attached dead leaves if they are freely exposed to the sky and effectively out of contact with the soil (which would act as a supplier of water from below). This will only happen on a large scale where the rate of leaf fall is great enough to lead to the formation of a multilayered A_0 horizon as in conifer and beech and oak woodland. In these niches the most recently fallen leaves often do not make effective hydraulic contact with the previous year's leaves and so dry out rapidly if they receive either energy from the sky or advection by unsaturated air. Leaves on woodland floors receive much less energy from the sky than leaves in the open so the potential rate of drying is much less, as are the temperature fluctuations. However, because of the interception of rain by the tree canopy and because of the much lower frequency of dew under trees (Made, 1956), there are not nearly so many wetting events as in the open. As a consequence the more slowly decomposed tree leaves, such as those of oak, beech and conifers, spend much of their first year after death in a dry state with relatively little microbial activity.

Once the dead leaves in a tussock or on the woodland floor have been covered by the following year's leaf fall, the niches and the organisms in them are protected to a great extent from the extremes of temperature and of water status which characterised their first year. Even if the hydraulic contact with the soil is not good, leaves in the subsurface position can receive water by condensation. When the surface leaves are wet and are warmer than the lower layers not all of the water will evaporate upwards into the air; some will evaporate into the spaces under the surface and will condense on lower layers of leaves. There will be a daily flux of water downwards from the litter surface caused by the daily flux of heat (Rose, 1968). At night the converse will tend to happen: the surface layers will cool and will attract not only water vapour from the air and from live plant material overhead but also water from below as the surface becomes cooler than deeper layers.

The further from the surface the niche is, the smaller is the daily flux in temperature and therefore in water status. Shameemullah, Parkinson & Burges (1971) have shown that the drying and rewetting of the H layer of mor humus are biphasic processes. At high water contents most of the water is in 'pores', the water potential is high and there is a rapid loss of water from the 'pore' phase at potentials above -0.1 MPa. The remainder of the water is held by 'adsorption' and is progressively removed as the potential of the surrounding medium decreases. As far as I am aware there have been no studies reported in which the diurnal changes in water status of the litter and humus horizons have been measured in detail equivalent to the temperature measurements reported above. The nearest information is that of Rose (1968).

The fungal floras of particular niches, their environments and adaptations

Phylloplane

The phylloplane environment is characterised by large fluctuations in water status and in energy balance, but because of the transpirational cooling of the leaves its temperature fluctuations are damped. However, within the phylloplane environment there exist more protected niches, e.g. the valleys overlying the junctions between epidermal cells and over leaf veins, and the depressions associated with stomata in many plants. The environments of the two sides of the leaf differ greatly in the *input* of energy and of water but not in their temperature and humidity (unless one surface is hairier than the other (Burrage, 1971)). The inclination, aspect and position of a leaf within the canopy or stand all affect its thermal and hydric environment in terms both of average and of extreme daily values. It is therefore difficult to generalise about a particular environment in numerical terms, but in biological terms the strongest selection pressures on the phylloplane organisms are the rapid fluctuations in temperature, water status and nutrient supply/loss. The last two factors are linked in that water is the medium for nutrient transfer and more importantly for the competition between microorganisms since there is an inevitable leakage of all solutes for a short while after rewetting (Simon, 1973). Blakeman (1973), and Stzejnberg & Blakeman (1973) considered that the leakage of soluble materials from hydrating spores of *Botrytis cinerea*, and the 'capture' of these materials by the pre-existing phylloplane residents, inhibited germination and was an important factor determining success or failure *in vivo*. Allen (1965) reported that urediniospores of *Puccinia graminis* germinated much more effectively when they were allowed to rehydrate slowly by exposure to a moist atmosphere than when they were immersed in water in the dry state. In the absence of liquid water around the hydrating spores there was minimal loss of soluble compounds whereas if they were immersed in water these compounds would be leached out. If the volume of water was large in relation to the spores the leached compounds would have to be taken up against a concentration gradient with the expenditure of precious energy reserves.

Most of the normal inhabitants of the phylloplane must survive and grow in an environment where the transition from dry conditions to those where free water is abundant is often very rapid and occurs frequently. One of the more striking adaptations of phylloplane organisms revealed by the scanning electron microscope is the production of large amounts of hydrophilic mucopolysaccharide (Blakeman, 1973; Bashi & Fokkema, 1976). Park (1982) has shown that the common phylloplane organisms stopped growing in culture at ψ below -15 MPa but survived well in the dried state. On rewetting they recommenced growth from the margins of the colonies. In contrast, nonphylloplane fungi were seriously affected by exposure to ψ below those at which they could grow (Park, 1982) and after drying the hyphal tips did not recover (Robertson & Rizvi, 1968). As the hyphal tips of a vegetative colony contain high concentrations of all important nutrients, such a loss would be a serious blow and would greatly weaken the competitive status of a mycelium. Daily repetition of such events would rapidly kill the mycelium. Northolt & Bullerman (1982) reported that the minimum ψ for *Cladosporium herbarum* was -20 MPa and that for *Alternaria tenuis* was -8.5 MPa, which demonstrated the better adaptation of the former to dry conditions. All the phylloplane organisms that have been studied are more sensitive to water stress in the reproductive mode than in the vegetative mode. *Cladosporium herbarum* spores are not dispersed in wet conditions (Ingold, 1971) and may require low humidity for the build-up of that electrostatic charge which may be necessary for spore discharge. Humid conditions are required for the formation and release of ballistospores by species of *Sporobolomyces* and of *Tilletiopsis* (Zoberi, 1964; Ingold, 1971; Pady, 1973). A low temperature requirement for sporulation would allow sporulation in the atmospheric conditions that cause dew formation in temperate climates.

Last (1955) and Carroll (1979) have reported that populations of phylloplane organisms increase with the age of a particular leaf. In the case of grasses, populations increase towards the oldest part of the leaf, the tip (Last, 1955). Carroll (1979) found a maximum epiphytic biomass on three-year-old Douglas Fir needles out of a range of ages 0–8 y. However, the biomass of epiphytes was lower at the top of the tree than in the middle on leaves of the same age, which he ascribed to the more stable environment of the mid-canopy (cf. Fig. 17.1). It is likely that the observed increase of phylloplane populations with leaf age on 'annual' leaves is in part due to the more stable thermal and hydric environments of most older leaves which are deeper within the canopy than younger leaves. Up to now the most plausible explanation has been that leaves become more 'leaky' as they age (Last, 1955). It is possible that the length of exposure to infection by spores and other propagules accounts for some of the observed delay in achievement of maximum populations of phylloplane organisms.

Leaf pathogens

The most critical stage in a leaf pathogen's life cycle is the germination/infection phase when it must temporarily establish itself in the phyllosphere before penetrating its host. The potential loss of essential solutes by leakage during imbibition of water has already been alluded to. It must be remembered however that the hydration process can begin before the spore reaches its destination leaf surface. Tommerup (1984) reported that spores can take up water from air in isothermal conditions to -2.5 MPa. Experimentation in this area is difficult as it is impossible to control water potentials in a biphasic vapour/liquid system in near-saturation conditions and one cannot be sure that conditions are sufficiently isothermal for condensation not to happen around the spores being investigated (Griffin, 1977). Higgins (1984) used KCl solutions for germination tests with *Pseudocercosporella herpotrichoides* and found that optimum germination occurred at -2.7 MPa with a minimum at -5 MPa. The results of experiments such as these must be interpreted with caution because liquid water surrounds the spores and they might behave differently in air of the same rh. Yarwood (1936) reported that the powdery mildews were an exception to this rule and could infect their host in the absence of liquid water. Somers & Horsfall (1966) reported that the highest water content of *Erysiphe* conidia at liberation was 69% and that the conidia were not only highly retentive of water but could 'sorb' it from the air. Schnathorst (1965) in a review on the powdery mildews concluded that this rather special group of organisms had members with very different water requirements for germination. Some could germinate and infect in the absence of liquid water but others could not. Ward & Manners (1974) and Harris & Manners (1983) have reported that the ability of *Erysiphe* conidia to germinate without water was related to the water stress at the time of conidial production which reduced the water content of the conidia. However, Ayres (1977) reported that conidial production by *Erysiphe* was lowered by water stress of infected pea plants, and Woolacott & Ayres (1984) found that conidial production by *Erysiphe* was lowered by water stress of seedling and young barley plants (−760 kPa *vs.* −320 kPa leaf water potential) but not by water stress of adult plants (−530 kPa *vs.* −370 kPa leaf water potential at the beginning of the day). Although Ayres & Paul (Ch. 15) cite evidence to show that stress during conidial production can increase subsequent infectivity, it must be concluded that *Erysiphe*, which was considered as an almost xerophilic genus because of the ability to germinate without liquid water, its position outside the host's cuticle and its greater prevalence in dry weather, is more sensitive to water stress in all its life phases than was suspected. It is likely that decreased competition from phylloplane saprotrophs on water-stressed hosts could partly explain both the enhanced infection rates and increased growth rates observed during dry weather.

Plant material at the soil surface (alive and dead)

The water relations of *Fusarium* diseases of the lower parts of the plant have been much studied and are discussed in detail by Duniway & Gordon (Ch. 7). The effects of low ψ and of regular diurnal excursions of water status and of temperature on leaf litter fungi have not received much attention. Mishra & Dickinson (1984) have recently reported on the effects of water stress on the growth and competitive abilities of a number of phylloplane and litter fungi inhabiting holly leaves. Dix (1984) found that the minimum ψ for growth of water stress-tolerant, litter-inhabiting Basidiomycotina was −6 to −7.2 MPa but three species of *Mycena* ceased growth at −1 to −3.6 MPa. Wilson & Griffin (1979) found that only one of several litter-decomposing Basidiomycotina tested could grow at ψ as low as −10 MPa. Tewary, Pandey & Singh (1982) established a significant correlation ($r^2 = 0.67$) between litter water content (%d.wt) and the respiration rate of litter in the field in deciduous woodland. Hintikka (1970) found that F and H layer humus inhabited by white rotting Basidiomycotina was drier than humus occupied by other organisms (187 *vs.* 253%) but it is not clear whether this difference was a cause or an effect, and in any case the ψ at all the water contents reported would be nearly zero.

Cellulolytic, lignolytic and pectolytic fungi create new voids in plant material and so increase its water-holding capacity. As the voids become larger their ability to retain water decreases. It is likely therefore that as decay proceeds the fungal community of large woody niches especially becomes more vulnerable to dry conditions and less affected by wet conditions than at the outset of decay.

Fungi can also alter the water potential of their environment by translocation of water (see Eamus & Jennings, Ch. 2) and by their metabolism (see Boddy, Ch. 21). Water status in large bulks (of cereal grains, hay, wood chips, milled fen peat) is never completely uniform and fungal activity can start in the damper pockets. This activity generates heat which in a normal environment would be dissipated to the surroundings. However in bulk stores heat cannot escape, and it warms the substrate in the damp pocket. Water vapour then moves from the warm volume to surrounding cooler substrata, where it condenses on and warms the surfaces and permits initiation of fungal activity (see also Ayerst, Ch. 20).

Conclusions

In most terrestrial niches fungal communities can only respond to changes in their water status that are dictated by the environment. There must be some terrestrial niches where water status is always optimal and where other factors such as substratum quality and saprotrophic grazing determine community composition. However, in most climatic regions of the world, water supply from the atmosphere is irregular in timing and in amount. Many of the niches occupied by terrestrial fungi experience changes in water status that range from the supra-optimal to the sub-minimal on a time scale that is too short to allow for reproductive reorientation of the community. Under such conditions (as would be the rule in most of Europe for dead plant material above the soil) different members of a community cannot all be active simultaneously, especially at the extremes of water status and of temperature. The relative duration and direction of extreme periods must select the balance of active fungal species. Unfortunately, most methods commonly used for isolation and identification do not distinguish well between resting-but-viable and active species, and the effects of fluctuating water status on the *active* members of a community must often go unnoticed unless they are extreme. Plant and animal physiologists now routinely take account of diurnal fluctuations in important physical parameters in their laboratory studies. It is time that mycologists escaped from the womb of the Petri plate/culture flask in an incubator, into the real world.

318   *P. Dowding*

References

Allen, P. J. (1965). Metabolic aspects of spore germination in fungi. *Annual Review of Phytopathology*, **3**, 313–21.

Ashton, D. H. & Macauley, B. J. (1972). Winter leaf spot disease of seedlings of *Eucalyptus regnans* and its relation to forest litter. *Transactions of the British Mycological Society*, **58**, 377–86.

Ayres, P. G. (1977). Effect of water potential of pea leaves on spore production by *Erysiphe pisi* (powdery mildew). *Transactions of the British Mycological Society*, **68**, 97–100.

Bashi, E. & Fokkema, N. J. (1976). Scanning electron microscopy of *Sporobolomyces roseus* on wheat leaves. *Transactions of the British Mycological Society*, **67**, 500–5.

Blakeman, J. P. (1973). The chemical environment of leaf surfaces with special reference to spore germination of pathogenic fungi. *Pesticide Science*, **4**, 578–85.

Bruehl, G. W. & Manadhar, J. B. (1972). Some water relations of *Cercosporella herpotrichoides*. *Plant Disease Reporter*, **56**, 594–6.

Burrage, F. (1971). The leaf surface environment. In *Leaf Surface Organisms*, ed. T. F. Preece & C. H. Dickinson, pp. 91–101. London: Academic Press.

Carroll, G. C. (1979). Needle epiphytes in a Douglas fir canopy: biomass and distribution patterns. *Canadian Journal of Botany*, **57**, 1000–7.

Charlson, R. J. & Rodhe, H. (1982). Factors controlling the acidity of natural rainwater. *Nature*, **295**, 683–5.

Clark, F. E. (1965). The concept of competition in microbial ecology. In *Ecology of Soilborne Plant Pathogens*, ed. K. F. Baker & W. C. Snyder, pp. 339–45. Los Angeles:California University Press.

Cook, R. J. & Papendick, R. I. (1970). Soil water potential as a factor in the ecology of *Fusarium roseum* f. sp. *cerealis culmorum*. *Plant and Soil*, **32**, 131–45.

Dix, N. J. (1984). Minimum water potentials for growth of some litter-decomposing agarics and other basidiomycetes. *Transactions of the British Mycological Society*, **83**, 152–3.

Dowding, P. (1969). The dispersal and survival of fungi causing bluestain of pine. *Transactions of the British Mycological Society*, **54**, 169–81.

Eaton, J. S., Likens, G. E. & Bormann, F. H. (1973). Throughfall and stem flow chemistry in a northern hardwood forest. *Journal of Ecology*, **61**, 495–508.

Geiger, R. (1965). *The Climate Near the Ground*. Boston: Harvard University Press.

Griffin, D. M. (1977). Water potential and wood-decay fungi. *Annual Review of Phytopathology*, **15**, 319–29.

Harris, J. G. & Manners, J. G. (1983). Influence of relative humidity on germination and disease development in *Erysiphe graminis*. *Transactions of the British Mycological Society*, **81**, 605–11.

Hart, G. E. & Parent, D. R. (1974). Chemistry of throughfall under Douglas fir and Rocky Mountain juniper. *American Midwestern Naturalist*, **92**, 191–201.

Henderson, G. S., Harris, W. F., Todd, D. E. & Grizzard, T. (1977). Quantity and chemistry of throughfall as influenced by forest type and season. *Journal of Ecology*, **65**, 365–74.

Higgins, S. (1984). Factors affecting spore germination and initial growth of *Pseudocercosporella herpotrichoides* on wheat. *Transactions of the British Mycological Society*, **82**, 443–8.

Hintikka, V. (1970). Studies on white-rot humus formed by higher fungi in forest soils. *Communications Instituti Forestalis Fenniae*, **67**, 1–68.

Ingold, C. T. (1971). *Fungal Spores: Their Liberation and Dispersal.* Oxford: Clarendon Press.

Kerr, A. (1964). The influence of soil moisture on infection of peas by *Pythium ultimum. Australian Journal of Biological Science,* **17**, 676–85.

Kohnke, H. (1975). *Soil Physics.* New York: McGraw-Hill.

Lapple, C. E. (1961). Characteristics of particles and particle dispersoids. *Stanford Research Institute Journal,* **5**, 95.

Last, F. T. (1955). Seasonal incidence of *Sporobolomyces* on cereal leaves. *Transactions of the British Mycological Society,* **38**, 221–39.

Made, R. (1956). Zur Methodik der Taumessung. *Wissenschaftliches Zeitschrift der Martin Luther University,* **5**, 483–512.

Magan, N., Cayley, G. R. & Lacey, J. (1984). Effect of water activity and of temperature on mycotoxin production by *Alternaria alternata* in culture and on wheat grain. *Applied and Environmental Microbiology,* **47**, 1113–17.

Mishra, R. R. & Dickinson, C. H. (1984). Experimental studies of phylloplane and litter fungi on *Ilex aquifolium. Transactions of the British Mycological Society,* **82**, 595–604.

Nihlgaard, B. (1970). Precipitation, its chemical composition and effect on soil water in a beech and spruce forest in south Sweden. *Oikos,* **21**, 208–17.

Northolt, M. D. & Bullerman, L. D. (1982). Prevention of mold growth and toxin production through control of environmental conditions. *Journal of Food Protection,* **45**, 519–26.

Oke, T. R. (1978). *Boundary Layer Climates.* Methuen: London.

Ottar, B. (1978). An assessment of the OECD study on the long range transport of air pollutants. *Atmospheric Environment,* **12**, 445–54.

Ovington, J. D. (1954). A comparison of rainfall in different woodlands. *Forestry London,* **27**, 41–53.

Pady, S. M. (1973). Ballistospore discharge in *Tilletiopsis minor. Canadian Journal of Botany,* **51**, 589–93.

Park, D. (1982). Phylloplane fungi: tolerance of hyphal tips to drying. *Transactions of the British Mycological Society,* **79**, 174–8.

Robertson, N. F. & Rizvi, S. R. H. (1968). Some observations on the water relations of the hyphae of *Neurospora crassa. Annals of Botany, N.S.,* **32**, 279–91.

Rose, C. W. (1968). Water transport in soil with a daily temperature wave I. Theory and experiment. *Australian Journal of Soil Research,* **6**, 31–44.

Schnathorst, W. C. (1965). Environmental relationships in the powdery mildews. *Annual Reviews of Phytopathology,* **3**, 343–66.

Shameemullah, M., Parkinson, D. & Burges, D. (1971). The influence of soil moisture tensions on the fungal population of a pinewood soil. *Canadian Journal of Microbiology,* **17**, 975–86.

Simon, E. W. (1973). Phospholipids and plant membrane permeability. *New Phytologist,* **73**, 377–420.

Somers, E. & Horsfall, J. G. (1966). The water content of powdery mildew conidia. *Phytopathology,* **56**, 1031–5.

Stzejberg, A. & Blakeman, J. P. (1973). Studies on the leaching of *Botrytis cinerea* conidia and dye absorption by bacteria in relation to competition for nutrients on leaves. *Journal of General Microbiology,* **78**, 15–21.

Tewary, C. K. Pandey, U. & Singh, J. S. (1982). Soil and litter respiration rates in different microhabitats of a mixed oak–conifer forest and their control by edaphic conditions and substrate quality. *Plant and Soil,* **65**, 233–8.

Tommerup, I. C. (1984). Effect of soil water potential on spore germination by vesicular-arbuscular mycorrhizal fungi. *Transactions of the British Mycological Society,* **83,** 193–202.

Ward, S V. & Manners, J. G. (1974). Environmental effects on the quantity and viability of conidia produced by *Erysiphe graminis. Transactions of the British Mycological Society,* **62,** 119–28.

Wearing, A. H. & Burgess, L. W. (1979). Water potential and the saprophytic growth of *Fusarium roseum "graminearum". Soil Biology and Biochemistry,* **11,** 661–7.

Wilson, J. M. & Griffin, D. M. (1979). The effects of water potential on the growth of some soil basidiomycetes. *Soil Biology and Biochemistry,* **11,** 211–2.

Woolacott, B. & Ayres, P. G. (1984). Effects of plant age and water stress on production of conidia by *Erysiphe graminis* f. sp. *hordei* examined by non-destructive sampling. *Transactions of the British Mycological Society,* **82,** 449–54.

Yarwood, C. E. (1936). The tolerance of *Erysiphe polygoni* and certain other powdery mildews to low humidity. *Phytopathology,* **26,** 845–59.

Zoberi, M. H. (1964). Effect of temperature and humidity on ballistospore discharge. *Transactions of the British Mycological Society,* **47,** 109–14.

18
Water and the origins of decay in trees

A. D. M. RAYNER

School of Biological Sciences, University of Bath, Claverton Down, Bath, BA2 7AY, UK

Attenuation of the activity of decay fungi by both high and low moisture contents of wood is a long established fact, finding practical application in the seasoning and ponding of timber as means of preventing deterioration (e.g. Cartwright & Findlay, 1958). In dead timber, moisture, or lack of it, affects fungal metabolism directly, whilst in living trees there is the additional possibility of indirect action via effects on the metabolism and morphogenesis of living plant cells.

Most recent emphasis has been placed on allelopathic chemicals and direct responsiveness to infection of living wood tissues as major, primary determinants of decay distribution in standing trees (Shigo & Marx, 1977; Shortle, 1979; Mercer, 1982; Shigo, 1984). However, the very essence of the standing tree is that it is spatiotemporally markedly heterogenous with respect to moisture distribution, with different parts having high, low, intermediate or fluctuating water contents and Boddy & Rayner (1983a) have argued that *this* may have primary significance in the aetiology of tree decay. There is therefore a need to integrate knowledge of decay-patterns with what is known (or needs to be known) about water distribution and transport in trees, and how this is affected by injury, stress and pathogenesis. Accordingly, this chapter seeks to examine the extent to which observed patterns of decay in standing trees can be related to the way in which problems imposed by differing moisture regimes might be countered by decay fungi with distinctive colonisation strategies. First it is necessary to examine the distribution of moisture in trees and how this affects potential fungal microenvironments.

Water in the standing tree
Sapwood, heartwood and wetwood

In intact, unstressed trunks and branches, the most obvious differences in wood moisture content usually occur between the saturated outermost growth ring or rings, which are functional in water conduction, and progressively older xylem. These differences may develop either abruptly or in, a continuous gradient, and are usually most clear-cut in coniferous trees with a morphologically distinctive heartwood, where a relatively dry transition zone marks the boundary between moist sapwood and much drier heartwood. Sapwood–heartwood differences in angiosperms are less clear-cut and data for some species even indicate higher moisture contents exist in heartwood than sapwood (Skaar, 1972). This may sometimes be due to the greater frequency of the condition known as 'wetwood' in angiosperms than in conifers (see below). However, it should also be appreciated that it is difficult to obtain reliable measurements of sapwood moisture in angiosperms where severance of functional vessels under conditions of hydrostatic tension leads to the immediate withdrawal of water from the damaged elements and concomitant replacement with air; this problem is less acute in conifers with their much shorter conducting elements (i.e. tracheids).

Variations in the degree of hydrostatic tension or pressure may also account for the marked changes in sapwood moisture content data which have been reported to occur with season, especially in deciduous angiosperms. Commonly these data indicate a high moisture content in spring, a lowering during summer and a return to high levels following leaf fall in autumn (Gibbs, 1958). Such data clearly accord with alternating imposition and relaxation of hydrostatic tension and Zimmermann (1971) insisted that they should not be interpreted as resulting from the generation and refilling of gas-filled space. However, Milburn (1979) concluded that cavitation (the introduction of gas-filled space into water columns under tension) occurred rather readily in trees, and could be reversed by redissolution when hydrostatic tension is lowered by absorption from external sources including rainwater. Positive hydrostatic pressures developed during spring could also help to reverse cavitation, at least in lower parts of the stem. Cavitation is discussed further by Jones (Ch. 13).

Another methodological problem affecting interpretation of comparative data is the general use of percentage moisture per unit dry weight as the criterion for ascertaining moisture levels. As indicated by Boddy (Ch. 21) this measure is affected by the density of the wood, which of course can vary considerably both within and between trees. For the microenvironment of fungi, the most significant feature of moisture in living trees is its reciprocal relationship with the gaseous phase, so a more useful way of describing differences between tissues is in terms of the per cent volume of gases, free water and water within and between the cell wall and woody cell-wall substance. This is illustrated in Fig. 18.1 which demonstrates the replacement of the liquid by a gaseous phase in progressively older wood of a typical coniferous tree.

The gaseous phase in intact wood originates from a variety of sources, of which cavitation has already been mentioned. Freezing of xylem sap in winter represents another source, since the formation of ice brings gases out of solution, hence producing bubbles (Scholander, 1958). The drawing of gas through pores in conduit walls, a process known as nuclea-tion, might also contribute, but this is not clearly established in intact wood (Milburn, 1979). Due to the presence of pit tori in conifers, and small pit membrane apertures in angiosperms, which resist passage of an air–water interface by surface tension, movement of air from one

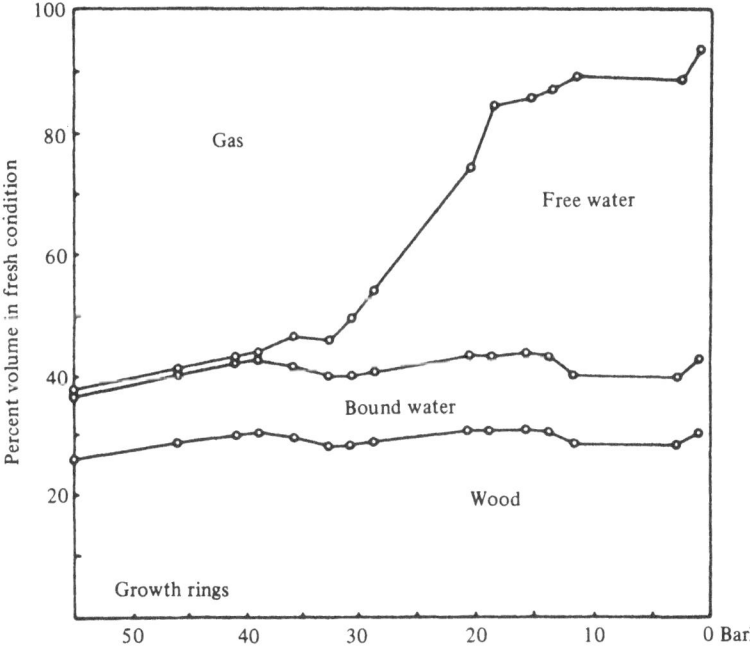

Fig. 18.1. Distribution of water and gas in xylem of a spruce trunk (after Trendelenburg, 1939).

conduit to another is not believed to occur readily in intact xylem (see Jones: Ch. 17).

In the relatively dry heartwood transition zone the presence of a gaseous phase is associated with the disappearance of nuclei and starch from ray parenchyma cells and may be due to a shift in metabolism, perhaps induced by ethylene, to the acetate and shikimic acid pathways (Hillis, 1977; Coutts, 1977; Shain, 1979). This in turn can lead to the accumulation of fungitoxic or fungistatic phenolic, tropolone or polyphenolic extractives, and in angiosperms tyloses occlude the vessels (if they have not done so already), completing the transition to heartwood.

Whilst the formation of a gaseous phase during heartwood formation might thus be thought to alleviate stresses resulting from lack of aeration in functionally intact xylem, the ameliorative effects may be countered at least partially by accumulation of allelopathic chemicals and formation of physical obstacles to fluid (and hyphal) migration. It is also important to note that the gas is not the same as ambient air since it is enriched with CO_2 and depleted of O_2. Thus CO_2 concentrations of 10–20% have commonly been reported, and rigorous studies using mass spectrometry indicated a composition approaching 100% CO_2 in living stems of *Acacia* (Hintikka, 1982; Carrodus & Triffett, 1975).

As already indicated, central wood need by no means always be drier than outer wood in standing trees, and a particular case is provided by the formation of wetwood. The term 'wetwood' has been applied to any non-living wood with a water-soaked appearance in living trees, and its aetiology is probably correspondingly diverse (Worrall & Parmeter, 1982). Thus it has been found in association with mechanical wounds, frost cracks, insect attacks, branch stubs, butt rot and mistletoe cankers (Hartley, Davidson & Crandall, 1961; Etheridge & Morin, 1962) as well as in intact and apparently healthy trees (Wilcox, 1968; Coutts & Rishbeth, 1977). It can contain dense bacterial populations, which in some cases have been attributed a causal role in its formation (Carter, 1945; Seliskar, 1952; Preece, 1977). Such populations include obligately anaerobic forms indicating highly anoxic conditions, as in cottonwood wetwood where methane is formed (Zeikus & Ward, 1974).

In several fir (*Abies*) species wetwood is usually present in the heartwood; these include *A. grandis* (Coutts & Rishbeth, 1977) and *A. concolor* (Worrall & Parmeter, 1982, 1983). In *A. concolor* hypoxic conditions in wetwood were demonstsrated by measurement of oxygen diffusion rates using platinum microelectrodes. Despite the fact that small amounts of air were probably introduced during their insertion, these still gave an average value of 35 ng cm^{-2} min^{-1} which is similar to data from water-logged soils and sediments (Armstrong, 1967; Worrall & Parmeter, 1983). Evidence that physiological activity is required for wetwood production in *Abies* species has been obtained by Coutts & Rishbeth (1977), Bauch, Holl & Endeward (1975) and Worrall & Parmeter (1982). In each case the formation of a dry zone, perhaps homologous with the heartwood transition zone (see above), has been implicated, in which nuclear disintegration and disappearance of starch occur. Worrall & Parmeter (1982) have also demonstrated more negative osmotic potentials in wetwood than in sapwood, associated with selective accumulation of potassium and calcium ions, and organic acids. This would provide a mechanism leading to water accumulation in wetwood.

Effects of wounding, colonisation and stress

Establishment of decay fungi in standing trees is commonly associated with mechanical injury, physiological stress affecting the whole or part of the tree, or pathogenic effects either of the fungi themselves or of other organisms. It is therefore necessary to understand how these factors change the moisture conditions prevailing in the intact, unstressed tree as well as their relationship to observed decay patterns.

Effects of wounding. The most immediate consequence of mechanical injury extending deeper than the vascular cambium is that the underlying xylem will be brought into direct contact with the exterior and hence, unless the structure is submersed, with relatively dry air. The extent to which this air actually enters will depend on many variable factors, but in particular on whether water columns in the xylem are under positive or negative hydrostatic pressure at the time of injury (Zimmermann, 1983; Leben, 1985). If the water columns are under tension (i.e. negative pressure) and the wound is deep, there will be an immediate withdrawal of water from severed conduits and concomitant entry of air, potentially as far as the connecting pit membranes of adjacent water-filled conduits where surface tension or aspiration will tend to prevent further passage of menisci. Lateral connections may also allow net flow of water away from inner severed elements, which may already be partially aerated prior to injury, towards outer, more actively conductive tissue. Lateral connections may also maintain a supply to the latter from adjacent undamaged elements. Thus, two inwardly tapered regions of air-accessed tissue occur, one being above and the other below a wound made into the main stem, the one above being of somewhat greater extent because

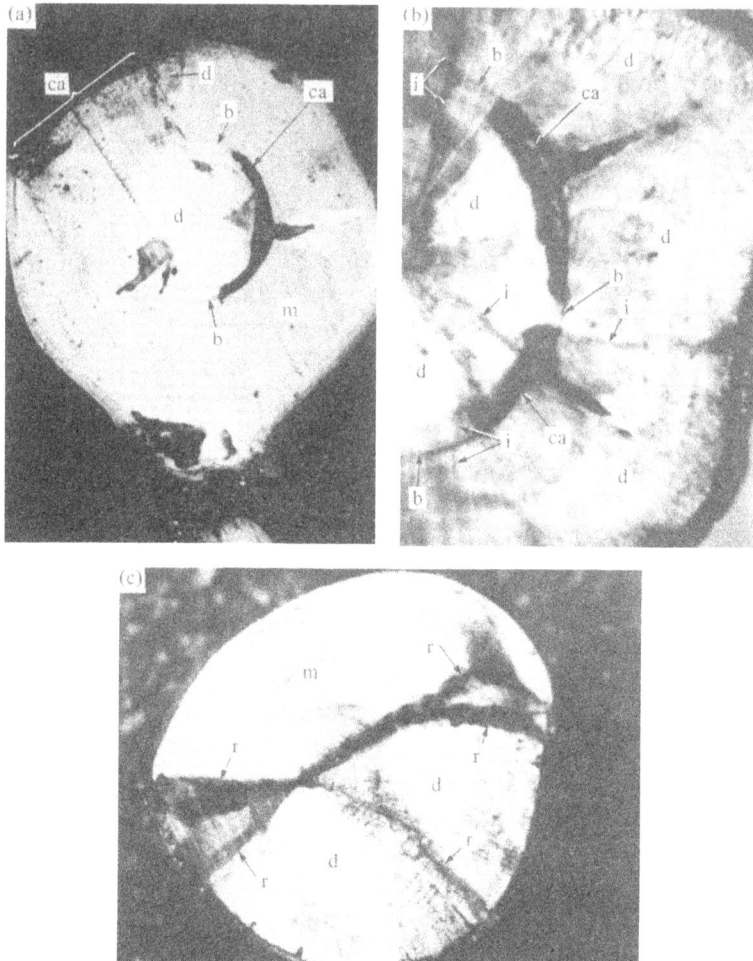

Fig. 18.2. Formation and breaching of 'barrier' and 'reaction' zones. (a) Transverse section through a cankered sycamore (*Acer pseudoplatanus*) trunk. A paler, drier central cylinder of wood is delimited from moist outer sapwood by a barrier zone produced in association with canker formation some years prior to felling. A newly formed canker on the outside of the trunk is associated with a wedge-shaped region of drying and discoloration/ decay. (b) Transverse section through a beech (*Fagus sylvatica*) branch which had been subject to water stress over a period of several years. As in (a), cankers formed several years previously are associated with a 'barrier' zone, but in this case the subsequently formed sapwood has been accessed by decay fungi in wedged-shaped regions separated by interaction zone lines which directly traverse the barrier zone. (c) Transverse section through an attached beech branch undergoing decay. Dark-coloured reaction zones delimit the decayed from the moist functionally intact sapwood,

of the upward flow of water (cf. Fig. 18.3). Where branches are severed, access into the main stem will tend to be below and not above the branch junction.

The extent of these air-accessed tissues will also be determined by the degree of tension and its duration in water columns, and by size and distribution of conducting elements – obviously angiosperms with very long vessels will be much more accessible than conifers. Eventually the accessed tissues will become sealed off from the remainder of the xylem by formation of tyloses and/or deposition of gums, resins or suberised layers and often by accumulation of polyphenolic extractives.

Because limitation of the spread of decay and discoloration in trees generally tends to coincide with deposition of these sealant layers they have commonly been interpreted as defensive barriers produced in direct response to and primarily against invasive microorganisms; the presence in them, in some instances, of allelopathic chemicals has strengthened this view (Mercer, 1982; Shigo, 1984; cf. Shain, 1979). However, although these barriers undoubtedly do contain fungal invasion which would otherwise probably continue unimpeded (for reasons to be discussed shortly), they do so non-specifically (Mullick, 1977; Shain, 1979) and it may be better to regard them as repair mechanisms which protect the plumbing of the tree from damage of either biotic or abiotic origin. They may help to arrest the spread of cavitation from air-accessed tissue as it dries and cracks and undergoes microbial invasion. In these terms, sealant zones represent the divide between microenvironmental conditions inhospitable to fungi in water-filled xylem and the more favourable conditions in aerated tissue, rather than defensive barriers *per se*. This view is supported by the fact that the chemicals they contain vary considerably in toxicity and that, as shown in Fig. 18.2, they are readily breached if the wood behind them becomes dried.

Although it is difficult to distinguish between primary and secondary effects, numerous observations of colonisation patterns from both inoculated and uninoculated wounds provide evidence that fungal invasion coincides rather precisely with expected patterns of access of air into

Fig. 18.2. (*cont.*)

and there are also older reaction zones undergoing degradation which have been breached as decay has spread tangentially. There is a close correlation between the position in which reaction zones have formed, and death and loss of bark cover. b, barrier zones; ca, cankers; d, dry, decayed or discoloured regions; i, interaction zone lines; m, moist regions of functionally intact sapwood; r, reaction zones.

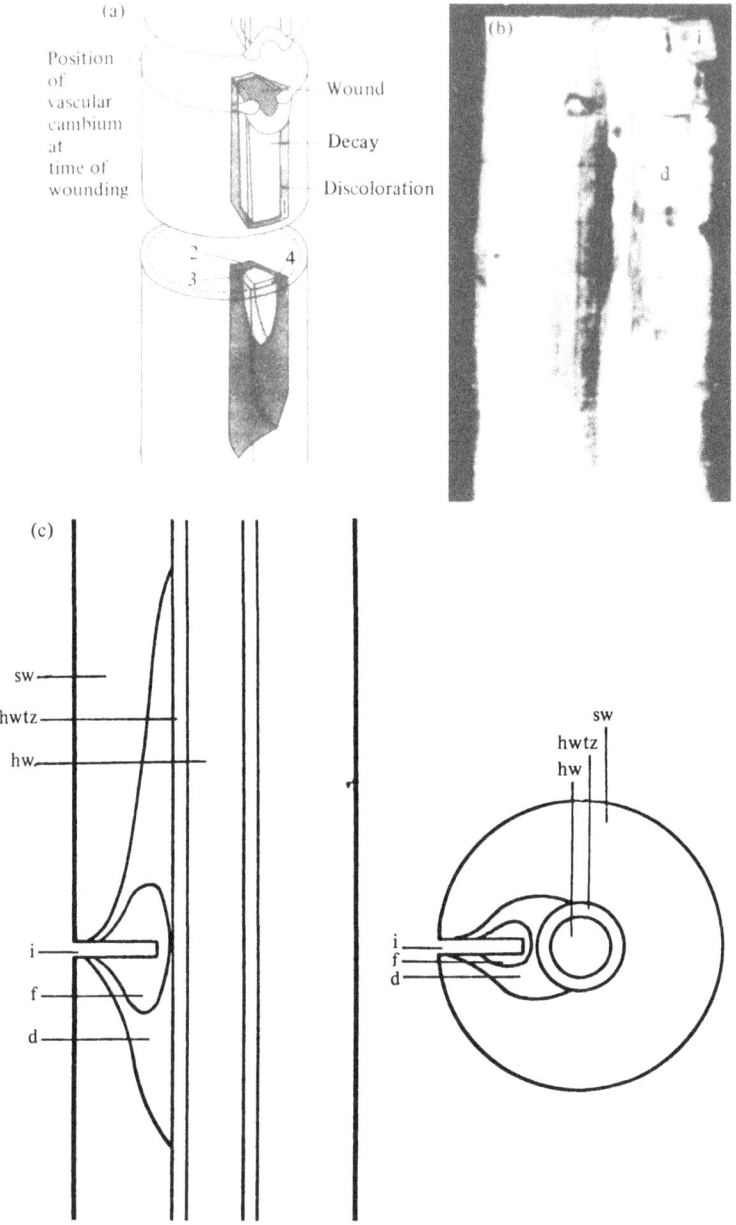

Fig. 18.3. Development of decay, discoloration and dry zones associated with deep wounds. (a) Idealised diagram illustrating typical pattern of colonisation. 1, 2, 3 and 4 represent the interfaces which develop between colonised and uncolonised wood, and have been interpreted by some (e.g. Shigo, 1984) as defensive 'walls'. (From Boddy & Rayner, 1983*a*.) (b) Decay

wood. Thus, as illustrated in Fig. 18.3, where deep wounds are made, it is common to find that colonisation, resulting in decay and discoloration, tapers away from outermost tissues and is more extensive longitudinally above than below the wound. (An interesting exception, perhaps reflecting a different mode of colonisation, has been found with *Cryptostroma corticale* in sycamore; see Dickenson & Wheeler, 1981). Several inoculation studies have also shown that growth of decay fungi into tissues adjacent to wounds becomes curtailed after 3–6 months (e.g. Rayner, 1979; Boddy & Rayner, 1984). Occasionally, however, as in oak, such curtailment is 'escaped', whence a long narrow decay column is formed – perhaps as a result of growth down a cavitated vessel (Boddy & Rayner, 1983*a*).

Several inoculation studies have also demonstrated the possibility of dynamic interplay between a decay fungus and the conducting system of the tree, such that air access is extended *beyond* that initially caused by wounding: this interplay may continue as long as the tissues do not become permanently sealed off and in this case the sealant zones could perhaps justifiably be regarded as primarily defensive in their effects. A very clear example has been provided by Coutts (1976) who demonstrated the formation of extensive dry zones in advance of infection by *Heterobasidion annosum* growing from wood dowel inocula inserted into the trunks of *Abies grandis* and several other conifer species. Two months after inoculation *H. annosum* had grown to an equal extent (30–50 mm) upwards and downwards. Tapered dry zones extended considerably in advance of infection: like mycelial extension, these were maximal in the inner sapwood, but they progressed further upwards than downwards (Fig. 18.3*c*). Similar observations were obtained by inoculating freshly cut pine logs, except that in this case there was no difference between the extension of dry zones towards the originally upper or lower portions of the trunk: hence the asymmetry was a feature of the standing tree – as would be expected if the directionality of water flow was an important factor. Further experiments with the logs demonstrated convincingly that formation of dry zones was at least partly due to withdrawal

Fig. 18.3. (*cont.*)

column (d) produced from a wood plug inoculum (i) in an attached oak (*Quercus robur*) branch, showing preferential colonisation of inner sapwood. (c) Longitudinal and transverse sections showing dry zones (d) and fungal colonisation (f) spreading from dowel inocula (i) containing *Heterobasidion annosum* into sapwood (sw) of conifer trunks. hw, heartwood; hwtz, heartwood transition zone. (From Coutts, 1976.)

of water from damaged tracheids by hydrostatic tension, and that the effect of infection was, via lysis of bordered pit tori, to extend the range over which water was withdrawn. The fact that the dry zones extended well beyond the region of lysis of pit tori indicated that an additional factor was responsible for the drying phenomenon in living wood. Further experiments indicated an active role for slowly dying ray parenchyma – a feature strongly reminiscent of the formation of dry heartwood transition zones previously mentioned.

If functional conduits are cut when under positive hydrostatic pressures the sap flows outwards and little air is introduced into the system before the damaged tissue is sealed off. Under these circumstances one would expect fungal colonisation to be limited. Evidence for this idea has recently been obtained by Leben (1985) who studied the occurrence of bark die-back and formation of discolored wood columns in trunks of red maple, *Acer rubrum*, wounded with horizontal chain-saw cuts 1 cm wide, 4–6 cm deep and 10–16 cm long. There was a marked difference between the extent of discolored wood columns resulting from wounds inflicted in March–May and ones inflicted in September–November, the latter giving rise to longer columns. Furthermore, the extent of discoloration was greater upwards than downwards, and there was a positive correlation between the extent of discoloration and bark die-back, which occupied approximately triangular areas above and below the wounds.

The pivotal significance of bark death. Xylem function and the maintenance of a living bark layer (taken here to mean all tissues outside and including the vascular cambium) are interdependent. Living bark protects the underlying xylem from the vagaries of the external environment and supplies nutrients to living medullary ray cells, whilst the vascular cambium is responsible for the supply of water, mineral nutrients and – at times of high physiological activity at the beginning of the growing season – organic nutrients to the living bark and cambial cells.

It is therefore not surprising that there is an intimate relationship between bark death and development of decay in xylem. This is illustrated by the association between beech bark disease caused by *Nectria coccinea* and development within the trunk of white rot, caused in particular by *Bjerkandera adusta*, rendering the tree susceptible to wind-snap (Parker, 1974, 1983; Houston, Parker, Perrin & Lang, 1979). However, this relationship has been obscured by overemphasis on the role of major mechanical wounds and direct pathogenesis *to xylem* – themselves something of an artefact of forest management practices – in the aetiology of decay in trees (Shigo & Marx, 1977; Shigo, 1979; Merrill & Shigo, 1979; Mercer, 1982). Natural decay processes are often initiated neither by obvious mechanical injury nor by decay fungi acting directly as pathogens. In these cases the interrelationship with bark death becomes very obvious, distinctive decay patterns occur and, although primary and secondary effects are hard to disentangle, the water status of the tree appears to be a fundamental underlying factor.

Bark death in unwounded trees can follow two basic patterns. Either it occurs in localised patches (cankers) – which may be close enough together to coalesce in extreme cases – or in continuous lengths. Whereas the former is probably normally due to growth of pathogens in the bark itself, the latter may more usually be associated with colonisation of, or lack of supply from, the underlying xylem. Both conditions commonly occur under conditions of water stress, and both offer a chance to understand decay aetiology in relation to moisture status.

The importance of water in relation to canker development has been recognised for some time. For example, Bier (1959, 1961) showed that development of cankers caused by *Cryptodiaporthe salicella* and *C. salicina* on *Salix*, and *Fusarium lateritium* on *Populus trichocarpa* depended on the relative turgidity of the bark being below a critical level (80% in the former case). More recently, Bedker & Blanchette (1983) have described the development of cankers caused by *Nectria cinnabarina* on honey locusts after root pruning. Other examples have been discussed by Schoeneweiss (1981; see also Ch. 9).

A very striking example of a drought-related canker disease is the diamond bark disease of sycamore (*Acer pseudoplatanus*) which occurred in Britain following the drought years of 1975 and 1976 (Bevercombe & Rayner, 1980). The disease, which is often associated with *Dichomera saubinetii*, results in formation on the main trunk of diamond-shaped necrotic areas of bark which generally increase in frequency upwards so that the tops of trees often died and lost their entire bark cover (Bevercombe & Rayner, 1980; Rayner, Bevercombe, Brown & Robinson, 1981). Examination of the wood underlying diamond cankers revealed the presence of relatively dry discolored and/or decayed regions often delimited by a greenish- or brownish-black 'reaction zone'. The position of the colonised regions and reaction zones varied and correlated with the degree of access of air into the tissues (Fig. 18.4). For example, on the lower parts of trunks colonisation was commonly limited to the outermost annual ring or rings so that only a peripheral zone of decay

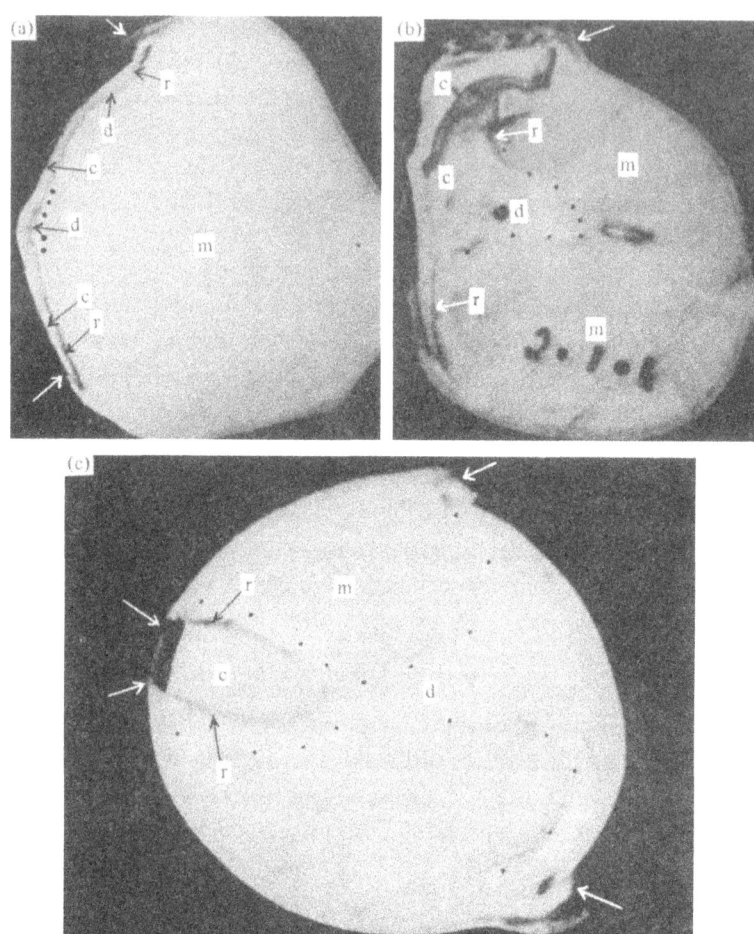

Fig. 18.4. Correlation between colonisation, drying and reaction zone formation underlying diamond cankers on sycamore. (a) Transverse section
through a cankered region from the lower part of a trunk showing limitation
of colonisation and drying to an outer peripheral zone. (b) Section through
a multiply cankered region showing drying and colonisation progressing
in a wedge-shaped zone to the centre of the trunk. (c) Section from the
upper part of a trunk with a large and much smaller canker. A region
of drying progressing from the large canker on the right is associated with
development of a wedge-shaped zone of colonisation, delimited by a reaction zone, progressing from the small canker on the left. c, colonised
regions; d, dry regions; m, moist regions; r, reaction zones; dotted lines,
junctions between pale dry and darker, moist regions; arrows, limits of
cankered regions.

or discoloration developed. On severely affected trees and on the upper parts of trunks other reaction zones formed along radii, delimiting wedge-shaped regions of decay or discoloration. Furthermore, series of transitory reaction zones were occasionally observed, illustrating the readiness with which they could be breached (Boddy & Rayner, 1983*a*; Cooke & Rayner, 1984; see also Fig. 18.2). The switch from peripheral to wedge-shaped colonisation patterns is consistent with the expected changes in moisture status up the tree, as well as providing a useful illustration of the two alternative consequences for fungal colonisation of localised bark death or removal.

The formation of elongated strips of dead bark also clearly indicates a connection with moisture distribution, not least because these strips seem to reflect the pattern of the underlying xylem, forming spiralling tongues up the trunks, particularly of angiospermous trees, as well as commonly being associated with drought stress. A good example is provided by a *Diatrype* sp. on beech (*Fagus sylvatica*) (Lonsdale, 1983). Formation of these strip cankers, which is often associated with branch or root dieback (Lonsdale, 1983), renders the underlying wood accessible to decay fungi; indeed in the case of certain *Hypoxylon* spp., for example, the 'causative' organisms (but see discussion of primary and secondary effects below) may themselves be decay species. Another possible example of a strip canker fungus is provided by the sooty bark fungus *Cryptostroma corticale* which formed elongated stromata up to several metres long, often spiralling up the trunk from sites of squirrel damage, on sycamore following the 1975 and 1976 droughts in southern Britain (Abbott, Bevercombe & Rayner, 1977; Young, 1978; Bevercombe, 1980; Cooke & Rayner, 1984). Interestingly, there is experimental evidence that water stress promotes spread of this fungus in xylem (Dickenson & Wheeler, 1981).

The foregoing examples relate to damage to the main trunk, particularly under conditions of drought predisposition (see Schoeneweiss, Ch. 9). However, bark death also occurs regularly in trees not obviously subject to extreme water stress, where it is associated with the decay and ultimate shedding of attached branches by a process which has sometimes been referred to as natural pruning (Millington & Chaney, 1973). As the latter term implies, this may be at least partly a beneficial process, providing a means for redistribution of resources from and disposal of branches which lose their usefulness during the upward and outward expansion of the canopy. As indicated by Kramer & Kozlowski (1979) one likely underlying mechanism is internal competition for water: those branches unable to compete become senescent and dehydrated and hence accessible to decay fungi.

Remarkably, development of decay in attached tree branches has been neglected. However, Boddy & Rayner (1981, 1982, 1983*b,c*, 1984) have recently made a detailed study of decay of branches of oak (*Quercus robur*) and, to a lesser extent, of beech (*Fagus sylvatica*) and ash (*Fraxinus excelsior*) (L. Boddy & A. D. M. Rayner, unpubl.; Boddy, Gibbon & Grundy, 1985). In all cases a close relationship between development of decay and bark death was found, associated with the delimitation of wedge-shaped decay regions by dark-coloured 'reaction zones' extending radially inwards from the junctions between living and dead cambial regions. In oak these zones were filled by gum deposits and tylosis-occluded vessels, and in branches large enough to contain heartwood they extended from and were morphologically similar to the latter. Hence the zones were termed 'heartwood wings'. The position in which they formed suggested that they were triggered by access of air through, and necrosis of, medullary ray parenchyma, recalling the mechanisms of heartwood formation in relatively dry transition zones (see above).

Heartwood wings can be several metres long. They delimit decay columns occupied by extensive individuals of Basidiomycotina such as *Stereum gausapatum, Phlebia rufa, Vuilleminia comedens* and *Peniophora quercina*. The wings persist long after the sapwood and bark have disappeared, giving rise to the characteristic 'ridged and grooved' appearance of long-dead oak branches. Similar decay patterns occur in ash, which like oak is ring-porous and has thick bark, the decay in this case commonly being caused by the Ascomycotina *Daldinia concentrica* and *Hypoxylon rubiginosum*, as well as Basidiomycotina such as *Peniophora limitata*, but durable heartwood wings of the type seen in oak do not occur.

In beech, which is diffuse-porous and has thin, smooth bark, rather more complex decay patterns occur, and it is common to find several foci of bark death, as well as the continuous columns found in oak and ash. *Oudemansiella mucida* is commonly associated with these foci (Rayner & Boddy, 1986). Decay in beech branches is also commonly less intense than that developing in oak, and this seems to be related to the greater rate at which the wood dries out and the protective bark layer is shed: in oak, intact, living bark and functional sapwood can persist, probably for many years, adjacent to decay columns.

This last point introduces the final variable into the equation between bark death and decay development. Whereas death of bark allows access of decay fungi, rapid loss of the tissue results in exposure of the wood to the atmosphere, so that its moisture content is no longer determined primarily by supply from within the tree, but fluctuates with the external environment.

Water and colonisation strategies

A coherent model relating decay patterns to moisture conditions may be developed in terms of the distinctive colonisation strategies adopted by decay fungi to counter the various types of stress that may be imposed on them. Five behavioural classes are recognised: unspecialised opportunism; specialised opportunism; heartrot; pathogenesis; and desiccation tolerance. Individual decay fungi may adopt any one of these behaviour patterns, or may combine two or more of them.

Heartrot

An obvious means of circumventing the inaccessibility of living sapwood is to grow in central wood where problems of aeration and active host response are minimised. Hence it is not surprising that so many fungi causing decay of standing trees produce heartrot. However, growth in heartwood often creates its own problems, due to the accumulation therein of allelopathic extractives, and many heartrot fungi exhibit slow growth, lack of combative ability and show a high degree of selectivity for particular tree species, indicating adaptive specialisation for this habitat (Cooke & Rayner, 1984). In a few cases, as with *Heterobasidion annosum* and certain *Phellinus* spp., these fungi may be capable of slow encroachment from the heartwood into sapwood, ultimately resulting in depletion of the latter until it becomes insufficient to meet the needs of the tree; these fungi may therefore also be regarded as truly pathogenic (Shain, 1979, J. R. Parmeter, personal communication).

Where wetwood occurs, heartwood loses its value as a haven from water saturation and anoxic or hypoxic conditions strongly inhibit decay fungi; indeed the condition has been regarded as a mechanism of decay resistance (Van der Kamp, Gokhale & Smith, 1979; Worrall & Parmeter, 1983). Here it is of interest that heartwood of *Abies* spp. is normally resistant to decay in the standing tree, where wetwood is present (see above), but in service is regarded as one of the least durable of conifer heartwoods (Brown, Panshin & Forsaith, 1949).

Unspecialised opportunism

Contrasting markedly with the behaviour of heartrot fungi is that of fungi which colonise sapwood made accessible by wounding or rapid death of bark. As has been described above, wounding or rapid death of bark renders accessible a region of aerated wood in which fungi can grow until their progress is arrested at the junction between accessed and functionally intact tissues. Wounding or bark death can thus be regarded in ecological terms as a form of enrichment disturbance wherby a virgin resource is suddenly made available for colonisation, which, under the relatively non-selective conditions, can be effected by a wide range of organisms (Cooke & Rayner, 1984; Coates & Rayner, 1985). The typical pattern of colonisation of wounds, with non-decay organisms dominant at early stages and decay fungi dominant later (Shigo, 1979; Mercer, 1982), follows the pathway from ruderal to combative communities which, according to a schema proposed by Cooke & Rayner (1984), is expected following enrichment disturbance. Similar community development patterns occur in dead seasoned wood placed under suitable conditions for colonisation (Levy, 1982). Therefore, it may well be a mistake to suggest that the observed changes in dominance are due to an ordered succession dictated by, and overcoming, host responses, albeit that detoxification of extractives is integral in the community development patterns (see Shigo, 1979). It may be better, therefore, to regard wound colonisation as being effected primarily by recruitment of fungi and other microorganisms which happen to be advantageously placed to capitalise on the opportunity. Although some adaptive specialisation to the microenvironmental conditions is to be expected, this will be less marked than with other colonisation strategies.

Specialised opportunism

A rather different type of opportunism is exhibited by decay fungi colonising intact sapwood which becomes predisposed to colonisation by water stress resulting from drought, root damage, internal competition or other factors (see Schoeneweiss, Ch. 9).

A common feature of such fungi appears to be that they very rapidly produce extensive individual genotypes within the tree and that, unlike the situation in wounded wood, there may be little or no participation of non-decay-causing microorganisms in the colonisation process. Thus in attached oak branches, lack of additional annual rings in living sapwood adjacent to decay columns several metres long and occupied by single genotypes, of for example *Stereum gausapatum*, *Phlebia rufa* and *Vuilleminia comedens*, indicated that these columns were formed within a single growing season (Boddy & Rayner, 1982, 1983b,c). This type of situation, which probably also applies to many host-specialised xylariaceous Ascomycotina such as *Daldinia concentrica* in ash and *Hypoxylon nummularium* in beech (Boddy *et al.*, 1985; Rayner & Boddy, 1986) as well as the basidiomycete *Piptoporus betulinus* in birch (Adams, 1982), suggests a mechanism of colonisation other than mycelial extension from a colonisation focus. This is supported further by the development of extensive decay columns of *Stereum gausapatum* colonising naturally compared with much smaller columns arising from inoculation into wounds in oak branches (Boddy & Rayner, 1984).

One mechanism which has been proposed to account for the development of extensive decay columns is prior colonisation and/or dissemination of propagules within xylem fluids, allowing extensive mycelial development to proceed rapidly as soon as microenvironmental conditions in sapwood are ameliorated by predisposition (Boddy & Rayner, 1982, 1983*a*; Rayner & Boddy, 1985). Such a 'latent invasion' strategy would parallel closely the behaviour of certain vascular wilt pathogens, and dissemination of propagules represents yet another way in which the presence of moving water columns could influence decay aetiology.

As mentioned above, colonisation under predisposing conditions is often associated with death of bark. A major issue which then arises is whether bark death precedes colonisation, is the result of it, or whether there is some interaction between the two.

Active pathogenesis

The last point introduces the fact that rather than awaiting injury or predisposing conditions, a fungus may invade unstressed sapwood directly by virtue of active pathogenesis. This is, of course, a large and complex issue and here it will only be pointed out that sapwood can be made accessible for both fluids and hyphae by active destruction of pit membranes, vascular and cork cambium and medullary ray cells. Perhaps such features as the ectotrophic infection habit of root-infecting decay species, whereby bark is colonised and vascular cambium killed well in advance of colonisation of the wood (Garrett, 1970), should be reexamined in this light.

Desiccation tolerance

Desiccating conditions may obtain within the tree with loss of sapwood function and of bark cover. There is a further group of fungi which seem to be selectively favoured by these conditions. At least some of these, such as *Rhodotus palmatus* which commonly inhabits the desiccated upper portions of standing dead elm trunks (Gibbs & Gulliver,

1977; Rayner & Hedges, 1982), and *Schizopora paradoxa* in decorticated oak branches, produce chlamydospores. For *R. palmatus* and several xylariaceous species there is the further possibility that they may themselves create or maintain dry conditions, and that this allows them to persist in the presence of potential competitors (A. D. M. Rayner & L. Boddy, unpubl.; see also Boddy, Ch. 21).

Concluding comments

Complex interplay may occur between the activities of decay fungi and the moisture conditions in trees. Only by coming to terms with and attempting to understand the nature of this interplay will it be possible to obtain a proper picture of the complex aetiology of decay in trees.

Acknowledgements. I would like to thank Dr Lynne Boddy and Dr David Lonsdale for their careful reading and helpful comments on the first draft of this article.

References

Abbott, R. J., Bevercombe, G. P. & Rayner, A. D. M. (1977). Sooty bark disease of sycamore and the grey squirrel. *Transactions of the British Mycological Society*, **69**, 507–8.

Adams, T. J. H. (1982). *Piptoporus betulinus*, some aspects of population biology. Ph.D. Thesis, University of Exeter.

Armstrong, W. (1967). The relationship between oxidation-reduction potentials and oxygen-diffusion levels in some waterlogged organic soils. *Journal of Soil Science*, **18**, 27–34.

Bauch, J., Höll, W. & Endeward, R. (1975). Some aspects of wetwood formation in fir. *Holzforschung*, **29**, 198–205.

Bedker, P. J. & Blanchette, R. A. (1983). Development of cankers caused by *Nectria cinnabarina* on honey locusts after root pruning. *Plant Disease*, **67**, 1010–13.

Bevercombe, G. P. (1980). Diseases affecting sycamore bark. Ph.D. Thesis, University of Exeter.

Bevercombe, G. P. & Rayner, A. D. M. (1980). Diamond-bark diseases of sycamore in Britain. *New Phytologist*, **86**, 379–92.

Bier, J. E. (1959). The relation of bark moisture to the development of canker diseases caused by native, facultative parasites. I. *Cryptodiaporthe* canker on willow. *Canadian Journal of Botany*, **37**, 229–38.

Bier, J. E. (1961). The relation of bark moisture to the development of canker diseases caused by native, facultative parasites. IV. Pathogenicity studies of *Cryptodiaporthe salicella* (Fr.) Petrak and *Fusarium lateritium* Nees, on *Populus trichocarpa* Torrey and Gray, *P. 'Robusta'*, *P. tremuloides* Michx., and *Salix* sp. *Canadian Journal of Botany*, **39**, 139–44.

Boddy, L., Gibbon, O. M. & Grundy, M. (1985). Ecology of *Daldinia concentrica* from ash: effect of abiotic variables on mycelial extension and interspecific interactions. *Transactions of the British Mycological Society*, **85**, 201–11.

Boddy, L. & Rayner, A. D. M. (1981). Fungal communities and formation of heartwood wings in attached oak branches undergoing decay. *Annals of Botany*, **47**, 271–4.

Boddy, L. & Rayner, A. D. M. (1982). Population structure, intermycelial interactions and infection biology of *Stereum gausapatum*. *Transactions of the British Mycological Society*, **78**, 337–51.

Boddy, L. & Rayner, A. D. M. (1983a). Origins of decay in living deciduous trees: the role of moisture content and a reappraisal of the expanded concept of tree decay. *New Phytologist*, **94**, 623–41.

Boddy, L. & Rayner, A. D. M. (1983b). Ecological roles of basidiomycetes forming decay communities in attached oak branches. *New Phytologist*, **93**, 77–88.

Boddy, L. & Rayner, A. D. M. (1983c). Mycelial interactions, morphogenesis and ecology of *Phlebia radiata* and *Phlebia rufa* from oak. *Transactions of the British Mycological Society*, **80**, 437–48.

Boddy, L. & Rayner, A. D. M. (1984). Internal spread of fungi inoculated into attached oak branches. *New Phytologist*, **98**, 155–64.

Brown, H. P., Panshin, A. J. & Forsaith, C. C. (1949). *Textbook of Wood Technology*. New York: McGraw-Hill.

Carrodus, B. B. & Triffett, A. C. K. (1975). Analysis of composition of respiratory gases in woody stems by mass spectrometry. *New Phytologist*, **74**, 243–6.

Carter, J. C. (1945). Wetwood of elms. *Illinois Natural History Survey Bulletin*, **23**, 405–48.

Cartwright, K. St G. & Findlay, W. P. K. (1958). *Decay of Timber and its Prevention*, 2nd edn. London: HMSO.

Coates, D. & Rayner, A. D. M. (1985). Fungal population and community development in cut beech logs. III. Spatial dynamics, interactions and strategies. *New Phytologist*, **101**, 183–98.

Cooke, R. C. & Rayner, A. D. M. (1984). *Ecology of Saprotrophic Fungi*. London, New York: Longman.

Coutts, M. P. (1976). The formation of dry zones in the sapwood of conifers. I. Induction of drying in standing trees and logs by *Fomes annosus* and extracts of infected wood. *European Journal of Forest Pathology*, **6**, 372–81.

Coutts, M. P. (1977). The formation of dry zones in the sapwood of conifers. II. The role of living cells in the release of water. *European Journal of Forest Pathology*, **7**, 6–12.

Coutts, M. P. & Rishbeth, J. (1977). The formation of wetwood in grand fir. *European Journal of Forest Pathology*, **7**, 13–22.

Dickenson, S. & Wheeler, B. E. J. (1981). Effect of temperature and water stress in sycamore on growth of *Cryptostroma corticale*. *Transactions of the British Mycological Society*, **76**, 181–5.

Etheridge, D. E. & Morin, L. A. (1962). Wetwood formation in balsam fir. *Canadian Journal of Botany*, **40**, 1336–45.

Garrett, S. D. (1970). *Pathogenic Root-Infecting Fungi*. Cambridge: Cambridge University Press.

Gibbs, J. N. & Gulliver, C. C. (1977). Fungal decay of dead elms. *European Journal of Forest Pathology*, **7**, 193–200.

Gibbs, R. D. (1958). Patterns in the seasonal water content of trees. In *The Physiology of Forest Trees*, ed. K. V. Thimann, pp. 19–41. New York: Ronald Press.

Hartley, C., Davidson, R. W. & Crandall, B. S. (1961). Wetwood, bacteria and increased pH in trees. *U.S. Forest Service, Forest Products Laboratory Report* **2215**, 1–34.

Hillis, W. E. (1977). Secondary changes in wood. In *The Structure, Biosynthesis and Degradation of Wood*, ed. F. A. Loewus & V. C. Runeckles, pp. 247–309. New York, London: Plenum Press.

Hintikka, V. (1982). The colonization of litter and wood by basidiomycetes in Finnish forests. In *Decomposer Basidiomycetes: Their Biology and Ecology*, ed. J. C. Frankland, J. N. Hedger & M. J. Swift, pp. 227–39. Cambridge: Cambridge University Press.

Houston, D. R., Parker, E. J., Perrin, R. & Lang, K. J. (1979). Beech bark disease: a comparison of the disease in North America, Great Britain, France and Germany. *European Journal of Forest Pathology*, **9**, 199–211.

Kramer, P. J. & Kozlowski, T. T. (1979). *Physiology of Woody Plants*. New York, San Francisco, London; Academic Press.

Leben, C. (1985). Wound occlusion and discolouration columns in red maple. *New Phytologist*, **99**, 485–90.

Levy, J. F. (1982). The place of basidomycetes in the decay of wood in contact with the ground. In *Decomposer Basidiomycetes: Their Biology and Ecology*, ed. J. C. Frankland, J. N. Hedger & M. J. Swift, pp. 161–78. Cambridge: Cambridge University Press.

Lonsdale, D. (1983). Some aspects of the pathology of environmentally stressed trees. *Yearbook International Dendrology Society*, **1982**, 90–7.

Mercer, P. C. (1982). Basidiomycete decay of standing trees. In *Decomposer Basidiomycetes: Their Biology and Ecology*, ed. J. L. Frankland, J. N. Hedger & M. J. Swift, pp. 143–60. Cambridge: Cambridge University Press.

Merrill, W. & Shigo, A. L. (1979). An expanded concept of tree decay. *Phytopathology*, **69**, 1158–60.

Milburn, J. A. (1979). *Water Flow in Plants*. London, New York: Longman.

Millington, W. F. & Chaney, W. R. (1973). Shedding of roots and branches. In *Shedding of Plant Parts*, ed. T. T. Kozlowski, pp. 149–204. New York, London: Academic Press.

Mullick, D. B. (1977). The non-specific nature of defense in bark and wood during wounding, insect and pathogen attack. In *The Structure, Biosynthesis and Degradation of Wood*, ed. F. A. Lowus & V. C. Runeckles, pp. 395–441. New York, London: Plenum Press.

Parker, E. J. (1974). Beech bark disease. *Forestry Commission Forest Record*, **96**, 1–15.

Parker, E. J. (1983). Beech bark disease in Great Britain. In *Proceedings, I.U.F.R.O. Beech Bark Disease Working Party Conference*, pp. 1–6. Hamden, Connecticut: USDA Forest Service.

Preece, T. F. (1977). Watermark Disease of the cricket bat willow. *Forestry Commission Leaflet*, **20**, 1–9.

Rayner, A. D. M. (1979). Internal spread of fungi inoculated into hardwood stumps. *New Phytologist*, **82**, 505–17.

Rayner, A. D. M. & Boddy, L. (1986). Population structure and the infection biology of wood decay fungi in living trees. *Advances in Plant Pathology* (in press).

Rayner, A. D. M. & Hedges, M. J. (1982). Observations on the specificity and ecological role of basidiomycetes colonizing dead elm wood. *Transactions of the British Mycological Society*, **78**, 370–3.

Rayner, A. D. M., Bevercombe, G. P., Brown, T. C. & Robinson, A. (1981). Fungal growth in a lattice: a tentative explanation for the shape of diamond-cankers in sycamore. *New Phytologist*, **87**, 383–93.

Schoeneweiss, D. F. (1981). The role of environmental stress in diseases of woody plants. *Plant Disease*, **65**, 308–14.

Scholander, P. F. (1958). The rise in sap in lianas. In *The Physiology of Forest Trees*, ed. K. V. Thimann, pp. 3–17. New York: Ronald Press.

Seliskar, C. E. (1952). Wetwood organism in aspen, poplar is isolated. *Colorado Farm Home Research*, **2**, 6–11, 19–20.

Shain, L. (1979). Dynamic responses of differentiated sapwood to injury and infection. *Phytopathology*, **69**, 1143–7.

Shigo, A. L. (1979). Tree decay: an expanded concept. *United States Department of Agriculture Forest Service Agricultural Information Bulletin* No. **405**.

Shigo, A. L. (1984). Compartmentalization: a conceptual framework for understanding how trees grow and defend themselves. *Annual Review of Phytopathology*, **22**, 189–214.

Shigo, A. L. & Marx, H. G. (1977). Compartmentalization of decay in trees. *United States Department of Agriculture Forest Service Agricultural Information Bulletin* No. **405**.

Shortle, W. C. (1979). Mechanisms of compartmentalization of decay in living trees. *Phytopathology*, **69**, 1147–51.

Skaar, C. (1972). *Water in Wood*. Syracuse University Press.

Trendelenburg, R. (1939). *Das Holz also Rohstoff.* Munich, Berlin: Lehmann.

Van der Kamp, B., Gokhale, A. A. & Smith, R. S. (1979). Decay resistance owing to near-anaerobic conditions in black cottonwood wetwood. *Canadian Journal of Forest Research*, **9**, 39–44.

Wilcox, W. W. (1968). Some physical and mechanical properties of wetwood in white fir. *Forest Products Journal*, **18**, 27–31.

Worrall, J. J. & Parmeter, J. R. (1982). Formation and properties of wetwood in white fir. *Phytopathology*, **72**, 1209–12.

Worrall, J. J. & Parmeter, J. R. (1983). Inhibition of wood-decay fungi by wetwood of white fir. *Phytopathology*, **73**, 1140–5.

Young, C. W. T. (1978). Sooty bark disease of sycamore. *Arboricultural Leaflet* No 3, London: HMSO.

Zeikus, J. G. & Ward, J. C. (1974). Methane formation in living trees: a microbial origin. *Science*, **184**, 1181–3.

Zimmermann, M. H. (1971). Transport in the xylem. In *Trees Structure and Function*, ed. M. H. Zimmermann & C. L. Brown, pp. 169–220. New York, Heidelberg, Berlin: Springer-Verlag.

Zimmermann, M. H. (1983). *Xylem Structure and the Ascent of Sap*. New York: Springer-Verlag.

19
Post-harvest spoilage of fruits and vegetables

C. DENNIS

Campden Food Preservation Research Association, Chipping Campden, Gloucestershire GL55 6LD, UK

Introduction: General aspects of post-harvest fungal disease

Post-harvest diseases of fruits and vegetables represent one of the most severe sources of loss of production (Table 19.1; Harvey, 1978; Coursey, 1967, 1971; Thorne, 1972; Eckert, 1977; Derbyshire & Shipway, 1978). The economic cost of post-harvest losses is proportionally greater than for losses in the field because costs of harvesting, transport and storage must be added to those of production.

The susceptibility of fruits and vegetables to infection by fungi varies considerably between different crops which is to be expected for a group which includes such diverse botanical structures as storage organs, such as tubers, tap roots and bulbs, leafy tissues, buds, inflorescences and fruits. Although fruits have the same basic reproductive function they are very varied in structure and hence in susceptibility to fungal spoilage. The occurrence and progress of fungal disease after harvest depends on the properties both of the fungus and of the fruit or vegetable concerned together with the interaction of these properties with growing, harvesting and storage conditions (Eckert & Ratnayake, 1983; Dennis, 1985). The most important considerations are the ability of the fungus to produce appropriate enzymes and toxins, especially pectolytic enzymes, the susceptibility of the host to these enzymes and/or toxins, the availability of natural openings or wounds for entry of the fungi and the susceptibility of invading fungi to preformed or postinfectional inhibitors in the host. The ability of wounded vegetative tissue to heal obviously will also influence susceptibility to rotting and, since healing is substantially influenced by storage conditions including relative humidity it is discussed later in the chapter.

The major post-harvest disease of fruits and vegetables and their time

343

Table 19.1 *Reported losses of fruits and vegetables caused by post-harvest diseases*

Fruit or vegetable	Fungus	% Loss	Reference
Apple	*Phytophthora syringae*	30% or more	Edney, 1978 Upstone, 1978
Apricot	*Monilinia* *Rhizopus* *Botrytis cinerea*	Up to 19%	Anon., 1965
Cherry	*Rhizopus* *Penicillium* *Monilinia fructicola* *Botrytis cinerea*	Up to 50%	Anon., 1965
Citrus	*Penicillium italicum* *Diplodia natalensis* *Phomopsis citri*	Up to 25%	Anon., 1965
Peach	*Monilinia fructicola* *Rhizopus*	Up to 35%	Anon., 1965
Pear	*Penicillium* *Botrytis cinerea*	Up to 25%	Anon., 1965
Strawberry	*Botrytis cinerea* *Rhizopus*	Up to 50%	Anon., 1965
Cabbage	*Phytophthora porri*	25–50%	Geeson, 1976
Carrot	*Mycocentrospora acerina*	Up to 38%	Derbyshire, 1973
Celery	*Mycocentrospora acerina*	30–100%	Derbyshire, 1973 Derbyshire & Crisp, 1971
Onion	*Botrytis allii*	15–60%	Maude & Presly, 1979 Derbyshire & Shipway, 1978 Maude *et al.*, 1984
Tomato	*Geotrichum candidum* *Rhizoctonia solani* *Sclerotium rolfsii*	5–20%	Onesirosan & Fatunla, 1976

of initiation are listed in Table 19.2. Susceptibility of crops and relative importance of fungi in causing post-harvest rotting can vary according to site, season, variety, damage and harvesting conditions as well as by post-harvest treatments and storage conditions (Derbyshire & Shipway, 1978; Dennis, 1983*a*, 1985).

At the time of harvest it is not possible to predict the storage potential of different fruits and vegetables, partly because visual disease symptoms of 'latent' or 'quiescent' infections only become apparent after a period in store, e.g. after five months for anthracnose disease of bananas (Simmonds, 1941). Quiescent infections are more common in fruits than in vegetative tissues (Swinburne, 1983; Dennis, 1983a), but the latter may become infected at harvest or during storage (see Day, Lewis & Martin, 1972; Dennis, 1983a).

The conditions of temperature, relative humidity (rh) and gaseous composition of the atmosphere during storage profoundly affect the development of fungal rotting. Temperature is undoubtedly the most important parameter as it can have not only a direct effect on fungal growth but also an indirect effect via the physiological status of the fruit or vegetable, by influencing metabolic activity. In addition temperature determines the amount of water vapour required to saturate the atmosphere, therefore affecting the evaporative loss which in turn will also affect the physiological state of the commodity. Thus, for many fruits and vegetables, low-temperature storage (except for chill-sensitive crops (Dennis, 1985)) is the most effective and practical method for delaying the development of fungal decay (see Dennis, 1983a,b).

The rh of the storage environment will be discussed in relation to its effect on the water status of the host in the next section. In general, however, fruits and vegetables which can be stored at temperatures approaching 0 °C require a rh of 97 to 100% to prevent excessive water loss; notable exceptions are onions, garlic and ginger. Fruits and vegetables which have to be stored at higher temperature, in order to avoid chilling injury, require a lower rh to avoid excessive fungal growth. It is thus fortunate that many such commodities have a waxy cuticle and are much less susceptible to evaporative loss than are leafy tissues, inflorescences and roots (Robinson, Browne & Burton, 1975).

Influence of water on post-harvest diseases

Water can affect the initiation and development of post-harvest diseases in various ways and at various stages of the disease cycle both before and after harvest. As already indicated, many important post-harvest diseases are initiated prior to harvest and, even with those that are not the level of inoculum on the commodity at harvest is related to the activity of the fungi in the field, which in turn can influence the severity of the disease in store. Water exerts its influence on post-harvest diseases largely by its effects on: pre-harvest sporulation; dispersal of inoculum; germination of inoculum and infection of the host; the physiological status of the host; post-harvest inoculum germination, infection and development of rot; and, finally, post-harvest physiological status

Table 19.2 *Major post-harvest diseases of fruits and vegetables*

Crop	Disease	Pathogen
(A) *Deciduous tree fruits*		
Apples and pears	Lenticel rot	*Gloeosporium perennans*[a]
	Eye rot	*Nectria galligena*[a]
	Blue mould rot	*Penicillium expansum*[b]
	Brown rot	*Monilinia fructigena*
		M. fructicola
	Phytophthora rot	*Phytophthora syringae*[a]
	Grey mould	*Botrytis cinerea*
Stone fruits (apricots, cherries, nectarines, peaches, plums)	Brown rot	*Monilinia fructicola*
	Rhizopus rot	*Rhizopus stolonifer*[b]
		R. arrhizus[b]
(B) *Soft fruits*		
Cane fruits (raspberries, blackberries, loganberries)	Grey mould	*Botrytis cinerea*[a]
	Soft rot	*Rhizopus stolonifer*[b]
		Mucor piriformis[b]
	Cladosporium rot	*Cladosporium herbarum*
		C. cladosporioides
Strawberries	Grey mould	*Botrytis cinerea*[a]
	Leak disease	*Rhizopus stolonifer*[b]
		R. sexualis[b]
		Mucor piriformis[b]
Grapes	Grey mould	*Botrytis cinerea*[a]
(C) *Subtropical fruits*		
Avocado	Anthracnose	*Collectotrichum gloeosporioides*[a]
Banana	Anthracnose	*Gloeosporium musarum*[a]
	Botryodiplodia fruit rot	*Botryodiplodia theobromae*[b]
		B. theobromae[b]
	Crown rot	*Fusarium roseum*[b]
		Verticillium theobromae[b]
		Ceratocystis paradoxa[b]

Pineapple	Black rot	*Ceratocystis paradoxa*
Citrus fruits	Stem-end rot	*Phomopsis citri*[a] *Diplodia natalensis*[a] *Alternaria citri*
	Brown rot	*Phytophthora citrophthora* and other *Phytophthora* species[a]
	Green mould rot	*Penicillium digitatum*[b]
	Blue mould rot	*Penicillium italicum*[b]
	Sour rot	*Geotrichum candidum*[b]
(D) Vegetable fruits		
Tomatoes, peppers, aubergines	Grey mould	*Botrytis cinerea*[a]
	Alternaria rot	*Alternaria alternata*
	Cladosporium rot	*Cladosporium herbarum*
	Rhizopus rot	*Rhizopus stolonifer*[b]
Cucumbers	Grey mould	*Botrytis cinerea*[a]
	Black rot	*Mycosphaerella melonis*
(E) Vegetables		
Beans	Grey mould	*Botrytis cinerea*
Celery	Grey mould	*Botrytis cinerea*[b]
	Liquorice rot	*Mycocentrospora acerina*[b]
Carrots	Grey mould	*Botrytis cinerea*
	Liquorice rot	*Mycocentrospora acerina*[b]
	Watery rot	*Sclerotinia sclerotiorum*
	Crater rot	*Rhizoctonia carotae*
	Soft rot	*Rhizopus stolonifer*[b]
Cabbage	Grey mould	*Botrytis cinerea*[b]
	Alternaria rot	*Alternaria brassisicola*
	Phytophthora rot	*Phytophthora porri*[b]

Table 19.2 (*cont.*)

Crop	Disease	Pathogen
Cauliflower	Brown rot	*Alternaria brassicae*
	Downy mildew	*Peronospora parasitica*
Lettuce	Grey mould	*Botrytis cinerea*
	Water soft rot	*Sclerotinia* spp.
Onion	Neck rot	*Botrytis allii*
	Black mould	*Aspergillus niger*
Potatoes	Blight	*Phytophthora infestans*
	Gangrene	*Phoma exigua* var. *foveata*[b]
	Dry rot	{ *Fusarium coeruleum*[b] { *Fusarium* spp.[b]
Sweet potato	Black rot	*Ceratocystis fimbriata*[b]
	Rhizopus rot	*Rhizopus stolonifer*[b]
Yam	Soft rot	*Aspergillus niger*[b]
		Botryodiplodia theobromae[b]
		Fusarium moniliforme[b]
		Penicillium sclerotigenum[b]

[a] Diseases resulting from pre-harvest infections, usually quiescent infections.
[b] Diseases resulting from infection through wounds caused during harvesting and subsequent handling.
N.B. Many of the fungi listed can also cause field disease, indeed, their activity during growing of the crop can influence their importance after harvest (Dennis. 1983a).

of host, including the process of wound healing. These are considered in turn below.

Sporulation and dispersal of inoculum

Sporulation and dispersal of spores in the field can have a profound effect on subsequent development of diseases after harvest. For most important post-harvest disease-causing fungi, rain or high rh conditions are conducive to both sporulation and dispersal (see also Lacey, Ch. 4). For example, rain encourages the production of conidia by *Gloeosporium album* and *G. perennans* which form quiescent infections on apples (Edney, Tan & Burchill, 1977), while long periods with a rh above 80% are conducive to sporulation of *Botrytis allii* on infected necrotic tissue of onions (Maude & Presly, 1977). Similarly, the sporulation of *B. cinerea* is enhanced by moist conditions (Jarvis, 1962*a*; Dennis *et al.*, 1979). Sporulation of *Phytophthora infestans* on potato tubers has been reported to occur when soil moisture exceeds 20% (dry weight basis) (Logan, 1983), while activity of other *Phytophthora* species important in fruit and vegetable diseases (e.g. *P. syringae*, *P. citrophthora*, *P. porri*) is undoubtedly enhanced by heavy rainfall (Edney & Chambers, 1981; Edney, 1983). In contrast *Rhizopus stolonifer* and *R. sexualis* only produce asexual sporangiospores at rh below 80% and, their incidence in the field is favoured by a long period of dry hot weather (Harris & Dennis, 1980). Under humid conditions *R. stolonifer* assumes a mycelial habit while *R. sexualis* produces abundant zygospores which are unimportant in spread of the disease during flowering and fruiting.

Rainfall is particularly important in the dispersal of conidia from acervuli of *Gloeosporium* spp. and *Colletotrichum* sp. (Edney, 1983; Edney *et al.*, 1977) and from pycnidia of Sphaeropsidaceous fungi, such as *Phoma exigua* var. *foveata* on potato (Logan, 1983) and *Diplodia natalensis* and *Phomopsis citri* on citrus (Eckert, 1978). Frequent periods of rain also encourage the release of conidia from *Botrytis allii* (Maude & Presly, 1977; Maude, 1983) while the hygroscopic movements of conidiophores in moist conditions are very important in conidia release of *B. cinerea* (Jarvis, 1962*b*).

Rain splash is important in dispersing soil fungi (see Ch. 5). For example, *P. syringae* on apple fruits is dispersed to a height of 0.45 m from the ground (Edney, 1978), while the incidence of Phytophthora rot is associated with rain during the harvest period (Upstone & Gunn, 1978). Rainstorms are also important in the dispersal of the wet spored *M. piriformis* in strawberry crops (Harris & Dennis, 1980).

In the case of Phytophthora rot of cabbage caused by *P. porri* (Semb, 1971; Geeson, 1976) it is wet conditions at the time of harvest which are important in affecting the severity of the disease in store. Both Semb (1971) and Geeson (1976) concluded that the fungus, surviving in the soil as oospores or chlamydospores, enters the cut stem of the cabbage during harvesting operations by soil splash or mud on the cutting knife. This is consistent with the prevalence of the disease when cabbages are cut in very wet field conditions (Derbyshire & Shipway, 1978).

Germination of inoculum and infection of host
Once spores, of whatever kind, have been produced and disseminated, germination and infection of fruit and vegetable crops by all pathogens is in general favoured by a period of high rh and/or rainfall (Edney, 1983; Dennis, 1983*a*; Logan, 1983; Maude, 1983; Maude & Presly, 1977; Grove, Madden, Ellis & Schmitthenner, 1985). In some cases a minimum period of wetness is required for successful infection, for example, zoospores of *P. syringae* require a period of 24 h at 15 °C for consistent infection (Edney, 1978). Apart from having a direct effect on fungal development, rainfall encourages infection indirectly in certain cases. Thus, the susceptibility of apples to infection by conidia of *Gloeosporium album* and *G. perennans* was found to be markedly influenced by the amount of water falling on the fruit prior to inoculation; it was suggested that this may result from the removal of some inhibitor (chemical or microbiological) from the lenticels, enabling infection to take place (Edney *et al.*, 1977).

Pre-harvest physiological status of host
Rainfall may have wider indirect effects on the susceptibility of fruits and vegetables to infection, both pre- and post-harvest, by its effect on the physiological status of the crop. For example, extended dry periods may result in cessation of crop development leading to premature senescence, e.g. high temperatures and low soil moisture resulted in premature foliage senescence and immature roots in carrot crops in 1975 (Heale, Harding, Dodd & Gahan, 1977). Such carrots lost more water in store than in other years, which the authors suggested may have been due to relatively incomplete formation of periderm. As will be seen later, water loss profoundly affects invasion of carrots by certain post-harvest pathogens. It is also possible that growth conditions, including soil moisture, may affect the accumulation of naturally occurring antimicrobial compounds or those produced in response to infection.

Post-harvest germination, infection and development of rot

The high rh which is necessary in storage environments (bulk store or packaged produce) to prevent excessive water loss, and hence wilting or shrivelling, is often conducive to germination of fungal propagules and subsequent growth on the surface of the commodities. Even where comparatively low rh occurs the microclimate at the surface of fruits or vegetables, due to evaporative loss, may still be conducive to germination and infection by fungi. Once infection has occurred, the fungus is able to obtain an adequate water supply from the host provided that excessive evaporative loss does not occur. Extensive fungal spoilage of stored produce is, therefore, usually prevented by maintenance of low temperatures (Dennis, 1983*a*), except for chill-sensitive crops where a lower rh is necessary to prevent extensive surface fungal growth (Dennis, 1983*c*) which can in turn initiate rotting. Even where low-temperature storage is possible, additional procedures such as use of fungicides and drying of the outer layers are adopted. For example, some dehydration of the outer wrapper leaves of Dutch White cabbage considerably reduces attack by *Botrytis cinerea* (Bunnemann & Hansen, 1973; Geeson & Browne, 1979; Shipway, 1981), which is consistent with the reports of Yoder & Whalen (1975), Shipway (1976) and Geeson (1983) indicating that high rh enhanced the rotting of cabbages by this fungus.

The apparently conflicting reports by Van den Berg & Lentz (1973) and Pendergrass & Isenberg (1974), which indicated that turgid wrapper leaves were more resistant to rotting by *B. cinerea*, may have resulted from the indirect effect on the physiological status of the host similar to that reported for carrots (see next section).

The post-harvest drying of onions to reduce the incidence of neck rot is a well established practice (Anon., 1977; Maude, 1983). In warm dry climates the sun's radiant energy is adequate to dry harvested onions in the fields. However, in temperate climates artificial post-harvest drying is necessary. The established practice in the UK where sources of infection include contaminated soil, onion clamps or overlapping seed-production crops, is to force air, at 30 °C at a rate of 425 $m^3 h^{-1}$ tonne^{-1}, through bulk-stored onions for 3 to 5 days, which substantially reduces neck rot (Maude, Shipway, Presly & O'Connor, 1984). This treatment is most effective if the crop is removed from the field for drying within 48 h of topping, thus avoiding severe infection of the damaged green tissues of the necks of the onions. However, such drying systems do not control neck rot in crops grown from infected untreated seed (Maude *et al.*, 1984).

The use of a relatively high temperature (30 °C) favours the selection of high-temperature pathogens such as *Aspergillus niger*, *A. fumigatus* and *Penicillium* spp., as well as soft rotting bacteria (Anon., 1984). Infection and consequent rotting are encouraged when warm air is recirculated and rh exceeds 80%. Results of recent work (Anon., 1984) suggest that storage rots by fungi can be controlled by the balanced use of heating and ventilation to produce a rh of less than 80% in a second drying stage, with the introduction of cool air as soon as possible thereafter to reduce onion stack temperatures.

The high humidities which are required for many fruits and vegetables often favour the spread of rots once they have been initiated, resulting in considerable wastage. This is particularly important in diseases caused by *Sclerotinia sclerotiorum*, *Rhizoctonia carotae* and *Botrytis cinerea* on carrots, *B. cinerea* on many other vegetables (see Table 19.2), and *Phytophthora* rots of apples, where all of the fungi are able to grow at 0 °C.

In the case of wound pathogens of potatoes it is usually the presence and persistence of moisture in the wound which favour infection rather than rh alone (Logan, 1983). This is especially true for dry rot caused by *Fusarium avenaceum*, whereas *F. coeruleum* is more tolerant of dry conditions and lower rh. Infection by *Phoma exigua* var. *foveata* through unwounded potato tubers can only occur in saturated atmospheres (Logan, 1983). Similarly, continued progression of the potato skin spot fungus (*Polyscytalum pullulans*) requires damp conditions, which also enhance sporulation on the surface of stored tubers (Logan, 1983).

Post-harvest physiological status of host

Most fungi involved in post-harvest diseases grow well at water potential deficits larger than those that are likely to occur in the tissues of palatable fruits and vegetables. However, the rh of the atmosphere may influence rotting by its effect on the water status (physiological status) of the harvested fruit or vegetable. For example, Årsvoll (1969) considered loss of root turgor to be a predisposing factor to infection of carrots by *B. cinerea*, while Goodliffe & Heale (1977) reported that a fresh weight loss of over 5% led to increased invasion of autumn-lifted carrots by *B. cinerea*. Subsequently, Goodliffe & Heale (1978) demonstrated that carrot roots showing 0 to 5% weight loss accumulated higher concentrations of the antifungal compound 6-methoxymellein compared to roots which had lost 10 or 15% water. Thus the absence of chemical inhibition may also explain the enhanced infection of roots which have lost 5% or more of their water. In addition to this mechanism, a second may have been operative since Thorne (1972) suggested that invasion of carrot by another pathogen, *Rhizopus stolonifer*, was facilitated by strips of carrot parenchyma separating one from another at about 7% water loss, leaving continuous intercellular spaces. Thorne also produced convincing evidence that there was a constant and critical weight loss (8%) below which invasion of carrots by *R. stolonifer* did not occur. At weight losses above 8% carrot roots rapidly increased in their suscepti-bility to invasion by this fungus. However, the weight losses from carrots reported to be critical for invasion by *B. cinerea* (Goodliffe & Heale, 1977) and *R. stolonifer* (Thorne, 1972) were achieved over a relatively short time period and may not have allowed equilibration of water loss across the different root tissues, i.e. the water loss from surface layers may have been critical for invasion rather than the total weight loss.

More recent work on ice-bank storage of carrots has clearly demon-strated that water losses in excess of those reported by Goodliffe & Heale (1977) and Thorne (1972) can occur over very extended storage periods (up to 10% over a 10-month storage period) (Dennis, 1981; Geeson, Browne & Dennis, 1983). Such a relatively slow (0.6–1.0% per month) loss would be more likely to allow equilibration of water content across the root tissue and hence maintain turgid roots which showed very low infection by *B. cinerea* (Dennis, 1981; Geeson *et al.*, 1983). Thus, more research is required on variation of water loss, ana-tomy, chemical composition and susceptibility of the surface layer of carrot roots, in order fully to explain the reported observations. Reports by Lauritzen (1932) and Mukula (1957) which showed that a rh of 85% and below was unfavourable for infection of carrots by *B. cinerea* could be explained by reduced germination and growth of the fungus, thus preventing rot initiation (see previous section). By contrast, work by Van den Berg & Lentz (1966), indicating that storage of carrots at 98 to 100% rh resulted in less disease caused by *B. cinerea* than at 95% rh, can be explained by the effect of the conditions on root turgor.

Wound healing

The use of an appropriate combination of temperature and rh plays an important role in the wound healing of certain tuber and root crops. Thus, exposure of potato tubers to temperatures of 15°–16°C and approximately 80% rh for 10 to 15 days allows wounds to heal quickly and minimises infection caused by *Phoma exigua* var. *foveata* (Malcolm-son & Gray, 1968) and *Fusarium* sp. (Logan, 1983). Similarly, curing of sweet potatoes and yams for 5 to 6 d at 25°–40°C and 80–90% rh allows tubers to form a suberised periderm to protect those injuries incurred at harvest from subsequent invasion by pathogenic fungi (Lutz & Simons, 1958; Been, Perkins & Thompson, 1977).

A curing period also promotes healing of wounds in carrot roots (Davies, 1977; Garrod, Lewis, Brittain & Davies, 1982). Treatment at high temperature and high rh markedly reduced infection of wounds by *M. acerina*. A period as short as 12 h reduced infection to negligible levels and the beneficial effects were maintained for several months of storage (Lewis, Davies & Garrod, 1981). The resistance of repaired phloem parenchyma tissue partly involves killing of inoculum by the antifungal compound, falcarindiol, which accumulates on the surface of wounds when the oil ducts in the tissue are severed during wounding, whereas on areas of wound surfaces where no oil ducts are present, other compounds, probably including 6-methoxymellein, have a fungistatic effect (Lewis *et al.*, 1981). Accumulation of these antifungal compounds appears to be the main mechanism that reduces infection of wounds by *M. acerina* since lignification and suberisation account for comparatively little impedance to this fungus (Garrod *et al.*, 1982).

References

Anon. (1965). Losses in Agriculture, *USDA Agric. Handbook*, **291**.

Anon. (1977). Dry bulb onions: effects of various handling and drying practices on storage. *Kirton Experimental Horticultural Station 14th Annual Report*, pp. 53–7.

Anon. (1984). High temperature storage rots of onions. *Agricultural and Food Research Council Annual Report*, 1983–4, pp. 61–2.

Årsvoll, K. (1969). Pathogen on carrots in Norway. *Scientific Reports of the Agricultural College Norway*, **48**, No. 2.

Been, B. O., Perkins, C. & Thompson, A. K. (1977). Yam curing for storage. *Acta Horticulturae*, **62**, 311–16.

Bunnemann, G. & Hansen, H. (1973). *Frucht und Gemüselagerung. Eine Anleitung für den Lagerwart*. Stuttgart: Verlag Eugen Ulmer.

Coursey, D. G. (1967). Yam storage – I: a review of yam storage practices and of information on storage losses. *Journal Stored Products Research*, **2**, 229–44.

Coursey, D. G. (1971). Biodeteriorative losses in tropical horticultural produce. In *Biodeterioration of Materials*, vol. 2, ed. A. H. Walters & E. H. Hueck-van de Plas, pp. 464–71. London: Applied Science Publishers Ltd.

Davies, W. P. (1977). Infection of carrot roots by *Mycocentrospora acerina* (Hartig) Seighton. Ph.D. Thesis, University of East Anglia.

Day, J. R., Lewis, B. G. & Martin, S. (1972). Infection of stored celery by *Centrospora acerina*. *Annals of Applied Biology*, **71**, 201–10.

Dennis, C. (1981). Lagring av gulrøtter. *Norsk Institutt for Naeringsmiddel-Forskning Informasjon*, **5**, 213–17.

Dennis, C. (ed.) (1983*a*). *Post-Harvest Pathology of Fruits and Vegetables*. London: Academic Press.

Dennis, C. (1983*b*). Soft fruits. In *Post-Harvest Pathology of Fruits and Vegetables*, ed. C. Dennis, pp. 23–42. London: Academic Press.

Dennis, C. (1983*c*). Salad crops. In *Post Harvest Pathology of Fruits and Vegetables*, ed. C. Dennis, pp. 147–78. London: Academic Press.

Dennis, C. (1986). Fungi. In *Post-Harvest Physiology of Vegetables*, ed. J. Weichmann. New York: Marcel Dekker Inc. (in press).

Dennis, C., Davis, R. P., Harris, J. E., Calcutt, L. W. & Cross, D. (1979). The relative importance of fungi in the breakdown of commercial samples of sulphited strawberries. *Journal of Science of Food and Agriculture*, **30**, 959–73.

Derbyshire, D. M. (1973). Post-harvest deterioration of vegetables. *Chemistry & Industry*, **22**, 1052–4.

Derbyshire, D. M. & Crisp, A. F. (1971). Vegetable storage diseases in East Anglia. *Proceedings 6th British Insecticide & Fungicide Conference*, **1**, 167–72.

Derbyshire, D. M. & Shipway, M. R. (1978). Control of post-harvest deterioration in vegetables in the U.K. *Outlook on Agriculture*, **9**, 246–52.

Eckert, J. W. (1977). Control of post-harvest diseases. In *Antifungal Compounds*, vol. 1, ed. M. R. Siegel & D. H. Sisler, pp. 269–352. New York: Marcel Dekker Inc.

Eckert, J. W. (1978). Post-harvest diseases of citrus fruits. *Outlook on Agriculture*, **9**, 225–32.

Eckert, J. W. & Ratnayake, M. (1983). Host–pathogen interactions in post-harvest diseases. In *Post-Harvest Physiology and Crop Preservation*, ed. M. Lieberman, pp. 247–64. New York: Plenum Press.

Edney, K. L. (1978). The infection of apples by *Phytophthora syringae*. *Annals of Applied Biology*, **88**, 31–6.

Edney, K. L. (1983). Top fruit. In *Post-Harvest Pathology of Fruits and Vegetables*, ed. C. Dennis, pp. 43–71. London: Academic Press.

Edney, K. L. & Chambers, D. A. (1981). Post-harvest treatments for the control of *Phytophthora syringae* rot of apple fruits. *Annals of Applied Biology*, **97**, 237–41.

Edney, K. L., Tan, A. M. & Burchill, R. T. (1977). Susceptibility of apples to infection by *Gloeosporium album*. *Annals of Applied Biology*, **86**, 129–32.

Garrod, B., Lewis, B. G., Brittain, M. J. & Davies, W. P. (1982). Studies on the contribution of lignin and suberin to the impedance of wounded carrot root tissue to fungal invasion. *New Phytologist*, **90**, 99–108.

Geeson, J. D. (1976). Storage rot of white cabbage caused by *Phytophthora porri*. *Plant Pathology*, **25**, 115–16.

Geeson, J. D. (1983). Brassicas. In *Post-Harvest Pathology of Fruits and Vegetables*, ed. C. Dennis, pp. 125–56. London: Academic Press.

Geeson, J. D. & Browne, K. M. (1979). Effect of post-harvest fungicide drenches on stored winter white cabbage. *Plant Pathology*, **28**, 161–8.

Geeson, J. D., Browne, K. M. & Dennis, C. (1983). Cashing in on ice banks for summer supplies. *Grower*, **100**, 49–52.

Goodliffe, J. P. & Heale, J. B. (1977). Factors affecting the resistance of cold-stored carrots to *Botrytis cinerea*. *Annals of Applied Biology*, **87**, 17–28.

Goodliffe, J. P. & Heale, J. B. (1978). The role of 6-methoxymellein in the resistance and susceptibility of carrot root tissue to the cold-storage pathogen *Botrytis cinerea*. *Physiological Plant Pathology*, **12**, 27–43.

Grove, G. G., Madden, L. V., Ellis, M. A. & Schmitthenner, A. F. (1985). Influence of temperature and wetness duration on infection of immature strawberry fruit by *Phytophthora cactorum. Phytopathology*, **75**, 165–9.

Harris, J. E. & Dennis, C. (1980). Distribution of *Mucor piriformis*, *Rhizopus sexualis* and *R. stolonifer* in relation to their spoilage of strawberries. *Transactions of the British Mycological Society*, **75**, 445–50.

Harvey, J. M. (1978). Reduction of losses in fresh market fruits and vegetables. *Annual Review of Phytopathology*, **16**, 321–41.

Heale, J. B., Harding, V., Dodd, K. & Gahan, P. B. (1977). *Botrytis* infection of carrot in relation to the length of the cold storage period. *Annals of Applied Biology*, **85**, 453–7.

Jarvis, W. R. (1962a). The infection of strawberry and raspberry fruits by *Botrytis cinerea* Fr. *Annals of Applied Biology*, **50**, 569–75.

Jarvis, W. R. (1962b). The dispersal of spores of *Botrytis cinerea* Fr. in a raspberry plantation. *Transactions of the British Mycological Society*, **45**, 549–59.

Lauritzen, J. I. (1932). Development of certain storage and transit diseases of carrots. *Journal of Agricultural Research*, **44**, 861–912.

Lewis, B. G., Davies, W. P. & Garrod, B. (1981). Wound healing in carrot roots in relation to infection by *Mycocentrospora acerina. Annals of Applied Biology*, **99**, 35–42.

Logan, C. (1983). Potatoes. In *Post-Harvest Pathology of Fruits and Vegetables*, ed. C. Dennis, pp. 179–217. London: Academic Press.

Lutz, J. M. & Simons, J. W. (1958). Storage of sweet potatoes. *Farmers Bulletin* 1442, United States Department of Agriculture.

Malcolmson, J. F. & Gray, E. (1968). The incidence of gangrene of potatoes caused by *Phoma exigua* in relation to handling and storage. *Annals of Applied Biology*, **62**, 89–101.

Maude, R. B. (1983). Onions. In *Post-Harvest Pathology of Fruits and Vegetables*, ed. C. Dennis, pp. 74–101. London: Academic Press.

Maude, R. B. & Presly, A. M. (1973). Onion neck rot and its control. *Proceedings 7th British Insecticide & Fungicide Conference*, vol. 2, pp. 609–11.

Maude, R. B. & Presly, A. M. (1977). Neck rot (*Botrytis allii*) of bulb onions. II. Seed-borne infection in relationship to the disease in store and the effect of seed treatment. *Annals of Applied Biology*, **86**, 181–8.

Maude, R. B., Shipway, M. R., Presly, A. M. & O'Connor, D. (1984). The effects of direct harvesting and drying systems on the incidence and control of neck rot (*Botrytis allii*) in onions. *Plant Pathology*, **33**, 263–8.

Mukula, J. (1957). On the decay of stored carrots in Finland. *Acta Agriculturae Scandinavica Supplement*, **2**, 1–132.

Onesirosan, P. T. & Fatunla, T. (1976). Fungal rots of tomatoes in Southern Nigeria. *Journal of Horticultural Science*, **51**, 473–9.

Pendergrass, A. & Isenberg, F. M. (1974). The effect of relative humidity on the quality of stored cabbage. *Hortscience*, **9**, 226–7.

Robinson, J. E., Browne, K. M. & Burton, W. G. (1975). Storage characteristics of some vegetables and soft fruits. *Annals of Applied Biology*, **81**, 399–408.

Semb, L. (1971). A rot of stored cabbage by *Phytophthora* sp. *Acta Horticulturae*, **20**, 32–5.

Shipway, M. R. (1976). Winter white cabbage: use of fungicidal dusts and polythene lining of storage bins. *Annual Report of Kirton Experimental Horticultural Station* 1975, pp. 35–6.

Shipway, M. R. (1981). Refrigerated storage of winter white cabbage. *Annual Review of Kirton Experimental Horticultural Station* 1980, pp. 44–50.

Simmonds, J. H. (1941). Latent infection in tropical fruits discussed in relation to the part played by species of *Gloeosporium* and *Colletotrichum*. *Proceedings of the Royal Society Queensland*, **51**, 92–129.

Swinburne, T. R. (1983). Quiescent infections in post-harvest diseases. In *Post-Harvest Pathology of Fruits and Vegetables*, ed. C. Dennis, pp. 1–21. London: Academic Press.

Thorne, S. M. (1972). Studies on the behaviour of stored carrots with respect to their invasion by *Rhizopus stolonifer* Lind. *Journal of Food Technology*, **7**, 139–51.

Upstone, M. E. (1978). *Phytophthora syringae* fruit rot of apples. *Plant Pathology*, **27**, 24–30.

Upstone, M. E. & Gunn, E. (1978). Rainfall and the occurrence of *Phytophthora syringae* fruit rot of apples in Kent 1973–75. *Plant Pathology*, **27**, 30–5.

Van den Berg, L. & Lentz, C. P. (1966). The effect of temperature, relative humidity and atmospheric composition on changes in quality of carrots during storage. *Food Technology*, **20**, 104–7.

Van den Berg, L. & Lentz, C. P. (1973). High humidity storage of carrots, parsnips, rutabagas and cabbage. *Journal of the American Society of Horticultural Science*, **98**, 129–32.

Yoder, O. C. & Whalen, M. L. (1975). Factors affecting post-harvest infection of stored cabbage tissue by *Botrytis cinerea*. *Canadian Journal of Botany*, **53**, 691–9.

20
Water and the ecology of fungi in stored products

G. AYERST

School of Applied Science, The Polytechnic, Wolverhampton, WV1 1LY, UK

Introduction

Stored products comprise primary agricultural products such as cereal grains, hay and oilseeds, and those derivatives, such as bran and the residues left after oil extraction, that are stored dry and in bulk. Water availability is the main factor controlling microbial activity but its effects can only be realistically understood in conjunction with other factors in this environment, especially temperature.

This chapter overviews the characteristics of the habitat and the fungi that occupy it. Detailed information on particular situations is readily available in longer reviews (Wallace, 1973; Christensen & Kaufman, 1974; Tuite & Foster, 1979; Lacey, 1980; Christensen & Saur, 1982; FAO, 1984).

The exact physical attributes of stored products vary with type and moisture content but to illustrate general principles realistically, the following average values for cereal grain will be used in this chapter: specific heat, $1.8 \, kJ \, kg^{-1} \, K^{-1}$; bulk density, $700 \, kg \, m^{-3}$; latent heat of vaporisation of water from a product at high moisture content, $2500 \, kJ \, kg^{-1}$; air space in a regular granular material, 40%.

The stored product habitat

This habitat is unusual in that nutrient supply rarely limits the growth of microbes. Typically, the substrata cease to be stored products before the easily available nutrients are exhausted but exceptionally, for example in hay, slowing of microbial activity follows the utilisation of soluble nutrients (Lacey, 1979).

Water is mostly absorbed onto insoluble polymers and not obviously present as a liquid although very thin layers of highly concentrated solu-

359

tion must occur on surfaces and adjacent to growing hyphae. Water content is not a useful parameter of water availability except within a single product (see Boddy, Ch. 21) and water potential (ψ) is preferred as explained by Papendick & Mulla (Ch. 1), and in the Preface.

The relationship between water content and water potential, at a particular temperature, is called the water sorption isotherm and is of a similar form for most stored products. Starting from near saturation a large proportion of water has to be removed to cause a small reduction in vapour pressure but the effect of each successive removal of water becomes greater until about -70 MPa at 25 °C, which is slightly below the lower limit for any fungal growth (see also Ch. 21). The effects of further decrements of water content on vapour pressure are then nearly constant until about -126 MPa at 25 °C and from there until complete dryness each successive removal of water causes a smaller reduction in ψ. None of the numerous attempts to model this relationship algebraically have been completely successful over the whole range of water content or for all products (e.g. Pixton & Howe, 1983) and this reflects the complexity of forces binding the water (e.g. Griffin, 1981).

An additional complication is that the water vapour pressure of most stored products depends upon whether their moisture content has been reached by adsorption or desorption. This hysteresis in water sorption can be quite large, for example 5 MPa to 20 MPa ψ at the same moisture content in cereal grain (Hunt & Pixton, 1974). In practice, however, the actual hysteresis states within a bulk of a stored product vary considerably and are rarely known with any certainty. For example, after artificial drying, the outer layers of a grain of wheat take up water from the inner layers (Becker & Sallans, 1956) so there are two hysteresis states in one grain and, in bulks of products each passage of a temperature front, carried by air flow through the material, results in absorption followed by desorption of water, or *vice versa* (Sutherland, Banks & Griffiths, 1971).

In practical situations the properties of large bulks of products are important in determining whether conditions suitable for fungal growth develop. These bulks vary from quantities of a few tonnes, in a small bin or a floor store two or three metres deep, to over 1000 tonnes in a single hold of a ship or a medium sized bin and as much as 20 000 tonnes in a large bin for storing surplus capacity (Bailey, 1974). Air occupies a continuous three-dimensional network of narrow channels between the closely packed hygroscopic particles and it rapidly reaches moisture and temperature equilibrium with them. If there is a continuing flow of air in one direction, changes in temperature and moisture content are carried through the bulk as 'fronts' (Sutherland, Banks & Griffiths, 1971). Such transient changes are not easy to detect by normal temperature and moisture monitoring and no systematic analysis of their effects on the mycoflora seems to exist. However, Christensen & Drescher (1954) adduce mycological and other evidence of large increases and decreases of moisture content within a particular commercial storage and it seems likely that such effects are not uncommon.

Thermal conductivity within bulks of products is low, often about one eighth that of concrete and about three times that of standard insulating materials such as glass wool and cork (Burrell, 1982), so they can retain steep temperature gradients. This results in the commonly experienced difficulty of detecting pockets of high temperature within a store because of the localisation of the effect. For example Sinha & Wallace (1965) recorded a gradient of $10°$ to $60°C$ within $0.5\,m$ in a horizontal direction through a small mass of wheat of high moisture content heating in a store of dry grain in Canada. Temperature gradients cause redistribution of moisture and may thus create conditions favourable for fungal growth. For example, when bulks are exposed to declining ambient temperatures during autumn or during sea voyages from hot to cold climates the centres remain warm and air convects upwards within them, picking up heat and water which is released by equilibration in the cooler upper surface layers (Muir, 1973). Less importantly moisture can move by diffusion from warm to cool material in any direction (Thorpe, 1982).

Only under extreme conditions of high local water content and very rapid temperature decrease at the surface, will water condense as liquid on the walls of storage containers in contact with a product (Disney, 1969) but, if humid air is not removed from the space above the product, dew collects on the inner surface of the container and this often runs down onto the product causing high moisture content and rapid fungal growth (see, e.g. Sauerbier, 1956; Bailey, 1974).

The fungi of stored products

Most authors make a broad but not absolute distinction between 'field fungi' which grow on the plant before harvest and 'storage fungi' which grow during storage (see, e.g. Christensen, 1972). The latter consist predominantly of members of the anamorph genera *Penicillium* and *Aspergillus* but others such as *Wallemia sebi*, *Scopulariopsis* sp., *Paecilomyces* spp. and *Absidia* spp. also occur. In addition, *Fusarium* spp. may be important under high moisture/low temperature conditions, yeasts such as *Hansenula* spp. may grow to the virtual exclusion of other fungi in micro-aerophilic conditions, and a number of thermophilic species occur at high temperature.

In general, storage fungi are able to grow at lower levels of water potential than field fungi although their vegetative hyphae are less tolerant of desiccation (Park, 1982). It was shown by Christensen (1951) that individual hyphae of field fungi beneath the pericarps of wheat grains would resume growth after eight years of dry storage but hyphae of storage fungi, several species of which often occurred within a single grain, were not viable beyond five years. Storage fungi survive on the surface of grains and other particles as long-lived conidia or other spores. Pixton, Warburton & Hill (1975) stored samples of Manitoba wheat and of English wheat at about −75 MPa in unsealed capped test tubes buried in pilot-scale bins of wheat at ambient temperature. Mould counts at two-year intervals showed that survival on both wheats was similar and that although the total numbers of colony-forming units of *Penicillium* spp. and *Aspergillus* spp., mainly *A. flavus* and *A. candidus*, declined steadily during the period with a decimal reduction time of about five years, there were still significant numbers of survivors after ten years. *Eurotium* spp. survived undiminished in number for ten years (Fig. 20.1).

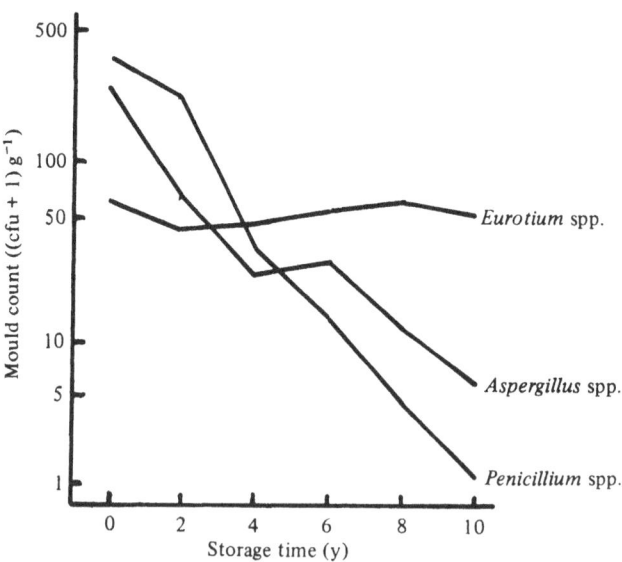

Fig. 20.1. Survival of colony-forming units (cfu) of fungi in stored wheat (after Pixton, Warburton & Hill, 1975).

In experiments to investigate the survival of *Aspergillus flavus* conidia on various natural substrata including soya, corn and cowpea flour, Beuchat (1979, 1984) found a good level of survival under normal storage conditions for up to 20 months. He showed, however, that the rate of decrease in viability depended on the substratum, temperature, and ψ. It was more rapid at higher temperatures up to 37 °C and at higher ψ up to about -34 MPa which is just below the lowest potential permitting germination of spores of this species.

There have been a number of investigations of the origins of fungi on stored products, particularly on cereal grains (Christensen & Kaufmann, 1974). In a careful study of British wheat and barley over several seasons, Flannigan (1978) found little penetration of sound grain by storage fungi at the time of harvest but consistent superficial contamination in the order of a hundred to ten thousand colony-forming units per gram. Pixton, Hyde & Ayerst (1964) found similar or slightly larger numbers of a range of species on good quality commercially harvested English and Canadian wheat.

Mycological examination of stored products

There have been various attempts to detect hyphae in products by direct examination and to identify them either by allowing them to grow in slide culture (Christensen, 1951) or by fluorescent-labelled antibody techniques (Warnock, 1971), but these methods have not been used for routine examination of samples. For culturing from stored products it is essential to include media of low ψ: many storage fungi fail to grow well on ordinary mycological media at ψ above about -1.5 MPa. A widely used medium is malt/salt agar containing 7.5% NaCl to give ψ of about -7 MPa at 25 °C, although, Pitt (1975) recommends a medium containing 40% glucose to give ψ of about -11.5 MPa. In addition the temperature requirements of the storage fungi vary widely and, although incubation at 20°–25 °C is satisfactory for most of them, temperatures in the range 40°–50 °C are needed for thermophilic and thermotolerant species (Lacey, Hill & Edwards, 1980).

Most investigators have either cultured from individual particles such as grains or have plated dilutions of suspensions. The latter 'mould counts' are essentially spore counts so their magnitude has only a tenuous association with biomass, but they are appropriate for investigating the rate of death of spores during storage (Pixton *et al.*, 1975; Beuchat, 1979, 1984). Spores released when samples are shaken in air have been counted and identified culturally by means of an Anderson sampler which

Table 20.1. *The effects of surface sterilisation, temperature of incubation, and culture medium on the detection of two fungi in naturally infected barley grains (data of Flannigan, 1969)*

			Grains with each fungus (%)	
Surface sterilised	Temperature (°C)	Culture medium	*Eurotium* spp.	*Aspergillus fumigatus*
+	25	MSA	28.7	0
+	25	PDA	<0.1	0.3
+	37	PDA	0	11.2
–	25	MSA	51.9	0
–	25	PDA	0.1	1.5
–	37	PDA	0	13.4

MSA = malt salt agar, −6.5 MPa water potential.
PDA = potato dextrose agar, −0.3 MPa water potential.

sorts them into sizes by differential impaction onto agar plates (Lacey & Dutkiewicz, 1976; Lacey *et al.*, 1980).

Plating individual grains after surface sterilisation should detect only those fungi that have actually grown in the grains and the method has been used as a basis for quality assessment (Lichtwardt & Barron, 1959; Christensen & Kaufmann, 1974). However, results may be misleading because even though a fungus is present in a grain, it may often not be detected if another species, better suited to the conditions of culture, is present (Table 20.1).

Pure culture studies on storage fungi

Several axenic studies have been made in order to understand the effects of moisture in stored products on the growth of individual fungi and on total fungal activity (e.g. Ayerst, 1969; Magan & Lacey, 1984*a,b,c*; Niles, Norman & Pimbley, 1985). In studying the effects of ψ on a storage fungus, one is in fact attempting to define one dimension of its 'fundamental niche', the range of environmental conditions within which it can survive and reproduce in the absence of competition (e.g. McNaughton, 1981; Cooke & Rayner, 1984). The term 'ecotope' used by Whittaker, Levin & Root (1973), appears to have the same connotation and these authors reserve the term niche for the position or role of a species within a community, pointing out that a species may occupy different niches in different communities. This distinction appears to be relevant to storage fungi which may occur with different roles in other habitats. For example, *Aspergillus flavus* exists in stored cereal grain where it is limited by low ψ and survives as dry spores. It is also a facultative plant parasite, sometimes limited by high ψ in the host (e.g. Austwick & Ayerst, 1963; Wotton & Strange, 1985), and a saprotroph in soil (Raper & Fennel, 1965), surviving as spores with dormancy induced by depletion of nutrients (Griffin & Roth, 1979).

In order to have a realistic understanding of the effects of ψ on fungi in stored products the interacting or direct effects of other environmental factors need to be considered. Temperature is always important and oxygen tension (Magan & Lacey, 1984c), presence of inhibitory chemicals, and possibly pH (Pitt & Hocking, 1977; Magan & Lacey, 1984a), become important under special circumstances. Antagonism (Magan & Lacey, 1984b) and competition for nutrients become relevant once colonisation is well advanced but the fungi seem to compete initially by the relative speed at which they utilise new resources in nutrient-rich substrata. Thus a suitable condition for laboratory studies is, conveniently, that of high nutrient supply.

To define the limits and the optimum of any environmental factor requires a measure of fungal growth or activity and most investigators have relied on speed of germination and subsequent rate of hyphal extension for this purpose. These are closely correlated when they are limited in the same fungus by ψ and temperature (see, e.g. Smith & Hill, 1982), although, exceptionally, at extreme conditions of high temperature or low ψ, growth may slow down and stop soon after germination (e.g. Ayerst, 1969; Magan & Lacey, 1984a).

Pure culture studies and the association of organisms with particular habitats and microclimatic conditions both assume that all members of a taxon will respond consistently. For this to be true the identification of taxa must be to an appropriate level. Thus the genus *Penicillium* has been recorded on stored products on many occasions but the information is of limited value when the species have not been identified because there are marked differences between them in their response to moisture and temperature (Ayerst, 1969; Mislivec & Tuite, 1970; Mislivec, Dieter & Bruce, 1975; Northolt & Bullerman, 1982; Magan & Lacey, 1984a). In particular Pitt (1973, 1979) found that growth rates of *Penicillium* spp. at different temperature and ψ were sufficiently diverse and stable to be used as taxonomic criteria. Similarly, although species of *Aspergillus* differ markedly in their temperature and ψ limits and optima, several authors employing different methods report only small differences within species if allowance is made for the fact that the minimum ψ for growth occurs near the optimum temperature and that the maximum temperature for growth often occurs at reduced ψ (Diener & Davis, 1967; Ayerst, 1969; Saez, 1975; Northolt, Verhulsdonk, Soentoro & Paulsch, 1976; Magan & Lacey, 1984a, Niles et al., 1985).

The response of fungi to ψ is so dependent on temperature that most recent investigations have included both factors as variables. It is convenient to summarise their combined effects in pure culture as growth or germination time isopleths on graphs whose axes are water activity or ψ and temperature (Fig. 20.2). These are effectively representations of two dimensions of the fundamental niche of each fungus at favourable states of other variables such as nutrient and oxygen supply.

Although these types of data are available for only a limited number of fungi, most of the species studied have been selected because of their importance on stored products, so they represent a high proportion of those of significance in spoilage or toxin production. Certain general points emerge from inspection of the diagrams. Firstly the tolerances of some species extend beyond the limits of any of the others and so they would be expected to be of major importance under these extreme conditions. For example, the ability of only P. martensii to grow at 5 °C at water potentials between 0 and −10 MPa fits well with the observation by Hill & Lacey (1983) that P. verrucosum, a species that Pitt (1979) combines with P. martensii in P. aurantiogriseum, dominated the microflora of barley stored at 5 °C and −6.6 MPa. Similarly the ability of Eurotium spp. and A. restrictus to grow at lower ψ than permit growth of nearly all other fungi explains their well known importance at levels below about −30 MPa (see, e.g. Christensen & Kaufmann, 1974). At less extreme conditions, above 15 °C and ψ above −30 MPa, several different species grow at similar rates and, although ψ is still a significant factor in selecting species, other factors such as the amount of initial inoculum of individual species modify the picture (e.g. Griffin, 1981).

Changes in the habitat caused by fungal activity

Fungi growing on stored products cause changes in the substratum such as death of seeds and chemical changes (e.g. Christensen, 1972), but there seems to have been little systematic investigation of the effects of these changes on subsequent fungal growth. It has been suggested that water produced by fungal respiration may lead to an ecological succession of species by causing an increase in moisture content (Christensen & Kaufmann, 1974; Cooke & Rayner, 1984). This change in water content and species composition may occur in containers that keep water in and let heat out. However, in bulk stores the situation is different because heat as well as water is retained, so the final result is a rise in temperature often accompanied by a reduction in water content (mc). For example, suppose that in each kilogram of a bulk of cereal grain at 16% mc (fresh weight basis) and 25 °C, 0.5 g each of fat and carbohydrate are respired aerobically and all the heat and water are retained. This would release about 0.74 g of water and 22 kJ of heat; sufficient to raise the moisture content to about 16.1% and the tempera-

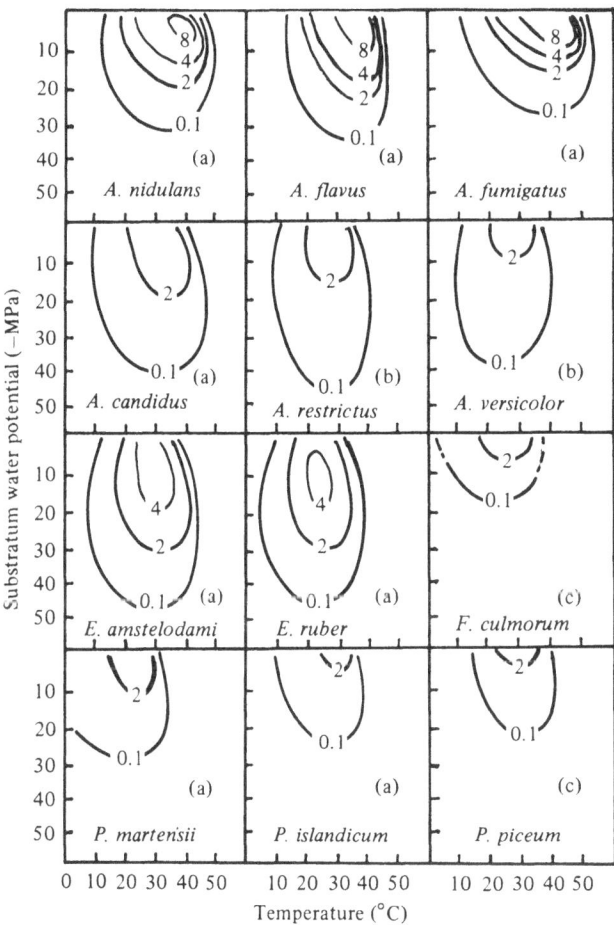

Fig. 20.2. Growth rate isopleths (mm d^{-1}) of *Aspergillus, Eurotium, Fusarium* and *Penicillium* species. (a) From Ayerst (1969); (b) from Smith & Hill (1982); (c) from Magan & Lacey, 1984*a*.

Table 20.2. *Predicted potential rates of heating in stored wheat at different levels of moisture (calculated from data of Oxley, 1948)*

Moisture content of wheat (% fresh weight)	<12	12–15	15–20	20+
Approx. water potential ($-$MPa)	>71	71–40	40–15	<15
Maximum rate (K d^{-1})	0.003	0.3	3.0	30.0

ture to over 37 °C. Alternatively the heat produced would be sufficient to evaporate nearly 9 g of water at constant temperature and so, allowing for the metabolic water, *reduce* the moisture content of the grain to about 15.2%. The practical consequences depend on the position, quantity and moisture content of the heating material and the moisture content, size and temperature of the rest of the bulk. Pockets of moist material may dry out before they cause further damage (Sinha & Wallace, 1965), but when the margin of safety is narrow, they may stimulate further fungal activity by warming the rest of the bulk or causing moisture migration.

A number of investigators have examined the effects of water content on the microflora of heating products (e.g. Lacey, 1979, 1980). Briefly, heating to about 45 °C can be caused by mesophilic, including xerophilic, fungi and so occurs at ψ below about -14.5 MPa (Fig. 20.2). Higher moisture contents are required for rapid growth of thermotolerant and thermophilic species of fungi and finally of prokaryotes, which are able to raise the temperature to over 60 °C. At high temperatures inorganic thermogenesis becomes important within a critical range of moisture content (Currie & Festenstein, 1971; Bowes, 1984). Early work indicating that maximum potential rates of heat production depend on moisture content was summarised by Oxley (1948) for wheat and can be expressed in terms of rate of temperature increase assuming perfectly adiabatic conditions (Table 20.2).

It is because potential rates of heating increase steeply with moisture content that uneven distribution of water causes increased total heat production (see Oxley, 1948; Christensen & Kaufmann, 1974). Rates of temperature increase have been observed in high-moisture wheat and soya beans in the range 3 to 17 K d^{-1} (Milner & Geddes, 1946; Sinha & Wallace, 1965) and in hay up to 40 K d^{-1} (Festenstein *et al.*, 1965; Lacey, 1979).

In 'hermetic storage', the decrease of O_2 and increase of CO_2, resulting from respiration, are used to control aerobic spoilage organisms and variations on this method have been reviewed by Hyde & Burrell (1982). Its consequences depend on ψ of the product and the efficiency of the seal. In containers that prevent any gaseous exchange the O_2 tension falls to very low levels and virtually all fungi are killed. If there is slight access of O_2 and ψ is too low for rapid bacterial growth, a characteristic micro-aerophilic mycoflora develops. It includes various myceliated yeasts such as *Hansenula* which may cause a dramatic whitening of the affected products, and often *Penicillium roquefortii* (see Nichols & Leaver, 1966; Clarke, Hill, Niles & Howard, 1969). Hermetically stored grain of high ψ is very susceptible to fungal attack when it is exposed to the air at the time of discharge, and serious heating can occur at this stage (Lacey, 1971).

Cooling bulk products to preserve them has been a widely used technique during the last two decades (Burrell, 1982). It is effective in low ψ products as a method of controlling insect pests, many of which are inactive at temperatures below 20°C, and it also prevents the growth of fungi as long as the temperature is low enough relative to ψ. Thus, the lowest ψ for growth of any fungi at 0°, 10°, and 20°C are about $-15\,\text{MPa}$, $-32\,\text{MPa}$, and $-48\,\text{MPa}$ respectively (see, e.g. Lacey *et al.*, 1980). If low temperature at high ψ is maintained by ventilation, *Penicillium* spp. and *Fusarium* spp. may be selected and this may result in the accumulation of mycotoxins (see Sherwood & Peberdy, 1974; Gross & Robb, 1975; Hill & Lacey, 1983).

References

Austwick, P. K. C. & Ayerst, G. (1963). Toxic products in groundnuts: groundnut microflora and toxicity. *Chemistry and Industry (London)*, 1963, 55–61.

Ayerst, G. (1969). The effects of moisture and temperature on growth and spore germination in some fungi. *Journal of Stored Products Research*, **5**, 127–41.

Bailey, J. E. (1974). Whole grain storage. In *Storage of Cereal Grains and their Products*, 2nd edn, ed. C. M. Christensen, pp. 333–60. St Paul, Minnesota: American Association of Cereal Chemists.

Becker, H. A. & Sallans, H. R. (1956). A study of the desorption isotherms of wheat at 25°C and 50°C. *Cereal Chemistry*, **33**, 79–91.

Beuchat, L. R. (1979). Survival of conidia of *Aspergillus flavus* in dried foods. *Journal of Stored Products Research*, **15**, 25–31.

Beuchat, L. R. (1984). Survival of *Aspergillus flavus* conidiospores and other fungi on cowpeas during long term storage under various environmental conditions. *Journal of Stored Products Research*, **20**, 119–23.

Bowes, P. C. (1984). *Self-Heating: Evaluating and Controlling the Hazards*. London: HMSO.

Burrell, N. J. (1982). Chilling. In *Storage of Cereal Grains and their Products*, 3rd edn, ed. C. M. Christensen, pp. 407–41. St Paul, Minnesota: The American Association of Cereal Chemists.

Christensen, C. M. (1951). Fungi on and in wheat seed. *Cereal Chemistry*, **28**, 408–15.

Christensen, C. M. (1972). Microflora and seed deterioration. In *Viability of Seeds*, ed. E. H. Roberts, pp. 59–93. London: Chapman & Hall.

Christensen, C. M. & Drescher, R. F. (1954). Grain storage studies. XIV. Changes in moisture content, germination percentage and moldiness of wheat samples stored in different portions of a wheat bulk in commercial bins. *Cereal Chemistry*, **31**, 206–16.

Christensen, C. M. & Kaufmann, H. A. (1974). Microflora. In *Storage of Cereal Grains and their Products*, 2nd edn, ed. C. M. Christensen, pp. 158–92. St Paul, Minnesota: American Association of Cereal Chemists.

Christensen, C. M. & Sauer, D. B. (1982). Microflora. In *Storage of Cereal Grains and their Products*, 3rd edn, ed. C. M. Christensen, pp. 219–40. St Paul, Minnesota: American Association of Cereal Chemists.

Clarke, J. H., Hill, S. T., Niles, E. V. & Howard, M. A. R. (1969). Ecology of the microflora of moist barley in 'sealed' silos on farms. *Pest Infestation Research for 1968*, p. 17. London: HMSO.

Cooke, R. C. & Rayner, A. D. M. (1984). *Ecology of Saprotrophic Fungi*. London: Longman.

Currie, J. A. & Festenstein, G. N. (1971). Factors defining spontaneous heating and ignition of hay. *Journal of the Science of Food and Agriculture*, **22**, 223–230.

Diener, U. L. & Davis, N. D. (1967). Limiting temperature and relative humidity for growth and production of aflatoxin and free fatty acids by *Aspergillus flavus* in sterile peanuts. *The Journal of the American Oil Chemists Society*. **44**, 259–63.

Disney, R. W. (1969). The formation of dew on a cooled surface in contact with wheat. *Journal of Stored Products Research*, **5**, 281–8.

FAO (1984). *Post-harvest losses of food grains*. *FAO Food and Nutrition Paper 29*. Rome: FAO.

Festenstein, G. N., Lacey, J., Skinner, F. A., Jenkins, P. A. & Pepys, J. (1965). Self heating of hay in Dewar flasks and the developments of farmer's lung antigens. *Journal of General Microbiology*, **41**, 389–407.

Flannigan, B. (1969). Microflora of dried barley grain. *Transactions of the British Mycological Society*, **53**, 371–9.

Flannigan, B. (1978). Primary contamination of barley and wheat grain by storage fungi. *Transactions of the British Mycological Society*, **71**, 37–42.

Griffin, D. M. (1981). Water and microbial stress. In *Advances in Microbial Ecology*, vol. 5, ed. M. Alexander, pp. 91–136. New York: Plenum Press.

Griffin, G. J. & Roth, D. A. (1979). Nutritional aspects of soil mycostasis. In *Soil Borne Plant Pathogens*, ed. B. Schippers & W. Gams, pp. 79–96. London: Academic Press.

Gross, V. J. & Robb, J. (1975). Zearalenone production in barley. *Annals of Applied Biology*, **80**, 211–16.

Hill, R. A. & Lacey, J. (1983). Factors determining the microflora of stored barley grain. *Annals of Applied Biology*, **102**, 467–83.

Hunt, W. H. & Pixton, S. W. (1974). Moisture – its significance, behaviour and measurement. In *Storage of Cereal Grains and their Products*, 2nd edn, ed. C. M. Christensen, pp. 1–55. St Paul, Minnesota: American Association of Cereal Chemists.

Hyde, M. B. & Burrell, N. J. (1982). Controlled atmosphere storage. In *Storage of Cereal Grains and their Products*, 3rd edn, ed. C. M. Christensen, pp. 443–78. St Paul, Minnesota: American Association of Cereal Chemists.

Lacey, J. (1971). The microbiology of moist barley stored in unsealed silos. *Annals of Applied Biology*, **69**, 187–212.

Lacey, J. (1979). Aerial dispersal and the development of microbial communities. In *Microbial Ecology, A Conceptual Approach*, ed. J. M. Lynch & N. J. Poole, pp. 40–70. Oxford: Blackwell Scientific Publications.

Lacey, J. (1980). Colonisation of damp organic substrates and spontaneous heating. In *Microbial Growth in Extremes of Environment*, vol. 5 of *Society of Applied Bacteriology Technical Series*, ed. G. H. Gould & J. E. L. Corry, pp. 53–70. London: Academic Press.

Lacey, J. & Dutkiewicz, J. (1976). Methods of examining the microflora of mouldy hay. *Journal of Applied Bacteriology*, **41**, 13–27.

Lacey, J., Hill, S. T. & Edwards, M. A. (1980). Microorganisms in stored grain: their enumeration and significance. *Tropical Stored Products Information*, **39**, 19–33.

Lichtwardt, R. W. & Barron, G. L. (1959). A quantitative deterioration rating scale for shelled corn. *Iowa State Journal of Science*, **34**, 139–46.

Magan, N. & Lacey, J. (1984a). Effect of temperature and pH on water relations of field and storage fungi. *Transactions of the British Mycological Society*, **82**, 71–81.

Magan, N. & Lacey, J. (1984b). Effect of water activity, temperature and substrate on interactions between field and storage fungi. *Transactions of the British Mycological Society*, **82**, 83–93.

Magan, N. & Lacey, J. (1984c). Effects of gas composition and water activity on growth of field and storage fungi and their interactions. *Transactions of the British Mycological Society*, **82**, 305–14.

McNaughton, S. J. (1981). Niche: definition and generalisations. In *The Fungal Community: Its Organisation and Role in the Ecosystem*, ed. D. T. Wicklow & G. C. Carroll, pp. 79–88. New York: Marcel Decker.

Milner, M. & Geddes, W. F. (1946). Grain storage studies. IV. Biological and chemical factors involved in the spontaneous heating of soybeans. *Cereal Chemistry*, **23**, 449–70.

Mislivec, P. B., Dieter, C. T. & Bruce, V. R. (1975). Effect of temperature and relative humidity on spore germination of mycotoxic species of *Aspergillus* and *Penicillium*. *Mycologia*, **67**, 1187–9.

Mislivec, P. B. & Tuite, J. (1970). Temperature and relative humidity requirements of species of *Penicillium* isolated from yellow dent corn kernels. *Mycologia*, **62**, 75–88.

Mossel, D. A. A. (1975). Water and microorganisms in foods – a synthesis. In *Water Relations of Foods*, ed. R. B. Duckworth, pp. 347–61. London: Academic Press.

Muir, W. E. (1973). Temperature and moisture in grain storages. In *Grain Storage: Part of a Systems*, ed. R. N. Sinha & W. E. Muir, pp. 49–70. Westport, Connecticut: The AVI Publishing Co.

Nichols, A. A. & Leaver, C. W. (1966). Methods of examining damp grain at harvest and after sealed and open storage: changes in the microflora of damp grain during sealed storage. *The Journal of Applied Bacteriology*, **29**, 566–81.

Niles, E. V., Norman, J. A. & Pimbley, D. (1985). Growth and aflatoxin production of *Aspergillus flavus* in wheat and barley. *Transactions of the British Mycological Society*, **84**, 259–66.

Northolt, M. D. & Bullerman, L. B. (1982). Prevention of mould growth and toxin production through control of environmental conditions. *Journal of Food Production*, **45**, 519–26.

Northolt, M. D., Verhulsdonk, C. A. H., Soentoro, P. S. S. & Paulsch, W. E. (1976). Effect of water activity and temperature on aflatoxin production by *Aspergillus parasiticus*. *Journal of Milk Technology*, **39**, 170–4.

Oxley, T. A. (1948). Movement of heat and water in stored grain. *Transactions of the American Association of Cereal Chemists*, **6**, 84–100.

Park, D. (1982). Phylloplane fungi: tolerance of hyphal tips to drying. *Transactions of the British Mycological Society*, **79**, 174–8.

Pitt, J. I. (1973). An appraisal of identification methods for *Penicillium* species: novel taxonomic criteria based on temperature and water relations. *Mycologia*, **65**, 1135–57.

Pitt, J. I. (1975). Xerophilic fungi and spoilage of food of plant origin. In *Water Relations of Foods*, ed. R. B. Duckworth, pp. 273–307. London: Academic Press.

Pitt, J. I. (1979). The genus *Penicillium* and its teleomorphic states *Eupenicillium* and *Talaromyces*. London: Academic Press.

Pitt, J. I. & Hocking, A. D. (1977). Influence of solute and hydrogen ion concentration on the water relations of some xerophilic fungi. *Journal of General Microbiology*, **101**, 35–40.

Pixton, S. W. & Howe, R. W. (1983). The suitability of various linear transformations to represent the sigmoid relationship of humidity and moisture content. *Journal of Stored Products Research*, **19**, 1–18.

Pixton, S. W., Hyde, M. B. & Ayerst, G. (1964). Long term storage of wheat. *Journal of the Science of Food and Agriculture*, **15**, 152–61.

Pixton, S. W., Warburton, S. & Hill, S. T. (1975). Long term storage of wheat – III: some changes in the quality of wheat observed during 16 years of storage. *Journal of Stored Products Research*, **11**, 177–85.

Raper, K. B. & Fennel, D. I. (1965). *The Genus Aspergillus*. Baltimore: Williams & Wilkins.

Saez, H. (1975). La thermotolerance des *Aspergillus*. Examen de 400 souches sauvages et de collection. *Revue de l'Institute Pasteur de Lyon*, **8**, 35–51.

Sauerbier, C. L. (1956). *Marine Cargo Operations*. New York: John Wiley & Sons.

Sherwood, R. F. & Peberdy, J. F. (1974). Production of the mycotoxin zearalenone by *Fusarium graminearum* growing on stored grain. 1. Grain stored at reduced temperatures. *Journal of the Science of Food and Agriculture*, **25**, 1081–7.

Sinha, R. N. & Wallace, H. A. H. (1965). Ecology of a fungus-induced hot spot in stored grain. *Canadian Journal of Plant Science*, **45**, 48–59.

Smith, S. L. & Hill, S. T. (1982). Influence of temperature and water activity on germination and growth of *Aspergillus restrictus* and *A. versicolor*. *Transactions of the British Mycological Society*, **79**, 558–60.

Sutherland, J. W., Banks, P. J. & Griffiths, H. J. (1971). Equilibrium heat and moisture transfer in air flow through grain. *Journals of Agricultural Engineering Research*, **16**, 368–86.

Thorpe, G. R. (1982). Moisture diffusion through bulk grain subjected to a temperature gradient. *Journal of Stored Products Research*, **18**, 9–12.

Tuite, J. & Foster, G. H. (1979). Control of storage diseases of grain. *Annual Review of Phytopathology*, **17**, 343–66.

Wallace, H. A. H. (1973). Fungi and other organisms associated with stored grain. In *Grain Storage: Part of a System*, ed. R. N. Sinha & W. E. Muir, pp. 71–98. Westport, Connecticut: AVI Publ. Co.

Warnock, D. W. (1971). Assay of fungal mycelium in grains of barley, including the use of the fluorescent antibody techniques for individual fungal species. *Journal of General Microbiology*, **67**, 197–205.

Whittaker, R. H., Levin, S. A. & Root, R. B. (1973). Niche habitat and ecotope. *American Naturalist*, **107**, 321–38.

Wotton, H. R. & Strange, R. N. (1985). Circumstantial evidence for phytoalexin involvement in the resistance of peanuts to *Aspergillus flavus*. *Journal of General Microbiology*, **131**, 487–94.

21
Water and decomposition processes in terrestrial ecosystems

LYNNE BODDY

Department of Microbiology, University College Cardiff, Newport Road, Cardiff, CF2 1TA, UK

Introduction: decomposition processes

Terrestrial ecosystems can be considered to consist of three sub-systems – the plant, herbivore plus carnivore, and decomposer subsystems (Swift, Heal & Anderson, 1979). These are linked together by, and functioning of the whole ecosystem depends upon, the transfer of energy and matter between and within subsystems. The major functions of the decomposer subsystem are the breakdown of organic detritus resulting in mineralisation of nutrients and the formation of soil organic matter. Decomposition of organic matter occurs as a result of interacting biotic and abiotic factors and can be attributed to the effect of four distinct processes: leaching and volatilisation, comminution, catabolism and non-enzymatic chemical reactions.

Water can affect decomposition processes crucially. Too much water can inhibit decomposition as demonstrated by the accumulation of organic matter in permanently saturated environments such as lake sediments, marshes and peat bogs (e.g. Sommers, Gilmour, Wildung & Beck, 1981; Clymo, 1983); indeed, wood is sometimes stored underwater to reduce decay rate (e.g. Cartwright & Findlay, 1958; Boutelje & Kiessling, 1964). Too little water can also inhibit decomposition. For example little or no decay of wood occurs when the water content is much below the fibre saturation point (e.g. Cartwright & Findlay, 1958). Water usually affects decomposition processes in less spectacular ways both by its direct influence on the four component processes mentioned above, and also by its indirect effects, via resource quality and microclimate, on catabolism by the decomposer organisms. It is however, often difficult to separate the various effects.

Organic matter undergoes decomposition in a wide range of ecosystem types and climates ranging from the often-frozen tundra, through boreal forest (taiga), temperate forest and grassland, to hot dry desert, tropical savanna, scrubland, rainforest and seasonal rainforest. There is a vast literature covering all of these biome types, reference to the effects of water being frequently made although not necessarily quantified, and comprehensive coverage here would obviously be difficult. Synthesis of the information is further hindered by the non-comparability of the various terms used to express moisture in different studies (Ch. 1; see also below).

Two types of decomposition-ecology studies can be recognised: those which take the 'black box' approach and consider the overall decay of an organic substratum, and those involving individual organisms (autecology) or the identification of community structure and development (synecology). A schism appears to have developed between these two approaches which is particularly unfortunate since the understanding of one is dependent upon knowledge of the other. Studies of the role of water in decomposition processes likewise fall under these two headings. My aim in this chapter is to overview the direct and indirect effects of water on component processes of decay and to indicate how, by considering the interrelationship between water, community structure and overall decomposition, a joint approach would allow better understanding of decomposition processes. Firstly, however, in order to understand the influence of water on decomposer organisms and decomposition processes, it is necessary to consider the water relations of decomposing organic substrata.

Water in decomposing organic substrata and its relation to fungi

Substrata, undergoing fungal or other microbial decomposition, may be regarded as matrices consisting of a solid phase comprised of variously sized organic and inorganic particles, and a system of fluid-filled voids. For a fungus growing within these substrata, it is the spatiotemporal distribution of water in relation to the location of hyphae which is important.

Traditionally, water content has been expressed gravimetrically either as a percentage of dry weight or, less frequently, as a percentage of fresh weight. Percentage saturation is a useful term that has sometimes been employed, particularly when substrata have a high moisture content and reduced aeration limits decay (see below). However, as with soils, such terms cannot be used for comparative purposes except with reference to exactly the same sample. This is because different substrata, or even the same sample at different stages of decay, cannot be expected necessarily to have the same densities. The need for more absolute measures is therefore clear. One such direct measurement of water availability, the theory of which is now well established, is water potential (ψ) (Ch. 1; Griffin, 1977, 1981; Harris, 1981; Papendick & Campbell, 1981). However, although it is possible to obtain values of substratum water potential and from these to predict, in a general way, spatiotemporal distribution of water under given microenvironmental conditions, progress may be limited by our inability to relate these to the location of fungal hyphae. Further, it must be emphasised that such values relate only to equilibrium conditions ultimately developing under constant external climatic regimes. These may not adequately predict more usual situations where there is flux of water into or out of substratum voids, whence temporary but nonetheless considerable spatial heterogeneity is likely.

Thus, although sophisticated methods for quantifying water relations exist, most only yield estimates of gross moisture content, and resulting data must be carefully interpreted with regard to their likely relationship to spatiotemporal distribution. In view of this, although ψ represents our most fundamental criterion of availability, moisture content expressed as a percentage of oven-dry weight probably remains, at present, as useful a term as any in studies of substratum decomposition. Indeed, since many of the published data concerning natural substrata are presented in these terms, it will largely be employed here.

Water potential and percentage moisture content are, of course, related. This relationship can be expressed graphically as the 'moisture characteristic' curve obtained by determining the ψ at a given moisture content. No single relationship exists because the forces involved during wetting (adsorption) and drying (desorption) differ, and temperature also affects the relationship. Boundary curves can be obtained within which all other curves, resulting from different histories of partial wetting and drying, will fall. However, the hysteresis can be quite considerable (see also Ayerst, Ch. 20). Further, the component of organic substrata exhibit different relationships, as is illustrated in Fig. 21.1 for the major constituents of wood (Christensen & Kelsey, 1959; Skaar, 1972). Clearly then, the moisture characteristic curve of one litter component, or even the same type of component but of a different species, will differ from that of another both because of differences in void size distribution and of the water relations of different components (Bartholomew & Norman, 1946; Dix, 1984). As decay proceeds with the removal of different chemical components and changes in the void size distribution, the relationship between ψ and percentage moisture content will also change.

Moisture contents in organic substrata are subject to variation in direct correlation with oscillations of the external environment, which imposes wetting and drying regimes, and contents are ultimately governed by capacity, input and output. Capacity is determined by the distribution and volume of available void space, which will in turn depend on the anatomy and extent of decay of individual substrata and on the degree of compaction and composition of aggregates such as leaf litter layer, animal frass etc. As a generalisation, the less dense the substratum the greater will be its water-holding capacity. This is counterbalanced, to a greater or lesser extent, by the reduced likelihood of saturation being maintained where large voids are present because these drain under the influence of gravity.

Input of water can be by four main routes: precipitation and adsorption from the air; capillarity from external reservoirs; translocation by fungal mycelia; production of 'metabolic' water as a result of respiration during

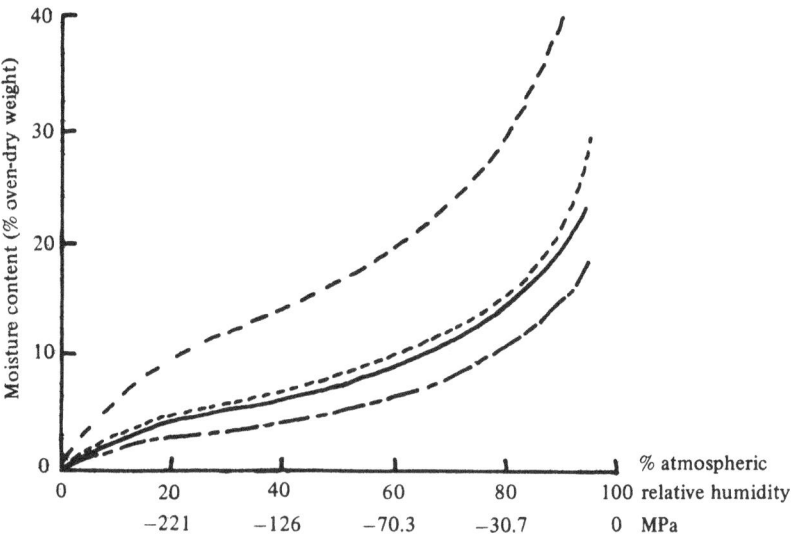

Fig. 21.1. Adsorption isotherms for wood hemicellulose (— — —), holocellulose (----), Klason lignin (— · —), and gross wood (———) of *Eucalyptus regnans*; (from Christen Skaar, *Water in Wood*. Syracuse, NY: Syracuse University Press, 1972. Adapted from Christensen & Kelsey, 1959). Note that the curves for chemically isolated constituents may be different from constituents in gross wood because of extraction effects and interrelationships among constituents in the wood cell wall.

decomposition. Input via precipitation, either as rain, snow, hail or mist, is of a sporadic nature (see also Dowding, Ch. 17). Also, the amount of water reaching organic substrata and its distribution within these will depend upon their location and other inherent features such as porosity, distance over which water has to travel (i.e. depth in soil or litter, size of wood, etc.), presence or absence of bark for wood, and vegetation cover. Indeed, interception by the canopy or litter can be quite considerable. Capillarity will operate wherever a substratum is in contact with a source of water of greater potential than in itself. This is probably especially important where the substratum is partially submerged below ground while the remainder is exposed to predominantly desiccating conditions, e.g. in tree stumps or fence posts.

The significance of metabolic water, and water translocated into substrata via fungal mycelium, is well demonstrated by the behaviour of the dry rot fungus, *Serpula lacrimans*, which is able to invade previously dry timber by translocating water from external sources (Ch. 2; Jennings, 1982, 1984). On complete microbial decomposition, over 50% of cellulose (the main plant cell-wall polymer) can be metabolised to water (Miller, 1932). Further examples of alteration of substratum water status as a result of microbial activity are given in Chapters 17 and 20. It should, however, be pointed out that fungal input of water will be advantageous only when water content is below optimum for growth and decomposition.

Loss of water occurs via three main routes: gravitational drainage; translocation via mycelium; evaporation – which is greatly influenced by temperature, humidity and inherent features of the substratum as already mentioned. It is possible that some fungi, whilst being unable to translocate water to any great extent, may be able to maintain a suitable environment by preventing, or at least reducing, input or loss of water. For instance, wood-rotting *Armillaria* and *Xylaria* spp. achieve this by forming relatively impermeable pseudosclerotial plates around occupied volumes of wood (Lopez-Real & Swift, 1975; I. Chapela & L. Boddy, unpubl.).

Given this variety of modes of water input and loss, it is clear that there will be spatiotemporal heterogeneity of water distribution within individual substrata, in substrata decomposing in different ecosystems and at different locations within ecosystems. For example, in litter decomposing above ground, such as standing dead plants, attached leaves and dead decorticated branches, fluctuating conditions occur which tend to be predominantly desiccating, whereas in buried substrata much more stable moisture regimes will occur (see also Ch. 17). Spatio-temporal distribution patterns of water within ecosystems, and on a smaller scale within substrata, are thus complex and rather unpredictable. However, temporal fluctuations reflecting seasonal and shorter-term differences can be monitored on a gross level (see Fig. 21.2). Indeed, the overall moisture relationships of the leaf litter layer have received considerable attention and accurate prediction of moisture content is now possible (e.g. Meentemeyer, 1974a; Moore & Swank, 1975).

A further complication to understanding water in relation to decomposition processes derives from the fact that hyphae of the same mycelium may be subject simultaneously to many different environmental regimes (Boddy, 1984). Unfavourable moisture conditions at one point may, to some extent, be offset by translocation from elsewhere. It is the influence of water at the mycelial level, resulting from the combined effects at the hyphal level, which is of relevance to resource capture, spread and overall substratum decay rate.

Components of decomposition processes
Leaching and volatilisation

Leaching influences decomposition processes by modifying resource quality and by contributing to weight loss. It begins prior to litter fall, and indeed even before death, with a loss of inorganic ions and to a lesser extent soluble organic materials (e.g. Frankland, 1966; Duvigneaud & Denaeyer de Smet, 1970; Tukey, 1970; Ch. 17). Such losses may be quite considerable, although there is wide variation in mobility of different ions. For instance, in a mixed temperate deciduous woodland, 2, 20, 45, 6 and 50% respectively of the total input of N, P, K, Ca and Mg at litter-fall are attributable to leachate (Duvigneaud & Denayer de Smet, 1970). Many plant litters contain inhibitors of microbial activity, such as polyphenols, and their levels may also be reduced as a result of leaching (e.g. Anderson, 1973a,b).

Leaching also occurs following fall to the floor and potentially may continue throughout the decomposition process. Decay rate curves frequently show a rapid initial weight loss, due partly to almost immediate uptake and decomposition of simple organic compounds but also to removal by leaching (Anderson, 1973a; Vossbrink, Coleman & Wooley, 1979; Seastedt, Crossley, Meentemeyer & Waide, 1983). Loss of compounds via leaching will obviously influence subsequent decomposition of an organic substratum. Loss of essential mineral nutrients and simple carbon compounds may hinder microbial activity (e.g. Platt, Hader &

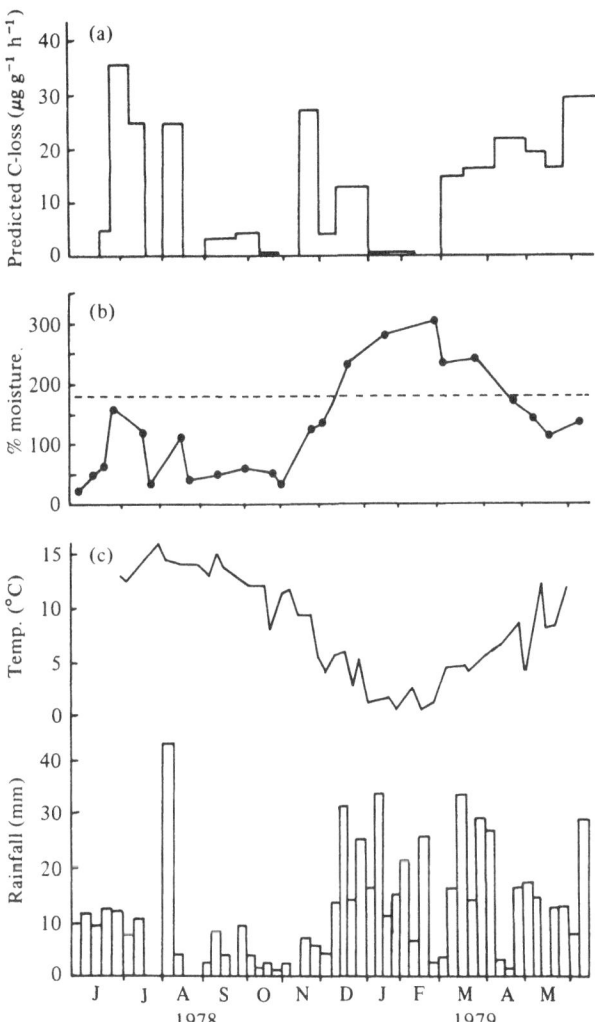

Fig. 21.2. Seasonal fluctuations in microclimate and decay rate of small beech (*Fagus sylvatica*) branches (0.3–0.4 g cm⁻³) decomposing on the floor of a temperate deciduous woodland, from June 1978 to June 1979. (a) Decay rate (y) predicted from $y = 13.796 + 0.025 \times$ water content $\times$ temperature. (b) Branch moisture content (% oven-dry weight); error bars have been omitted for clarity but variation was sometimes large; hatched line denotes moisture content at saturation after gravity drainage. (c) Air temperature (°C) and rainfall (mm). (From Boddy, 1980, 1983b, reprinted with permission of Pergamon Press. See text for details.)

Chet, 1984), whilst removal of inhibitory compounds may promote it and/or change the balance of organisms within the community. It must, however, be pointed out that these leachates are not necessarily lost from the decomposer subsystem, but are relocated and will subsequently influence microbial communities and be decomposed elsewhere. The relocation of leachates will depend on the water relations of the organic substratum, particularly on the mode of water loss. Thus in the leaf litter layer gravitational loss may predominate with concomitant relocation of leachates to lower horizons. By contrast, water loss from wood may often be predominantly by evaporation; thus solute loss from the substratum is less likely, although relocation within the wood may occur on drying (e.g. King, Oxley & Long, 1974).

Volatilisation of oils, waxes and resins will also result in dry weight loss from organic substrata. This is likely to be of particular significance in substrata which contain large quantities of such substances and which experience high temperatures. For example, the leaves of desert plants, e.g. creosote bush, have wax layers and resins on their stems and leaves, and eucalypts contain volatile oils (Whitford *et al.*, 1981; Nagy & Macauley, 1982).

The importance of leaching and volatilisation relative to other components of the decomposition process probably depends upon climate (particularly rainfall and temperature respectively), the soluble component of different organic substrata – the lowest values being found in wood and the highest in soft herbaceous tissue – physical factors including extent of decay and comminution, and on the techniques used for assessment. As Swift *et al.* (1979) point out, it is not surprising then that the literature shows inconsistency with some authors attributing almost all decomposition to microbial catabolism whilst others indicate that between 10 and 30%, and sometimes more, weight loss results from leaching.

Losses by leaching from undisturbed ecosystems are usually small although it might be expected that they would occur where rainfall is high, such as in tropical rainforest. However, streams issuing from these ecosystems are low in nutrients, losses apparently being largely prevented by a meshwork of mycorrhizal roots which constitute an absorptive barrier (Stark & Jordan, 1978). Large nutrient losses from ecosystems can occur following flooding, rapid run-off, or disturbance (e.g. Clark & Rosswall, 1981; Khanna, 1981). The latter is occasioned particularly by agricultural practices in which removal of extensive areas of vegetation occurs (e.g. Likens *et al.*, 1970; Vitousek, 1979).

Comminution

Comminution can be brought about by the direct abiotic effects of water, as the result of wetting/drying and freezing/thawing cycles and by other physical phenomena such as wind battering, whilst animals are usually the main biotic agents. Quantitatively, the contribution of animals to decomposition processes is usually (but not always cf. p. 394) relatively insignificant but, in addition to their role as agents of comminution, animals can also affect decomposition in a number of other ways. They catabolise both simple and complex organic molecules, either as a result of their own metabolism or in association with symbiotic microorganisms, facilitate microbial colonisation (e.g. Swift & Boddy, 1984) and stimulate or inhibit microbial activity by selective grazing (Anderson, Rayner & Walton, 1984). Water will have an indirect effect on decomposition processes, and the decomposer microorganisms, when it influences the presence and activity of animals.

Particle size influences the pattern of subsequent decomposition in two main ways: firstly by affecting the physical environment and thus its moisture regime, gaseous composition and susceptibility to leaching, and secondly by altering the pattern of microbial colonisation, small particles tending to select for unicellular as opposed to mycelial forms (Hanlon & Anderson, 1980; Swift & Boddy, 1984). Such effects have not been quantified, and indeed, surprisingly few studies have related to particle size in terrestrial ecosystems (see Swift *et al.*, 1979).

Catabolism

Catabolism is the component of decomposition processes which is largely mediated by microorganisms and is limited at both high and low moisture levels. Low ψ, due either to osmotic or matric forces, or freezing, reduce catabolism and growth rates. The two most significant factors affecting such reductions are problems associated with the need for osmoregulation (see Chs 1 and 2) and arising from lowered enzyme activity and movement. For example, cellulolytic activity may be reduced at ψ of about -2.8 MPa and probably somewhat lower (Meyers & Reynolds, 1959; Griffin, 1977). Kundu, Ghosh & Ghose (1983) have shown suboptimal rates of cellulose hydrolysis, by a cellulase of *Aspergillus terreus*, at both low and high water contents. The former was hypothesised as resulting from loss of active sites on the enzyme molecule and/or increased end-product repression, and the latter from greater diffusion of the enzyme from the substrate. Physical properties of the substratum are also important at low matric potentials (ψ_m) where water may be present only in micropores within the cell walls, whose dimensions are so small that enzyme protein molecules are unable to enter them (Cowling & Brown, 1969; Griffin, 1977).

At high moisture contents water will be limiting as a result of its influence on gaseous diffusion. Under aerobic conditions in organic substrata or in soils, microbial metabolism involves the oxidation of C compounds to CO_2, thus the main gases of consequence are O_2 and CO_2. The rates of diffusion to and from the decomposer organisms are determined by Fick's Law. Water content, along with size and porosity of the substratum, is the most significant factor which increases the effective path length and hence decreases the rate of gaseous exchange with the external atmosphere – diffusion of O_2 and CO_2 is slower in water than in air. Thus as water content increases gaseous diffusion will decrease and anaerobic conditions will develop. Under the latter, different terminal electron acceptors are used, e.g. NO_3^-, Mn^{4+}, Fe^{3+}, SO_4^{2-}, H_2 and CO_2, different gaseous end products are formed, e.g. CH_4, N_2, N_2O, H_2 and H_2S, and different microorganisms dominate the decomposer communities. It should also be remembered that enzymes vary in their sensitivity to O_2.

At this juncture it is worth re-emphasising that the spatial heterogeneity of decomposing substrata results in the possibility of different catabolic activities occurring at different locations. A striking example is provided by Greenwood & Goodman (1967) who demonstrated that even in the best aerated soils, anaeorbic conditions exist in water-saturated soil crumbs if their radii are greater than 3 mm.

Various groups of microorganisms have different abilities in respect of germination, survival and growth under different water regimes, details of which can be found elsewhere (e.g. Wilson & Griffin, 1979; Harris, 1981; Griffin, 1981; Boddy, 1983a; Chs. 3, 4, 7 and 20). An important point to note here is that these abilities under particular regimes are not necessarily related to an organism's rate of catabolism. For instance, the cord-forming Basidiomycotina *Phanerochaete velutina*, growing from wood blocks into soil at four different moisture levels at 20 °C, achieved maximum extension at a soil moisture content below field capacity (28–37%) whilst maximum decomposition occurred near or slightly above field capacity (C. G. Dowson, L. Boddy & A. D. M. Rayner, unpubl.).

Non-enzymatic chemical reactions

Molecular transformations in decomposing substrata can also be effected by non-enzymatic molecular reactions. For instance, cellulose degradation can be partly achieved by photosensitised oxidation by sunlight (Mark, Gaylord & Bikales, 1970) – which is greatly enhanced by the presence of moisture in the atmosphere (Ergeton, 1948) – by weakening or direct cleavage of chain bonds by ultraviolet light (Pillai & Rohatgi, 1981), and by hydrolytic degradation by weak acids (e.g. Stamm, 1964) such as might occur in rain water. Non-enzymatic reactions are probably of most significance in exposed situations, e.g. in the degradation of roof-thatching straw (Perry, 1981; Pillai & Rohatgi, 1981).

Water and overall substratum decomposition
Relationship between water and decomposition rate

Limitations due to high and low moisture content result in patterns of overall substratum decomposition and respiration typified by the model shown in Fig. 21.3 (Bunnell & Dowding, 1974). It can be seen that as moisture increases from a minimum so aerobic respiration increases towards a maximum rate which is set by the type of organisms present, temperature and resource quality (see below). Once the maximum rate has been attained a further increase in moisture does not affect the respiration rate further until the point at which there is a reduction in gaseous diffusion to and from the organisms, resulting in

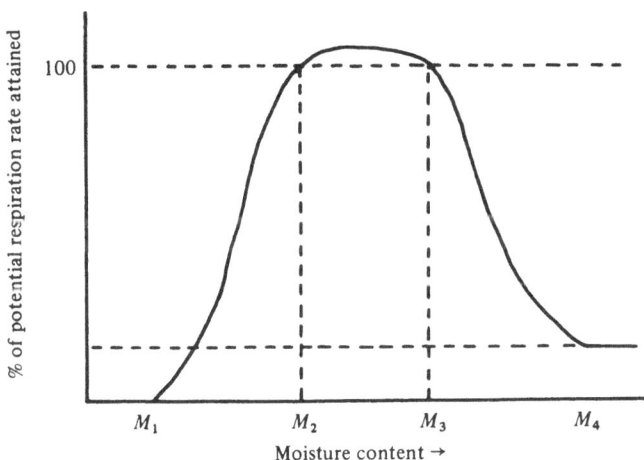

Fig. 21.3. Theoretical effect of moisture content on respiration rate expressed as a percentage of maximum possible rate. M_1, lowest moisture content at which respiration is possible; M_2, moisture content at which optimal conditions for respiration are obtained; M_3, moisture content at which conditions begin to become anaerobic; M_4 moisture content at which weight loss is attributable to facultative anaerobes. (From Bunnell & Dowding, 1974.)

a decline in respiration rate towards zero: Respiration often does not cease completely because although diffusion is slower some oxygen will probably be present, as may some facultative anaerobes.

The form of the model, minimum and maximum moisture contents at which respiration is possible, and the optimum substratum moisture content for respiration will vary both between different organic substrata, and at different stages of decay for the same substratum, depending upon the water and gas relations of the substratum (see earlier). The form of the relationship between moisture content and respiration may also vary according to temperature, various aspects of resource quality, etc. Thus the phase where increase in water content results in increase in respiration rate to a maximum may be lengthened or shortened, as may the phase where neither low aeration nor lack of water is limiting (Fig. 21.4). In fact, this middle phase may be non-existent with inhibition due to insufficient O_2 to meet demand occurring even before the maximum decay rate under non-gaseous limiting conditions is reached (Fig. 21.4a). Conversely a decline at high moisture contents may never occur.

Changes in the form of the model are well illustrated by the effect of moisture content and temperature on the fungal decomposition of wood (Fig. 21.5; Boddy, 1983b). At low temperatures (5 °C) increasing the moisture content to the maximum resulted in a concomitant increase in rate of decomposition. Raising the temperature to 15 °C, however, resulted in higher decay rates and a change in the relationship between water content and decay rate. Thus, as moisture content increased so did decay rate, but only to a point beyond which, as described above,

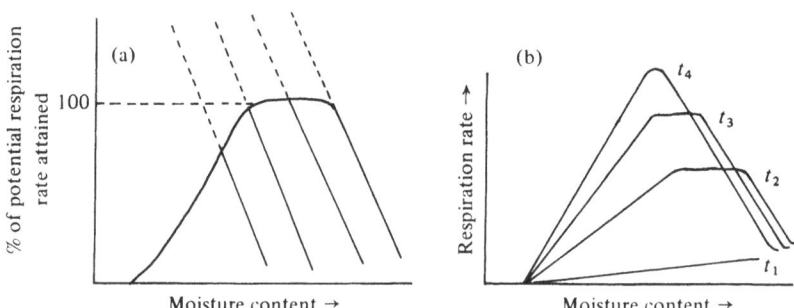

Fig. 21.4. Theoretical effect of moisture content on respiration rate, illustrating different possible forms of the model. (a) Indicates that the moisture level at which aeration starts to become limiting may vary between substrata or for a substratum at different stages of decay. (b) Indicates the effect of temperature (t) on the relationship. $t_4 > t_3 > t_2 > t_1$. See text for details.

further increases inhibited decay rate. Similar results have been found for other plant litter components, e.g. by Flanagan & Veum (1974) in tundra leaf litter, whereas Orchard & Cook (1983) found a continual increase in CO_2 evolution from loosely packed soil at 25 °C over a range of about -10 to higher than -0.01 MPa.

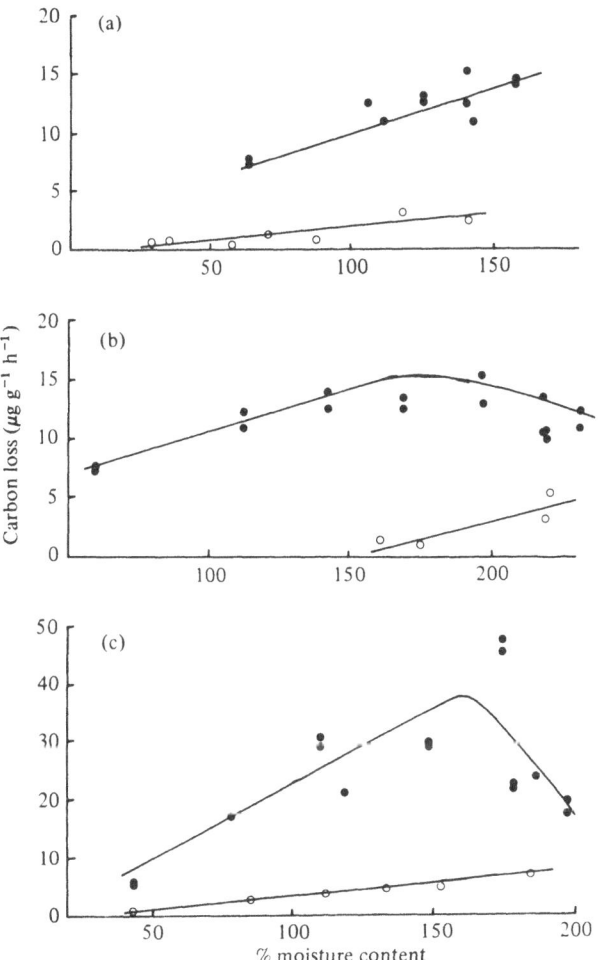

Fig. 21.5. Typical relationships between moisture content (% over-dry weight) and rate of carbon loss from small beech (*Fagus sylvatica*) branches (2–2.5 cm diam., 15 cm length) in the laboratory. Densities of (a) 0.35 g cm^{-3}; (b) 0.25 g cm^{-3}; and (c) 0.29 g cm^{-3} (density of undecayed wood approx. 0.6–0.65 g cm^{-3}); ●, 15 °C; ○, 5 °C (curves fitted by eye). (From Boddy, 1983*b*, reprinted with permission of Pergamon Press. See text for details.)

Increases in respiration and decay rates with increasing water content result from water becoming available to microorganisms in increasingly larger voids (Griffin, 1977, 1981). At higher temperatures, however, decay processes are more rapid, as is the utilisation of O_2 and production of CO_2. Since high water contents result in decreased rates of diffusion of O_2 to CO_2 from the organisms, high temperature may result in insufficient supply of O_2 and removal of CO_2 to meet demands. Consequently decay rate is reduced. This situation, modelled in Fig. 21.4b, may also occur as a result of driving variables other than temperature. For instance, high nutrient levels (or low C:nutrient ratio) may be analogous to the high-temperature situation. On the other hand, substrata with high levels of inhibitory compounds, or with decomposer organisms having inherently low rates of activity, or substrata in early stages of colonisation, may be more analogous to the low-temperature situation. Clearly other factors which influence gaseous composition will interact and affect decay rate in a similar manner. Thus, bulky, dense substrata will provide an additional constraint as a result of decreased rate of gaseous diffusion (see examples in Boddy, 1983b).

Modelling decay rate: relative effect of water, other climatic variables and resource quality

From the above it is also evident that the effect of water on decomposition processes is not independent of other climatic and resource quality factors, and therefore must to some extent be considered along with them. The relative importance of each of these factors can be explored further by fitting models and analysing the explained variation. An appropriate model has been derived whose component terms describing water are based on the above discussion (Bunnel & Dowding, 1974; Bunnel, Tait, Flanagan & Van Cleve, 1977a):

$$R(T,M) = [M/(a_1 + M)] \times [a_2/(a_2 + M)] \times a_3 \times a_4^{[(T-10)/10]}$$

where $R(T,M)$ is the respiration rate at T (°C) and moisture M (% oven-dry weight), a_1 is percentage moisture content at which activity is at half its optimal value (i.e. a measure of water availability), a_2 is percentage moisture content at which gas exchange is limited to half of its optimal value (i.e. a measure of limiting effects due to decreased gaseous exchange rates), a_3 is the respiration rate at 10 °C when neither oxygen or moisture is limiting, a_4 is the Q_{10} coefficient. The authors did consider using ψ as the term regulating availability to microorganisms but concluded that, except at extremely low water contents, percentage water content was an equally acceptable term for the model.

The influence of water on decomposition, and the 'goodness-of-fit' of the model, can be seen by examining actual and predicted response surfaces of various litters to moisture and temperature (Fig. 21.6). Even though the model ignores leaching and comminution, it gives a good fit to the observed data. It is difficult to quantify specific effects associated with the various coefficients because of gaps in the data set and lack of statistical independence, but Bunnel *et al.* (1977*a*) consider that the inclusion of the moisture content terms accounts for at least 23–31% of the variation. Other studies have also indicated a significant role for moisture (e.g. Witkamp, 1966; Wildung, Garland & Buschbom, 1975; Nyan, 1976; Boddy, 1980; Sommers *et al.*, 1981; Orchard & Cook, 1983).

The crucial role of water evident from laboratory studies is modified by prevailing microclimatic conditions in the field, where moisture can sometimes be of considerably less significance than temperature. For example, Witkamp (1966) estimated that moisture accounted for only about 5% of the variation, in his multiple regression model for field respiration data of four tree leaf-types, mulberry, redbud, oak and pine, compared with 64% of the variation being accounted for by temperature. This results from relatively constant field moisture contents which were probably never limiting (means never fell below 100%). When more critical moisture contents were employed in the laboratory, respiration was correlated with moisture content (Witkamp, 1963).

Similarly, in a black spruce taiga ecosystem, temperature was the dominant influence in control of forest floor respiration and biomass, where moisture contents fluctuated relatively little seasonally (from 80 to 140%)

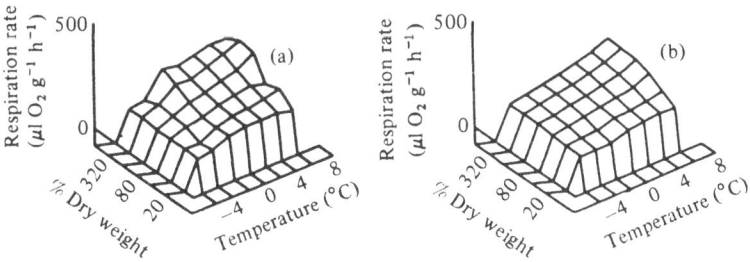

Fig. 21.6. (a) Actual and (b) predicted effect of moisture and temperature on respiration rate (uptake of O_2) of mixed Alaskan graminoid litter in the laboratory. (From Bunnell *et al.*, 1977*a*, reprinted with permission of Pergamon Press. See text for details.)

(Flanagan & Van Cleve, 1977). However, occasionally high moisture contents (>150%) were found in the L layer and these resulted in a decline in microbial respiration and biomass. Likewise, multiple regression models indicated that temperature was relatively more important than moisture in forest floor respiration in birch and aspen stands in interior Alaska (Van Cleve & Sprague, 1971).

On the other hand, moisture (quantified as precipitation or humidity, and not as soil or litter moisture content) was found to have a considerably greater influence than temperature on decomposition rate of litter in a tropical montane rain forest in Thailand (Thaiutsa & Granger, 1979). Similarly, in seasonal tropical rainforest in Panama, temperature was fairly constant (mean monthly range approx. 26–31 °C) and the litter decomposition rate was closely related to rainfall (Healey & Swift, unpubl., cited in Swift *et al.*, 1979).

The relative importance of water and temperature will often vary seasonally and on a shorter time scale. This is particularly evident in temperate ecosystems and is well illustrated by a study of wood decay rate (Fig. 21.2). Here a regression model was fitted to laboratory data of the effect of temperature and moisture on decay rate and, by incorporating appropriate field measurements, was used to predict decay rate throughout the year (Boddy, 1980, 1983*b*,*c*). From late October to February rainfall and low temperatures resulted in a trend for gradual increase in water content, until the maximum water-holding capacity was reached. With the advent of warmer, drier conditions a trend for gradual drying occurred. In the summer months precipitation was interspersed by strongly drying conditions resulting in rapid fluctuations in moisture content. It must be noted that the data points in the figure were not obtained daily, although they probably reflect closely actual conditions in the winter. In the summer, however, they represent only the conditions pertaining at the time of measurement, although they do model fluctuations that would be seen over a few days rather than weeks. During summer months it is moisture and not temperature that is often limiting, and accordingly decay rate fluctuates in response to moisture. Temperature tends to be the limiting factor as moisture increases during autumn and winter, and the low predicted decay rates correspond with low field temperatures.

Meentemeyer (Meentemeyer, 1974*a*,*b*; Seastedt *et al.*, 1983) also recognised the prime importance of both moisture and temperature when he predicted annual and seasonal variations in leaf litter breakdown, using actual evapotranspiration (AE) as a joint indicator of microclimate.

This variable takes into account temperature and precipitation inputs, latitude (to compute a day-length factor) and soil or litter storage capacity (Thornwaite & Mather, 1957; Meentemeyer, 1974*a,b*). AE increases as temperature increases, provided that water is available, and is reduced if energy or water becomes limiting. Meentemeyer (1977, 1978*a*) emphasised that litter decomposition rate is under climatic regulation, and predicted well by AE, not just in a particular ecosystem component but also on a global (excluding arid zones) scale. He obtained excellent correlation when log annual percentage decomposition was regressed on annual AE, obtained from data in stable terrestrial ecosystems covering many biome types.

It is evident then that moisture (and temperature) is a driving variable in decomposition processes but, in all of the studies mentioned, the gross climatic factors alone leave considerable variation unaccounted for. This is only to be expected in view of the earlier discussion of water relations in decomposing organic substrata and further insight will undoubtedly be gained from considering additional environmental variables (including spatial variation of water and changes in moisture-characteristic curves and water-holding capacity as decay proceeds) and also physical characteristics of litter (cf. Bunnell, Tait & Flanagan, 1977*b*; Meentemeyer, 1977*b*). In addition considerable variation is likely to result from factors associated with the decomposer organisms themselves.

Interrelationship between water, community structure and overall decomposition

Decomposition of organic substrata is brought about by, and the effect of water on decomposition processes is mediated via, a variety of organisms which include not only fungi but also bacteria, actinomycetes and various soil microfauna. The rate and pattern of substratum decomposition will depend on the community composition, and the functioning of component organisms under the prevailing moisture regime. Thus, in order to attain a complete understanding it is necessary to analyse the structure of the decomposer communities within organic substrata and to relate the effect of water to the functioning of the main members of the community.

There have been very few studies combining information derived from the two approaches. Thus, it is possible only to construct a framework of ideas which need subsequently to be tested. Important issues are the outcome of interactions between other members of the decomposer community, and the ability of individual organisms to germinate, grow, survive and catabolise, both in *absolute* terms and *relative* to other organisms, under relevant regimes. Apart from catabolism, these abilities are not necessarily directly relevant to decomposition rate. However, they are, together with other features including quantity and type of inoculum arriving, conditions pertaining at this time, and past history of abiotic and biotic regimes etc., of paramount importance in colonisation and subsequent development of microbial communities in organic substrata (see Griffin, 1981; Harris, 1981; Cooke & Rayner, 1984).

Reflecting this variety of factors influencing development, microbial communities show considerable variation both between similar substrata and with time within a particular substratum. The structure of these communities can obviously influence overall decomposition rate, since various organisms are able to catabolise different organic molecules at different rates and their abilities may vary according to the abiotic environmental regime. Further, factors allowing establishment of organisms may not necessarily be positively correlated with decomposer ability. Thus, organisms that are able to survive or establish themselves most rapidly under conditions adverse to the majority may be less-able agents of decomposition than others under the most favourable conditions. The former may remain *in situ*, at least temporarily, depressing overall decomposition rates. These features, and the need for a combination of holistic and community structure approaches, are well illustrated by recent studies on wood.

In a study of felled beech (*Fagus sylvatica*) logs, the Basidiomycotina *Bjerkandera adusta*, *Chondrostereum purpureum*, *Coriolus versicolor* and *Corticium evolvens*, and the Ascomycotina *Xylaria hypoxylon* regularly formed decay columns, originating from aerial surfaces, after three months exposure to natural/artificial inoculum; after six months *Stereum hirsutum* was also evident (Coates & Rayner, 1985a,b,c). *C. purpureum* rapidly occupied large volumes of wood but caused relatively little decomposition, and after about eighteen months had, however, been virtually eliminated. By contrast, after nearly five years *C. versicolor*, *S. hirsutum* and *X. hypoxylon*, together with some colonisers from log bases and other later additions, still remained in a large number of logs (I. Chapela, L. Boddy & A. D. M. Rayner, unpubl.). Blocks of wood removed from representative samples of the decay columns showed average weight losses of about 73, 58, and 30%, respectively. Thus, *X. hypoxylon* caused relatively little decay but remained in wood for considerable periods. It was not replaced by the more active decomposers *C. versicolor* and *S. hirsutum*; it is perhaps aided, at least in part, by its aforementioned ability to maintain low moisture regimes in the wood that it occupies. *X. hypoxylon* can, however, be replaced by cord-forming fungi. In view of the possibly important role of water in resistance of this species to replacement, it is tempting to speculate that the ability of the cord-forming species to translocate water may confer upon them an advantage in combat with *X. hypoxylon*. Further, it is now evident that the outcome of interactions between a range of fungi when paired in agar culture are sometimes modified by environmental conditions (Magan & Lacey, 1984; Boddy, Gibbon & Grundy, 1985).

Another important feature was noted when some beech logs were artificially inoculated with different spore concentrations of certain fungi. There was evidence that high spore loads inhibited decomposition (Coates & Rayner, 1985*a*), further emphasising the need for a joint holistic and synecological approach.

Xylariacious Ascomycotina are often found in dry, relatively unde-cayed wood and it is possible that the characteristics described for *X. hypoxylon* are found in a number of members of this group. For example they may be possessed by *Daldinia concentrica* which, like a number of other Ascomycotina and Basidiomycotina that are primary colonisers of attached branches, appears to adopt latent invasion strategies. It is suggested that such fungi are present in the living wood in a cryptic form, being able to develop substantially only after the stress of high water content and low aeration has been removed (Boddy & Rayner, 1983*a,b*; Cooke & Rayner, 1984; Boddy *et al.*, 1985; Rayner, Ch. 18). *D. concentrica* is able to defend against, or even replace, many other competitive decay fungi. Although it may ultimately cause considerable decay, it has been observed to persist in ash (*Fraxinus excelsior*) for probably over twenty years whilst causing only moderate decay (Boddy *et al.*, 1985; A. D. M. Rayner & L. Boddy, unpubl.).

As a generalisation, in early stages of development of decay communi-ties, different wood resource units show variation in extent of decay, volume of wood colonised and proportions of active wood-rotting species present. In intermediate stages, under favourable microclimatic condi-tions, although wood units will differ in species composition and activity, characteristic communities of *active* wood decomposers tend to predomi-nate. In terminal stages, both wood and community structure shows considerable variation as wood-rotting species are declining and the soil microflora and microfauna, with their different enzymatic capacities and water tolerances, enter. Interestingly, probably corresponding with these stages, although moisture and temperature explained significant amounts of variation in intermediate decay categories (RD 0.3–0.4 g cm^{-3}) in the study mentioned earlier (p. 390), very little was explained by these variables in late stages of decay (RD < 0.2 g cm^{-3}) (Boddy, 1980).

A further example, showing that knowledge of component organisms is essential before variation can be explained, is provided by unexpectedly high rates of decomposition occurring in some arid systems where AE is low (e.g. Meentemeyer, 1978a; Santos & Whitford, 1981; Whitford *et al.*, 1981; Elkins, Steinberger & Whitford, 1982). In these environments soil microarthropods and termites (Isoptera) play a major role in breakdown and removal of litter. For example, Whitford *et al.* (1981) cite a study in the Chihuahuan and Sonoran deserts in which percent weight losses of litter were 2.6- to 2.9-fold greater on plots where termites were not excluded compared with those from which they were excluded. Interestingly, these figures were very similar to the amount by which the AE model underestimated the decay rate in deserts. These authors emphasise that although the soil biota are affected by water regimes they can migrate vertically and go into anhydrobiosis to escape harsh surface conditions. In addition they make use of short periods, of a few hours duration, when moderate surface litter microclimatic regimes occur in order to feed and rapidly transport quantities of organic substrata to more favourable locations. Intense daily biological activity is thus possible for some, but not all, soil organisms.

Acknowledgements. I am grateful to Alan Rayner and Ignacio Chapela for comments on the draft manuscript.

References

Anderson, J. M. (1973a). The breakdown and decomposition of sweet chestnut (*Castanea sativa* Mill.) and beech (*Fagus sylvatica* L.) leaf litter in two deciduous woodland soils. I. Breakdown, leaching and decomposition. *Oecologia (Berl.),* **12,** 251–74.

Anderson, J. M. (1973b). The breakdown and decomposition of sweet chestnut (*Castanea sativa* Mill.) and beech (*Fagus sylvatica* L.) leaf litter in two deciduous woodland soils. II. Changes in carbon, hydrogen, nitrogen and polyphenol content. *Oecologia (Berl.),* **12,** 275–88.

Anderson, J. M., Rayner, A. D. M. & Walton, D. W. H. (ed.) (1984). *Invertebrate–Microbial Interactions.* Cambridge: Cambridge University Press.

Bartholomew, W. V. & Norman, A. G. (1946). The threshold moisture content for active decomposition of some mature plant materials. *Soil Science Society of America Proceedings*, **11**, 270–9.

Boddy, L. (1980). Decomposition ecology of fallen branch-wood. Ph.D. Thesis, University of London.

Boddy, L. (1983*a*). The effect of temperature and water potential on the growth rate of wood-rotting basidiomycetes. *Transactions of the British Mycological Society*, **80**, 141–9.

Boddy, L. (1983*b*). Carbon dioxide release from decomposing wood: effect of water content and temperature. *Soil Biology and Biochemistry*, **15**, 501–10.

Boddy, L. (1983*c*). Microclimate and moisture dynamics of wood decomposing in terrestrial ecosystems. *Soil Biology and Biochemistry*, **15**, 149–57.

Boddy, L. (1984). The micro-environment of basidiomyceyte mycelia in temperate deciduous woodlands. In *Ecology and Physiology of the Fungal Mycelium*, ed., D. H. Jennings & A. D. M. Rayner, pp. 261–89. Cambridge: Cambridge University Press.

Boddy, L., Gibbon, O. M. & Grundy, M. A. (1985). Ecology of *Daldinia concentrica* from ash: effect of abiotic variables on mycelial extension and interspecific interactions. *Transactions of the British Mycological Society*, **85**, 201–11.

Boddy, L. & Rayner, A. D. M. (1983*a*). Origins of decay in living deciduous trees: role of moisture content and a re-appraisal of the expanded concept of tree decay. *New Phytologist*, **94**, 623–41.

Boddy, L. & Rayner, A. D. M. (1983*b*). Ecological roles of basidiomycetes forming decay communities in attached oak branches. *New Phytologist*, **93**, 77–88.

Boutelje, J. B. & Kiessling, H. (1964). On water-stored oak timber and its decay by fungi and bacteria. *Archiv fur Mikrobiologie*, **49**, 305–14.

Bunnell, F. L. & Dowding, P. (1974). ABISKO – a generalised decomposition model for comparison between Tundra sites. In *Soil Organisms and Decomposition in Tundra*, ed. A. J. Holding, O. W. Heal, S. F. MacLean & P. W. Flanagan, pp. 227–48. Stockholm: Tundra Biome Steering Committee.

Bunnell, F. L., Tait, D. E. N., Flanagan, P. W. & Van Cleve, K. (1977*a*). Microbial respiration and substrate weight loss. II. A general model of the influences of abiotic variables. *Soil Biology and Biochemistry*, **9**, 33–40.

Bunnell, F. L., Tait, D. E. N. & Flanagan, P. W. (1977*b*). Microbial respiration and substrate weight loss. II. A model of the influences of chemical composition. *Soil Biology and Biochemistry*, **9**, 41–7.

Cartwright, K. St G. & Findlay, W. P. K. (1958). *Decay of Timber and Its Prevention*. London: HMSO.

Christensen, G. N. & Kelsey, K. E. (1959). The sorption of water vapour by the constituents of wood. *Holz als Roh- und Werkstoff*, **17**, 189–204.

Clark, F. E. & Rosswall, T. (ed.) (1981). *Terrestrial Nitrogen Cycles. Processes, Ecosystem Strategies and Management Impacts*. Stockholm: Ecological Bulletins.

Clymo, R. S. (1983). Peat. In *Ecosystems of the World, 4A Mires: Swamp, Bog, Fen and Moor*, ed. A. J. P. Gore, pp. 159–224. Amsterdam, Oxford, New York: Elsevier Scientific Publishing Co.

Coates, D. & Rayner, A. D. M. (1985*a*). Fungal population and community development in beech logs. I. Establishment via the aerial cut surface. *New Phytologist*, **101**, 153–71.

Coates, D. & Rayner, A. D. M. (1985*b*). Fungal population and community development in beech logs. II. Establishment via the buried cut surface. *New Phytologist*, **101**, 173–81.

Coates, D. & Rayner, A. D. M. (1985*c*). Fungal population and community development in beech logs. III. Spatial dynamics, interactions and strategies. *New Phytologist*, **101**, 183–98.

Cooke, R. C. & Rayner, A. D. M. (1984). *Ecology of Saprotrophic Fungi*. London & New York: Longmans.

Cowling, E. B. & Brown, W. (1969). Structural features of cellulosic materials in relation to enzymatic hydrolysis. In *Cellulases and Their Applications*, ed. G. J. Hajney & E. T. Reese, *Advances in Chemistry Series*, **95**, 152–87.

Dix, N. J. (1984). Moisture content and water potential of abscissed leaves in relation to decay. *Soil Biology and Biochemistry*, **16**, 367–70.

Duvigneaud, P. & Denaeyer de Smet, S. (1970). Biological cycling of minerals in temperate deciduous forests. In *Analysis of Temperate Forests Ecosystems*, ed. D. Reichle, pp. 199–225. London: Chapman & Hall.

Elkins, N. Z., Steinberger, V. & Whitford, W. G. (1982). Factors affecting the applicability of the AET model for decomposition in arid environments. *Ecology*, **63**, 579–80.

Ergeton, G. S. (1948). Some aspects of photochemical degradation of nylon, silk and viscose rayon. *Textile Research Journal*, **13**, 669–80.

Flanagan, P. W. & Veum, A. K. (1974). Relationships between respiration, weight loss, temperature and moisture in organic residues in tundra. In *Soil Organisms and Decomposition in Tundra*, ed. A. J. Holding, O. W. Heal, S. F. MacLean & P. W. Flanagan, pp. 249–78. Stockholm: Tundra Biome Steering Committee.

Flanagan, P. W. & Van Cleve, K. (1977). Microbial biomass, respiration, and nutrient cycling in a black spruce taiga ecosystem. *Ecological Bulletins*, **25**, 262–73.

Frankland, J. C. (1966). Succession of fungi on decaying petioles of *Pteridium aquilinum*. *Journal of Ecology*, **54**, 41–63.

Greenwood, D. J. & Goodman, D. (1967). Direct measurements of oxygen in soil aggregates and in columns of fine soil crumbs. *Journal of Soil Science*, **18**, 182–96.

Griffin, D. M. (1977). Water potential and wood-decay fungi. *Annual Review of Phytopathology*, **15**, 319–29.

Griffin, D. M. (1981). Water and microbial stress. *Advances in Microbial Ecology*, **5**, 91–136.

Hanlon, R. D. G. & Anderson, J. M. (1980). Influence of macroarthropod feeding activities on microflora in decomposing oak leaves. *Soil Biology and Biochemistry*, **12**, 255–61.

Harris, R. F. (1981). Effect of water potential on microbial growth and activity. In *Water Potential Relations in Soil Microbiology*, ed. J. F. Parr, W. R. Gardner & L. F. Elliott, pp. 23–95. Madison: Soil Science Society of America.

Jennings, D. H. (1982). The movement of *Serpula lacrimans* from substrate to substrate over nutritionally inert surfaces. In *Decomposer Basidiomycetes: Their Biology and Ecology*, ed. J. C. Frankland, J. N. Hedger & M. J. Swift, pp. 91–108. Cambridge: Cambridge University Press.

Jennings, D. H. (1984). Water flow through mycelia. In *The Ecology and Physiology of the Fungal Mycelium*, ed. D. H. Jennings & A. D. M. Rayner, pp. 143–64. Cambridge: Cambridge University Press.

Khanna, P. K. (1981). Leaching of nitrogen from terrestrial ecosystems – patterns, mechanisms and ecosystem responses. In *Terrestrial Nitrogen Cycles. Processes, Ecosystem Strategies and Management Impacts*, ed. F. E. Clark & T. Rosswall, pp. 343–52. Stockholm: Ecological Bulletins.

King, B., Oxley, T. A. & Long, K. D. (1974). Soluble nitrogen in wood and its redistribution on drying. *Material und Organismen*, **9**, 241–54.

Kundu, A. B., Ghosh, B. S. & Ghose, S. N. (1983). Rôle of water in the hydrolysis of cellulose in a solid state. *Journal of Fermentation Technology*, **61**, 185–8.

Likens, G. E., Bormann, F. H., Johnson, N. M., Fisher, D. W. & Pierce, R. S. (1970). Effects of forest cutting and herbicide treatment on nutrient budgets in the Hubbard Brook watershed-ecosystem. *Ecological Monographs*, **40**, 23–47.

Lopez-Real, J. M. & Swift, M. J. (1975). The formation of pseudosclerotia ('zone lines') in wood decayed by *Armillaria mellea* and *Stereum hirsutum*. II. Formation in relation to the moisture content of the wood. *Transactions of the British Mycological Society*, **64**, 465–71.

Magan, N. & Lacey, J. (1984). The effect of water activity, temperature and substrate on interactions between field and storage fungi. *Transactions of the British Mycological Society*, **82**, 83–93.

Mark, H. F., Gaylord, N. G. & Bikales, N. M. (ed.) (1970). *Encyclopedia of Polymer Science and Technology*, **3**, 117–19.

Meentemeyer, V. (1974a). Climatic water budget approach to forest problems. 1. The prediction of forest fire hazard through moisture budgeting. *Publications in Climatology*, **27**, 1–35.

Meentemeyer, V. (1974b). Climatic water budget approach to forest problems. II. The prediction of regional differences in decomposition rate of organic debris. *Publications in Climatology*, **27**, 35–74.

Meentemeyer, V. (1977). Climatic regulation of decomposition rates of organic matter in terrestrial ecosystems. In *Environmental Chemistry and Cycling Processes*, ed. D. C. Adriano & I. L. Brisbin, pp. 779–89. Washington: US Department of Energy Symposium Series CONF-760429.

Meentemeyer, V. (1978a). An approach to the biometeorology of decomposer organisms. *International Journal of Biometeorology*, **22**, 94–102.

Meentemeyer, V. (1978b). Macroclimate and lignin control of litter decomposition rates. *Ecology*, **59**, 465–72.

Meyers, D. P. & Reynolds, E. S. (1959). Growth and cellulolytic activity of lignicolous Deuteromycetes from marine localities. *Canadian Journal of Microbiology*, **5**, 493–503.

Miller, V. V. (1932). *Points in the Biological Diagnosis of House Fungi*. Leningrad: State Forestal Technical Publishing Office.

Moore, A. & Swank, W. T. (1975). A model of water content and evaporation for hardwood leaf litter. In *Mineral Cycling in Southeastern Ecosystems*, ed. F. G. Howell, J. B. Gentry & M. H. Smith, pp. 58–69. US Energy Research & Development Administration.

Nagy, L. A. & Macauley, B. J. (1982). *Eucalyptus* leaf litter decomposition: effects of relative humidity and substrate moisture content. *Soil Biology and Biochemistry*, **14**, 233–6.

Nyan, J. W. (1976). Influence of soil temperature and water tension on the decomposition rate of carbon-14 labelled herbage. *Soil Science*, **121**, 288–93.

Orchard, V. A. & Cook, F. J. (1983). Relationship between soil respiration and soil moisture. *Soil Biology and Biochemistry*, **15**, 447–53.

Ott, E., Spurlin, H. M. & Graffin, N. W. (ed.) (1959). *Cellulose and Cellulose Derivatives*. New York: Interscience Publishers Incorporated.

Papendick, R. I. & Campbell, G. S. (1981). Theory and measurement of water potential. In *Water Potential Relations in Soil Microbiology*, ed. J. F. Parr, W. R. Gardner & L. F. Elliott, pp. 1–22. Madison: Soil Science Society of America.

Perry, V. E. A. (1981). The suitability of wheat varieties for thatching. B.Sc. Honours Dissertation, University of Wales, Bangor.

Pillai, C. K. S. & Rohatgi, P. K. (1981). Mechanisms of deterioration of thatch in the tropics and its prevention. *Journal of Scientific and Industrial Research*, **40**, 363–72.

Platt, M. W., Hader, Y. & Chet, I. (1984). Fungal activities involved in lignocellulose degradation by *Pleurotus. Applied Microbiology and Biotechnology*, **20**, 150–4.

Santos, P. F. & Whitford, W. G. (1981). The effects of microarthropods on litter decomposition in a Chihuahuan desert ecosystem. *Ecology*, **62**, 654–63.

Seastedt, T. R., Crossley, D. A. Jr, Meentemeyer, V. & Waide, J. B. (1983). A two-year study of leaf litter decomposition as related to macroclimatic factors and microarthropod abundance in the southern Appalachians. *Holarctic Ecology*, **6**, 11–16.

Skaar, C. (1972). *Water in Wood*. Syracuse: Syracuse University Press.

Sommers, L. E., Gilmour, C. M., Wildung, R. E. & Beck, S. M. (1981). The effect of water potential on decomposition processes in soils. In *Water Potential Relations in Soil Microbiology*, ed. J. F. Parr, W. R. Gardner & L. F. Elliott, pp. 97–117. Madison: Soil Science Society of America.

Stamm, A. J. (1964). *Wood and Cellulose Science*. New York: The Ronald Press Co.

Stark, N. M. & Jordan, C. F. (1978). Nutrient retention by the root mat of an Amazonian rainforest. *Ecology*, **59**, 434–7.

Swift, L. & Boddy, L. (1984). Animal–microbe interactions in wood decomposition. In *Invertebrate–Microbial Interactions*, ed. J. M. Anderson, A. D. M. Rayner & D. W. H. Walton, pp. 89–131. Cambridge: Cambridge University Press.

Swift, M. J., Heal, O. W. & Anderson, J. M. (1979). *Decomposition in Terrestrial Ecosystems*. Oxford: Blackwell Scientific Publications.

Thaiutsa, B. & Granger, O. (1979). Climate and the decomposition rate of tropical forest litter. *Unasylva*, **31**, 28–35.

Thornwaite, C. W. & Mather, J. R. (1957). Instructions and tables for computing potential evapotranspiration and the water balance. *Publications in Climatology*, **10**, 185–311.

Tukey, H. B. (1970). The leaching of substances from plants. *Annual Review of Plant Physiology*, **21**, 305–24.

Van Cleve, K. & Sprague, D. (1971). Respiration rates in the floor of birch and aspen stands in interior Alaska. Arctic and Alpine Research, **3**, 17–26.

Vitousek, P. M. (1979). Clear-cutting and the nitrogen cycle. In *Terrestrial Nitrogen Cycles. Processes, Ecosystem Strategies and Management Impacts*, ed. F. E. Clark & T. Rosswall, pp. 343–52. Stockholm: Ecological Bulletins.

Vossbrink, C. R., Coleman, D. C. & Wooley, T. A. (1979). Abiotic and biotic factors in litter decomposition in a semi-arid grassland. *Ecology*, **60**, 265–71.

Whitford, W. G., Meentemeyer, V., Seastedt, T. R. & Cromack, K. Jr, Crossley, D. A. Jr, Santos, P., Todd, R. L. & Waide, J. B. (1981). Exceptions to the AET model: deserts and clear-cut forest. *Ecology*, **62**, 275–7.

Wildung, R. E., Garlund, T. R. & Buschbom, R. L. (1975). The interdependent effects of soil temperature and water content on soil respiration rate in plant root decomposition in arid grassland soil. *Soil Biology and Biochemistry*, **7**, 373–8.

Wilson, J. M. & Griffin, D. M. (1979). The effect of water potential on the growth of some soil basidiomycetes. *Soil Biology and Biochemistry*, **11**, 211–12.

Witkamp, M. (1963). Microbial populations of leaf litter in relation to environmental conditions and decomposition. *Ecology*, **44**, 370–7.

Witkamp, M. (1966). Decomposition of leaf litter in relation to environment, microflora and microbial respiration. *Ecology*, **47**, 194–201.

Index

www.ingramcontent.com/pod-product-compliance
Ingram Content Group UK Ltd.
Pitfield, Milton Keynes, MK11 3LW, UK
UKHW010853090126
466816UK00011B/205